Growing Up Caring

Growing Up Caring

FIRST EDITION

FRANCES SCHOONMAKER BOLIN

ASSOCIATE PROFESSOR OF EDUCATION
TEACHER'S COLLEGE, COLUMBIA UNIVERSITY
NEW YORK, NEW YORK

GLENCOE
Macmillan / McGraw-Hill

Lake Forest, Illinois
Columbus, Ohio
Mission Hills, California
Peoria, Illinois

Contributors

Roberta Larson Duyff
Nutrition Education Consultant
St. Louis, MO

William E. Bolin
Educational Writer and Consultant
New York, NY

Eddye Eubanks
Educational Writer and Consultant
Austin, TX

Gale Cornelia Flynn
Educational Writer and Consultant
Hockessein, DE

Herbert J. Kramer
Director of Communications
Joseph P. Kennedy, Jr. Foundation
Washington, DC

Eunice Kennedy Shriver
Executive Vice President
Joseph P. Kennedy, Jr. Foundation
Washington, DC

Copyright © 1990 by Glencoe/McGraw-Hill Educational Division.

All rights reserved. Printed in the United States of America. Except as permitted under the United States Copyright Act of 1976, no part of this publication may be reproduced or distributed in any form or by any means, or stored in a database or retrieval system, without prior permission of the publisher.

Send all inquiries to:
Glencoe Division, Macmillan/McGraw-Hill
15319 Chatsworth Street
P.O. Box 9609
Mission Hills, CA 91346-9609

ISBN 0-02-652401-5

2 3 4 5 6 7 8 9   96 95 94 93 92 91 90

Design
Design Associates, Inc.
Paul C. Uhl and William A. Seabright

Advisory Board

Eileen Crane
Home Economics Teacher
Northern Valley Regional
High School
Old Tappan, NJ

Joan Dietz
Home Economics Department Head
Central High School
Little Rock, AR

Eunice Hartmann
Health Education Coordinator
Penfield Central School
Penfield, NY

Juanita Mendenhall
Home Economics Department Head
Southside High School
Fort Wayne, IN

Linda Pagel
Home Economics Teacher
Sumner High School
Sumner, IA

Reviewers

Bonnie Adams
Houston Independent School District
Houston, TX

Father James DiGiacomo, S.J.
Teacher of Theology
Regis High School
New York, NY

Wilma Ferguson
Community of Caring Facilitator
Franklin Middle School
Long Beach, CA

Janet B. Hardy, M.D.
Professor of Pediatrics Emeritus
Johns Hopkins University
Baltimore, MD

James H. Hughes
Drug and Alcohol Educator
Tatnall Middle School
Wilmington, DE

Joseph Hunt
Assistant Director of Vocational
Education
Covenant House
New York, NY

Patty Jessen
Community of Caring Coordinating
Committee
Maple Park Middle School
Kansas City, MO

Donna Joannes
Community of Caring Coordinating
Committee
Maple Park Midddle School
Kansas City, MO

Rosemary E. Lassiter
Certified Alcoholism Counselor
Stroudsburg, PA

Eugertha Minnicks
Assistant Principal
Armstrong High School
Richmond, VA

Mario Pazzaglini, Ph. D.
Mental Health, Drug, and Alcohol
Consultant
Newark, DE

Elizabeth B. Pertzoff
Administrator
Addictions Coalition of Delaware, Inc.
Yorklyn, DE

Connie Pflug
Community of Caring Coordinating
Committee
Maple Park Middle School
Kansas City, MO

Christi Todd
Seventh Grader
Granville Middle School
Granville, OH

Gene V. Todd
Home Economics Consultant
Ohio Department of Education
Columbus, OH

Introduction

In this book one value stands above all others; it is the value of caring. All of us need to care for others and to be cared for throughout our lifetime. Caring is one of the most essential values in building strong families, communities, or nations.

One of the most powerful forms of caring is love. As you read this book, think about love as the overpowering feeling when someone falls in love or when someone does a heroic deed for another person. Think about love as experienced in celebrations; in family gatherings. Think about the love of a parent for a child, a child for a parent; think about the love of a grandparent.

Love can express itself in great works like the civil rights movement. Love of this earth has inspired dedicated people to give their time and energy to clean the environment.

There have been many examples of love in history, literature, and poetry. Stories about love are found in the Bible, the Koran, the great teachings of Buddha, and in the celebrations and rituals of every society. As you read this book, I ask you to think about what love means in your own life story, in your family, and in your relationships with others. When the year is over, I think your understanding will be different from what you think and feel today.

Martin Luther King, Jr., the great civil rights leader, never used violence or hatred. He used love. When he wrote his magnificent "Letter from Birmingham Jail" he said,

Let us all hope that the dark clouds of racial prejudice will soon pass away...and in some not too distant tomorrow the radiant stars of love and brotherhood will shine over our great nation with all their scintillating beauty.

Dr. Robert Coles, a famous psychiatrist at Harvard University, has worked long and hard on behalf of children. During the early days of the civil rights movement, Dr. Coles met six-year-old Ruby, who was being sent to an all-white school that did not want her. Every day for months when Ruby tried to enter the school, she was spit upon and called names, but she never hated. When Dr. Coles asked her why, she said, "I prayed for them and I asked God to love them."

When President John F. Kennedy said, "Ask not what your country can do for you — ask what you can do for your country," he was talking about love of country, of giving of oneself. This was the central idea behind the Peace Corps, which asks young people, and old, to work and sacrifice to help others in places where living conditions are sometimes very hard.

All of these examples of love reveal caring, sacrifice, responsibility, and respect. Most of all, they reveal that the values people live by are not accidental; they are freely chosen from among all other possible choices. In Charles Dickens' *A Tale of Two Cities*, Sidney Carton is a man who has led a selfish, wasted life.

During the French Revolution he discovers a cause greater than himself and deliberately chooses to sacrifice his life to save the life of another. Mother Teresa made a conscious decision to spend her life helping the poor and sick of Calcutta. Gandhi gave up a successful career as a lawyer to win justice for the "untouchables" of India. Once they made their choices, all these people chose to express their love by working for the greater good.

But what about love closer to home? What about the young man at school who says to his girlfriend, "I love you and want to have sex with you"? Isn't he confused about what love really is? He isn't expressing love. Where are the caring, the responsibility, the sacrifices? This sounds more like physical desire than love. Physical desire is a beautiful part of mature love, but alone it is only using someone for your own pleasure.

Then there is the love of marriage and family. In this kind of love, when two people say "I love you," they follow this by saying, "I want to commit myself to you for the rest of my life." They plan their wedding, a special celebration of their love, and vow that they will stay together "for better or for worse, in sickness and in health, until death do us part." In this love there is a commitment to stay together, to create a home and family, to want the best for each other. This isn't love that is here today, gone tomorrow.

The value of love lies in the traditions of the peoples of Asia, Africa, Europe, and the Americas; in the religions and philosophies of Judaism, Christianity, Hinduism, Buddhism, and Islam. So you see, our caring for self and others isn't something new; no, it has been a part of our common heritage and history. It is a part of us.

This book understands that much of our world falls far short of our ideals. It recognizes that there is very little love in the lives of many teenagers; it doesn't turn away from the dark side. After all, no one of us has perfect values, and no one of us has perfect love.

But this book does try to show that whatever your life is like now, you can begin to shape the kind of future you want and form the kind of person you want to be. You can decide to be healthy instead of unhealthy; educated instead of a dropout; sexually responsible instead of being a person who uses or is used by others; drug-free instead of hooked and out of control; caring and loving instead of selfish and cold.

We hope this book will help you sort out your own values and decide which ones you want to live by. As Martin Luther King, Jr., said, "Our goal is to create a beloved community and this will require a qualitative change in our souls as well as a quantitative change in our lives." This book is only a starting place. The rest is up to you.

Eunice Kennedy Shriver

Values

VALUE	WHAT IT MEANS	WHAT YOU MIGHT HEAR
Acceptance	Approving of yourself or others. Believing someone or something is okay.	"She wears really strange clothes, but that's just her way — she's all right!"
Caring	Being concerned about someone or something. Acting in a thoughtful way.	"I started supper because I knew you'd all be tired when you got home."
Citizenship	Being a member of a society or country. Practicing those things that make you a responsible member of society	"I don't litter — I put trash where it belongs because its my responsibility."
Commitment	Giving yourself to someone or something you have confidence in or believe is worthwhile.	"I'm going to get good grades this year, even if I have to miss out on some good times to do it!"
Compassion	Feeling sympathy for others' problems and troubles. Caring for those who are hurting and wanting to help them.	"I'm really sorry your boyfriend broke up with you — I'm here for you when you need me."
Confidence	Believing, trusting, and relying on someone or something. Knowing you can depend on yourself or others.	"Have a good time at the movies, Mom. I'll take good care of the baby."
Cooperation	Working with others to get something done. Having a common purpose and working for it.	"I knew if we all pitched in and helped we'd get done in time for the game!"
Courage	Facing things. Being willing to deal with difficult or painful things.	"I'm not going out with you if you're going to drink."
Courtesy	Being polite and helpful.	"Excuse me! I didn't mean to interrupt. What were you saying?"
Diligence	Working hard and carefully at something; sticking with it. Giving your best effort.	"I'm doing this report and doing it right — even if it takes me the rest of the night!"
Ecology	Interdependence between living things and their environment. Being concerned about the conditions, circumstances, influences, and surroundings that affect life and the growth of all living things.	"This is my dog and I take care of him. If he makes a mess, I clean it up."
Education	Learning those things a society believes to be most important. Learning from experience.	"I'm going to stay in school and learn everything I can!"
Family	A group of two or more people who live together and/or are related by blood or marriage.	"We stick together — we're a family!"

VALUE	WHAT IT MEANS	WHAT YOU MIGHT HEAR
Friendship	Being around people we know well and care about. Supporting and helping friends.	"Of course I'll write to you while you're away at camp."
Future Planning	Planning ahead. Making options for yourself.	"I'm getting some tutoring in math so I'll be prepared for next year's math course!"
Honesty	Telling the truth. Being trustworthy.	"I won't lie to you. I got home half an hour late because we were having so much fun I hated to leave."
Humor	Being able to see the funny side of things.	"I guess I did look pretty silly when I dropped all those books!"
Integrity	Having high ethical and moral standards. Being someone who is trustworthy.	"You can count on her to do what she believes is right!"
Interdependence	Depending on each other. Knowing you need and can count on others and they need and count on you.	"Mom and Dad both work, so on Saturdays we all pitch in to help clean and do the laundry."
Knowledge	Learning, knowing things. Having ideas, information, and skills.	"I'm staying off drugs. I've learned what they can do to people."
Respect	Being considerate. Feeling and acting in a thoughtful way toward others. Honoring others.	"I'm getting along better with my parents since I stopped talking back and started talking with them."
Responsibility	Being accountable and dependable. Knowing right from wrong and making wise choices about how you act.	"I'm calling home to let my folks know that we're going to be late."
Self-Worth (Self-Esteem)	Valuing yourself. Knowing you matter just because you're you.	"I'm okay!"
Self-Control (Self-Discipline)	Being in charge of yourself. Regulating your own emotions, desires, and actions in a healthy way.	"I am so angry at you I feel like slapping you, but I don't have to act that way!"
Self-Respect	Appreciating yourself. Having pride in yourself as a person.	"Maybe I would have passed the quiz if I'd let him give me the answers, but I have to feel good about myself."
Trust	Having faith in someone or something. Believing in the honesty and dependability of someone or something.	"I know you're afraid, but you can talk to the counselor at school. She has never let me down!"
Tolerance	Recognizing and respecting others' beliefs and actions that are different from your own.	"My best friend always celebrates Christmas, and I celebrate Hanukkah."

Contents

UNIT 1
WHY CARE

1. Your Values Are Showing
SECTION
1. VALUES AND WHERE THEY COME FROM 4
2. FINDING YOUR OWN VALUES 12

2. Who Cares?
SECTION
1. CARING IS A VALUE 24
2. CARING SHOWS WHAT YOU VALUE 31

3. Your Decisions Show You Care
SECTION
1. MAKING YOUR OWN DECISIONS 40
2. HOW PEOPLE MAKE DECISIONS 45
3. HOW PEOPLE MAKE WISE DECISIONS 50

UNIT 2
CARING FOR YOURSELF

4. Eating Smart, Staying Healthy
SECTION
1. FOOD CHOICES AND YOU 62
2. EATING FOR FITNESS 69

5. Exercising And Staying Fit
SECTION
1. VALUING EXERCISE 82
2. GETTING ENOUGH EXERCISE 92

6. Caring For Your Mental Health
SECTION
1. SPIRITUALITY AND SELF-WORTH 106
2. EMOTIONS AND SELF-WORTH 115

7. Saying No And Feeling Good About It
SECTION
1. WHY IT'S TOUGH TO SAY *NO* 128
2. DRAWING THE LINE: WHEN TO SAY *NO* 131
3. HOW TO SAY *NO* EFFECTIVELY 136
4. WHEN A SIMPLE *NO* WON'T DO 140

8. Staying Drug Free
SECTION
1. KNOW THE FACTS 150
2. FACE THE FACTS: WHAT DRUGS CAN DO TO YOU 155
3. A FEW MORE FACTS ABOUT DRUGS 163
4. UNDERSTANDING CHOICES 167

9. Recovering: Coming To Terms With Drugs
SECTION
1. OFF THE ESCALATOR: GETTING WELL AGAIN 180
2. WHEN FAMILIES OR FRIENDS ABUSE SUBSTANCES 187

10. Planning For Tomorrow
SECTION
1. FACING UP TO THE FUTURE 202
2. PREPARING FOR THE FUTURE 213

UNIT 3
CARING FOR OTHERS

11. Communicating With Others

SECTION
1. WHAT IS COMMUNICATION 228
2. COMMUNICATING IS MORE THAN WHAT YOU SAY 237

12. Understanding Conflict

SECTION
1. RECOGNIZING CONFLICT 250
2. WHERE DOES CONFLICT COME FROM? 255
3. DEALING WITH CONFLICT 259
4. USING CONFLICT FOR PEACE 263

13. Being A Friend

SECTION
1. WHAT IS FRIENDSHIP? 276
2. BUILDING FRIENDSHIPS 285

14. Understanding Your Sexual Identity

SECTION
1. SEXUAL IDENTITY 298
2. ADOLESCENT SEXUALITY 303
3. SEXUAL ROLES THROUGHOUT LIFE 312

15. Making Sexual Decisions

SECTION
1. SAYING *YES* AND *NO* TO SEX 324
2. THINKING ABOUT SEXUAL OPTIONS 331
3. FACING THE CONSEQUENCES 341

UNIT 4
CARING FOR YOUR FAMILY

16. Understanding Families

SECTION
1. ACCEPTING YOUR FAMILY 356
2. STRENGTHENING YOUR FAMILY 365

17. Help For Troubled Families

SECTION
1. ONE FOR ALL AND ALL FOR ONE 376
2. TOGETHER IN TROUBLE 385

18. Building A Strong Family For Tomorrow

SECTION
1. TRUE LOVE 398
2. PREPARING FOR MARRIAGE 406

UNIT 5
CARING FOR YOUR COMMUNITY

19. Protecting The Natural Environment

SECTION
1. UNDERSTANDING THE NATURAL ENVIRONMENT 422
2. CARING FOR THE NATURAL ENVIRONMENT 427

20. Building A Caring Community

SECTION
1. HABITS OF THE HEART 436
2. WHAT IS GOOD CITIZENSHIP? 439
3. WHY IS CITIZENSHIP IMPORTANT? 445
4. VOLUNTEERING: CARING ENOUGH TO SHARE YOURSELF FREELY 449

UNIT 1
Why Care

CHAPTER 1   **YOUR VALUES ARE SHOWING**
CHAPTER 2   **WHO CARES?**
CHAPTER 3   **YOUR DECISIONS SHOW YOU CARE**

CHAPTER 1

Your Values Are Showing

A TEEN SPEAKS

I know I spend a lot of time in front of the mirror. My mom is always on me about it. "Leslie, all you care about is how you look," she'll say.

That's not true. I care about a lot of things. I just don't talk about them. I'm thinking about different stuff I never thought of before, like what I'm going to do with my life. I think about getting my own car and if I should get a job so I'll have enough money for it. I think about my grades, too.

Then I keep thinking, will somebody blow up the planet before I have my turn? Or if somebody doesn't nuke us out, what about the ozone layer? The whole world is turning into some kind of garbage dump.

But I do care about myself. Sure, I think about how I look. What's wrong with that? At least I can do something about that—most of the time!

It's kind of funny, but from the minute I get up I know how the day is going to go. Sometimes I get up and everything's okay. I feel great, I look great—everything just sort of falls into place. Then other days nothing is together. Everything I put on is wrong.

It doesn't matter if I wore the same thing last week and it was the best, it feels wrong. So I keep putting on different stuff. Then I have to grab something fast because my dad is yelling at me to hurry up or I'll miss my breakfast. You can bet that's the time I'll spill or drop things. The whole day will be like that, from then on. I can tell. That's why I spend so much time in front of the mirror. But it's not like that's the only thing I care about.

Leslie

SECTION 1

VALUES AND WHERE THEY COME FROM

Every day you make choices. Sometimes even a choice as routine as deciding what to wear can become a big issue. If you spend too much time on how you look, your family may complain. But if you don't care about looks, members of your family are just as likely to complain.

Sometimes other people judge you by what they see you doing. They have no idea what you are really thinking. They imagine they know you, but they might be surprised to know who you really are and what you care about most. But when others base their opinion of you on the choices they see you make, they aren't always wrong.

What Are Values?

Every choice you make—no matter how important it is—is based on what you think is important or value. Your **values** are those things that you prize or cherish most. Values are

- **who and what you care about,**
- **what you believe to be true,**
- **who and what you pay attention to,**
- **what you consider most important.**

Your values show in
- **how you act,**
- **what you talk about,**
- **what you stand up or fight for,**
- **what you are willing to sacrifice for.**

Your values are always showing! Your behavior almost always shows what you value. Who you are is connected to what you value.

The teen years are a time when you become more aware of your own values. You begin to make some very important **decisions,** or choices, about who you will be. You also decide what and who are worth caring about. Some decisions are conscious because you take the time to think about them. Many are little choices. You may not think about the values your decisions reflect and the values they are building. Either way, the choices you make show others what you think is important. Your choices show what you value.

A **society** is a group of people living together in a community or country. Each society has rules of behavior that reflect the most important values of the people. Certain values are important in

Throughout history values have been a part of the political process.

almost every country—and have been throughout history. Examples of these values are goodness, beauty, and truth.

The writers of the Declaration of Independence thought that the values of life, liberty, and pursuit of happiness were especially important. The people who signed the Declaration pledged their lives, their fortunes, and their "sacred honor" to support their values. These American revolutionaries were willing to lose all their money and even to die. They considered these values that important. Other values of our democratic society are spelled out in our Constitution and Bill of Rights.

Are any of your values so important to you that you would risk everything you own? Do you have any values for which you would be willing to die?

This book focuses on the value of caring for yourself, your family, and others. It discusses how you can be a trustworthy and responsible member of your family, school, and community. Other values you will think and learn about in this course are listed in the chart on pages VIII-IX.

What do you care about? What are you willing to give or commit yourself to? Why care at all? These are questions you will answer as you read this book. You will also look at how to improve the way you care for yourself, for your family, and for your community. And you will learn how living out your values can make a difference in the world. The difference can be positive or negative, depending on what values you choose to live by.

Older friends can be good role models. What characteristics would make this man a positive role model for teens?

Where Do Values Come From?

Your values come from your family and the world around you through

- people you try to be like. This kind of learning is called **modeling.**
- what you have been taught. This kind of learning is called **direct teaching.**
- what you learn from experience. This kind of learning is called **trial and error.**
- **maturing or growing.**

Modeling — "I want to be just like you!"

Modeling or patterning yourself after someone you admire is one of the most powerful ways of establishing values. You may not realize it, but you model others even when you are not aware of it. Being like someone else gives you a chance to try out his or her values and see if they work for you. You are modeling when you identify with someone and then try to dress the way that person dresses, talk the way that person talks, and act the way that person acts.

But, what happens if you try to model the wrong person? In trying to be like the other person, you may find yourself acting in ways that put you in conflict with your family or friends.

Or, without knowing it, you may model someone for the wrong reasons. For example, you might want to be like a professional athlete because the athlete makes so much money.

It is true that you can develop some important values by trying to be like someone you admire. But you need to ask yourself if the person you admire shows the values of caring, respect, trust, family, and responsibility.

One of the exciting things about being a teenager is discovering who you really are. You can learn a lot from a good model.

Direct Teaching — "Do what I say."

You learn values from people you admire. You also learn values because someone taught you to have them.

Direct teaching began when you were very young. You were probably told to say "please" and "thank you." And you heard hundreds of direct messages like, "We don't hit each other," or "Don't take more than you can eat."

Direct teaching can take place when there is a crisis, or it may happen at times set aside for family talk. Some parents like to bring up important topics at the dinner table for example.

How might this be an example of direct teaching?

Your dad might say "One thing that's really important in life is telling the truth."

Your dad is directly teaching you that he values the truth. He has brought up the subject because he thinks it is too important to leave to chance. He wants you to know where he stands. He wants you to value telling the truth.

Values are taught in different ways, at different times, and in many different places—for example:
- in your family,
- at school,
- at church or synagogue,
- in organizations such as Girl Scouts and Boy Scouts,
- on television.

We have local, national, and international laws and rules that give us guidelines about what is and is not acceptable behavior. These laws are based on and teach us what our society values. For example, laws against speeding and murder both show that our society values life. Our Constitution and Bill of Rights are based on some of the values of a democratic society. These values include respect for individuals, each person's right to express political opinions, and each person's right to make free choices about religious beliefs.

Does direct teaching help you to have the right values? To answer this question, you might ask yourself if you always do what you are told is right. Probably not always! After thinking about values and looking at other people's behavior, you must decide for yourself what you will value most. You have to learn the **consequences** or results of your values, both for yourself and for others. For example, you will not become more caring just because you have been taught that caring is important. You must decide to be caring and then make this value your own by acting in caring ways.

Trial And Error — "I'll figure it out!"

Some families give their children very little guidance or help in knowing the right thing to do. The parents believe that there are no absolute right and wrong ways of living. They think each person must decide what is right or wrong.

Sometimes parents are uncomfortable talking about values. So values are never discussed in the family.

Still other families believe that experience is the best teacher. They think that children learn best those values they figure out on their own.

There are even families who tell their children one thing and do another. In these families children are really on their own. They have to try to figure out what is most important while they are getting mixed messages.

Does trial and error work? Sometimes you can learn a lot by trial and error. But you can also make a lot of mistakes and you can get hurt. Even though you may get tired of having your parents and teachers tell you what to do, some direction can be helpful. When you know what your parents and teachers value, you have something to guide you.

Growing — "Now I see what you mean!"

Do you remember how hard it was to learn to share when you were a little child? Sharing requires that you understand property rights. You have to respect and care for others enough to let them have or use something that is yours. When you were a child you saw things and wanted them for yourself. You didn't think about whether they belonged to somebody or if somebody else might want them. Sharing is a value that you are taught and learn over time.

Teens who respect others aren't hesitant to make friends with people who may be different or have different needs. What kinds of differences are present here? What can be learned from people who are different?

Suppose you have a two-year-old brother. When he sees something he wants, he is likely to take it. He may scream and cry when he is asked to give it back. You can tell your little brother it isn't right to take things that don't belong to him. But you won't make any sense to him. Until he is about 6, he won't really understand property rights.

At about 6 or 7 children begin to understand property rights and respect other people's things. If they take something, you can tell them that it is against the rules to take things that belong to someone else. They will understand following the rules. But it will do you little good to ask them how they would feel if someone stole from them.

To understand someone else's feelings, children need to put themselves in another person's place. This is called **empathy.** The ability to do this does not develop until much later, usually in the preteen or teenage years.

Cultural Values: The Individual or the Group?

"To each his own." "Do your own thing." These expressions reflect the value America places on the individual. Other cultures or countries put the group first.

Individualism, the belief that each person needs to look out for individual or self interests and well-being, is a strong cultural value in the United States. We're encouraged to develop our own unique talents and our own distinct identities.

In Japan, people learn to serve society first, not themselves. The family, not the individual, is the basic social unit. Public status depends on the groups—school, or profession—people belong to.

Decisions are made by groups, not individuals. Personal ideas and privacy are given less emphasis than in the West. Interestingly, there is no exact translation in Japanese for "individual privacy."

Growing Brings New Insights

As a teenager, you are capable of valuing in ways that you were not as a child. You understand laws and rules. You are beginning to be able to put yourself in someone else's place and consider how he or she might feel. You have the capacity to care about and respect others in ways that you could not before.

As a teenager, you are also ready to start evaluating the rules that you have taken for granted all your life. This is a time in your life when you often experience deep **conflict** or disagreement over what is right and wrong. You have the values you were taught at home, at school, in your church or synagogue, and in your community. You have the values you see in other people, especially in your friends, to draw upon. But sometimes the values you see are entirely different from those you've been taught. Sometimes, too, they are in conflict.

Growing Means New Challenges

As you try out the values of other people, you may begin to question rules and relationships. You may find yourself acting in ways that are unfamiliar and sometimes puzzling to your family and friends and even to yourself! Acting differently can cause others to have strong reactions to your behavior.

> Suppose you talk back to your dad. He reacts with, "So when did you get to be the wise guy?" and grounds you. His reaction seems harsh and unfair.

You may find that you, too, have strong reactions that are out of proportion to what has happened. These reactions are often as confusing to you as they are to other people. They seem to come from nowhere. Reactions that are inappropriate and out of proportion may leave you feeling foolish. And, they can get you into trouble.

> Suppose your mother sees you stretched out on the sofa reading a book and asks what you're doing. It would be reasonable to say, "Just reading." Instead, you ask why she always has to know exactly what you are doing every minute, and then you stomp out of the room in a huff.

Remember, you are still growing. Sometimes growth is painful. It takes time. But you don't have to let growth just happen. You can learn about how you are changing and what to expect. Learning about these things will help you to

- understand your own behavior and why people react to you as they do,
- bring your behavior more into line with your values,
- set realistic goals for yourself,
- make appropriate choices about who you are now and who you will be in the future.

SECTION REVIEW
STOP AND REFLECT

1. Describe someone you admire and want to be like.

2. Tell how the person you admire influences your values.

3. Make a list of your own values.

4. Which values on your list are values you didn't have three or four years ago? Write "new" next to these values.

5. For which values on your list would you be willing to risk your money and/or your life? Make a star next to these values. Choose one of the starred values and tell why it is so important to you.

6. Tell why you would be willing to risk so much for the other values you have starred.

7. Where did you get your values? Go back over your list and see if you can determine where each value came from.

SECTION 2

FINDING YOUR OWN VALUES

Your values come from experiences you have with family, with others, and with things in the world as you grow and mature. From the moment you were born, things began happening to you. You began trying to figure those things out. Parents, grandparents, neighbors, and people you met told you what to make of your experiences. They also told you what value you should attach to these experiences. Sometimes they told you directly. More often, however, they told you indirectly, through the ways they acted or the choices they made.

When Is A Value Really Your Own?

Patty really throws a great party, especially when her folks are gone. The other night we were partying and it was great. We were dancing and lying around listening to the music. Then somebody said, "Let's break out the booze!" I was real confused. I don't think kids my age ought to be drinking. But what was I supposed to do? These are my friends.

No matter who your parents are, or how you have been taught, you are still responsible for your own behavior. As you try to make sense of things, you have to think about your own experiences. You also have to think about what you have been taught. Your own ideas about life and what is important become your own values. This happens as you deliberately choose to act in ways that show your values.

Choosing — "Is this right?"

Choosing what you will value is the first step in valuing. To make a value truly your own, to make it personal rather than simply a family value or a community value, you have to know

Responsible teens stay in control. They don't leave their futures to chance.

enough about it to prize and cherish it. You can learn about values by testing them. This can be very risky business.

You need to make wise choices about how you will test values. Part of your growing self-awareness is a feeling of **invincibility.** This is the feeling you have when you believe "It can't happen to me." This, in turn, can lead you to use poor judgment in making choices that may involve risks.

Maybe you have thought "It's okay to smoke pot." You feel that you are invincible, that getting hooked can't happen to you. You can't help the fact that as a teenager you feel invincible. But you can act wisely. You can realize that many teenagers who have said, "I've got it under control," became addicted. They, too, felt invincible.

All teens share feelings of invincibility. Knowing this may help you to act in ways that reflect values of responsibility, self-respect, and caring.

Telling — "This is how I see it!"

Telling others what you value is a second step in claiming values for yourself. Telling other people lets them know what to expect from you and helps you to take a stand.

It is not always easy to tell people what you value. People may not understand what you say, or they may not be interested. For example, when your friends are eating fries and drinking sodas, they may not want to hear about the dangers of poor food choices. They might not want to see the connection between responsible eating and caring for themselves. On the other hand your friends might respect you for standing up for your values and for caring about them enough to take a healthy risk. You have to choose when to take a public stand. But you can act in ways that show your values whether or not you tell anyone.

Acting — "Just watch me!"

Your behavior shows what you value. For example, you may believe it is wrong to cheat on tests. But if you're taking a very difficult test and your friend has all the answers, it may be hard not to cheat. The clearer you are about what you value, the more certain you are about what your actions ought to be.

Regardless of what you say, it's your actions that show your values. What values are present here?

When trust is a basic value, you know that you must be trustworthy, even when faced with a hard test. And if you have told your friend that you do not think people should cheat, it will be easier to resist the temptation.

Values guide behavior. Values help you to choose appropriate actions from a long list of alternatives. Values help you to consider the consequences or results of your actions. For example, if you care about yourself and others you will refuse to ride with a friend who has been drinking. And you'll try to keep him or her from driving. Your responsible actions show your values.

Repeating — "Practice makes perfect!"

Repeatedly choosing to act in ways that show your values is still another step in valuing. The more you practice your values, the easier it is to practice them. However, you may sometimes encounter a conflict in acting on your values. You might hear several voices within your head. One says, "Do this." Another says, "No! Whatever you do, don't do that!"

For example, you may have been thinking about having sex with your girlfriend. You have always been taught

CHAPTER 1 YOUR VALUES ARE SHOWING

Money is a positive value when it's used to provide for your family. It's a negative value when it becomes all-important and is used selfishly or at the expense of others.

that you should wait until you are married to have sex. You agree, even though your friends make fun of you for being old-fashioned. Something inside tells you that you really should wait until you are older. At the same time, there is another side of you that urges you to go ahead. A third voice says, "Do it, but tell everyone you didn't! Then it will look like you are waiting." But another voice says, "Don't do it, but tell everyone you did. All of your friends who are putting pressure on you to do it will get off your back."

Inner conflict makes choosing difficult. But when you practice choosing and acting to show your values, you establish patterns of behavior. That is, you become used to behaving in the same way. Patterns of behavior make choosing easier. This is why it is so important to make appropriate choices. You can practice the wrong things!

Acting in ways that show positive values will help you to become a moral and ethical person. A moral and ethical person knows the difference between right and wrong. A moral and ethical person acts in ways that demonstrate the right choices. Positive values reflect who you really are as a person.

Good And Bad Values

Not all values are equal. Positive values are those that help you to experience
- beauty,
- truth,
- friendship,
- justice,
- freedom,
- health,
- self-expression.

Recognizing Positive and Negative Values

Values that build trust are positive. Values that destroy trust and lead to suspicion are negative.

Values that contribute to the quality of life (mental, physical, emotional, and spiritual health) are positive. Values that destroy the quality of life are negative.

Values that help you to reach your potential are positive. Values that destroy your potential are negative.

Values that increase your knowledge and capacity to learn are positive. Values that keep you from knowing and learning are negative.

Caring, family, trust, respect, and responsibility are positive values.

Negative values keep you from having positive values. They destroy your character and can harm your health and ruin your future. Caring only for yourself is a negative value. Money is a negative value when it becomes more important than truthfulness or family or respect.

Whether a value is positive or negative may depend upon how it contributes to some purpose. For example, health is a value because life is an even greater value. By practicing good health habits you contribute to your high quality of life. In this sense, valuing health is positive. But suppose you were on a life raft in the middle of the ocean with another person. You might drink all the fresh water in order to maintain your health. Would health still be a positive value? What would this action say about your value of respect for others?

Generally, values that build your **self-worth,** the importance you place on yourself, and the worth of other people are positive. Values that harm or destroy people are negative values even though they may contribute to your pleasure. You should always consider the worth of your values in relation to other people and the effect your values have on them.

Friendship is another value that is usually seen as positive. Having friends helps you feel self-worth. But a friendship might lead you into activities that threaten your health and life or the health and life of others.

Can you value your friends even when you do not accept all of their values? You probably can't unless your friends respect your values too. For example, you may keep seeing a friend who uses crack. But if your friend urges you to use crack or wants to borrow money to buy crack, she is placing her negative values before your values. But you may not be able to continue the friendship if it requires you to compromise what you believe.

Friendship can be a negative or positive value. Can you give an example that describes when this friendship might be positive? And when it might be negative?

Values Can Change

You are not stuck with the values you now hold. The very fact that your teen years are a time of testing values should suggest that it is possible to change them. Some of your values will naturally change as you grow older and mature. Others will not change unless you take action.

Should you change your values? Which values should you want to change? To answer these questions you should ask yourself two others:
- What kind of person do I want to be?
- Will my present values allow me to be the kind of person I want to be?

If values need to be changed, you can change them in much the same way you learn them. You must take action. One way to do this is to set goals.

Setting Goals — "I really want to do this."

Suppose that you want to be an adult who is generous and caring. You want to be the kind of person that other people will seek out for a friend. Yet you know you are often selfish and uncaring to your family—particularly your sister. You want to be a caring person, but you see that you are practicing just the opposite behavior. So you decide to start by being more caring at home. You set a **goal,** something you want to do or be, for yourself. You are more likely to succeed if your goal is

- something you can put into words for yourself,
- appropriate,
- possible,
- realistic,
- visible,
- desired,
- serious,
- positive.

Sometimes you set goals for yourself quietly. At other times you may want to make your goal public. When you set a goal, carry out your plan. When you achieve it, you will feel satisfaction and you'll be ready for a new goal.

GOALS IMPROVE YOUR SELF-CONCEPT

GOALS ALLOW YOU TO DEVELOP DREAMS

GOALS PUT YOU IN CHARGE OF YOUR LIFE

GOALS ARE SOMETHING TO GET EXCITED ABOUT

GOALS CAN HELP YOU IN PLANNING FOR THE FUTURE

Setting goals and accomplishing them is a lot like climbing a mountain — one step at a time.

The Best Values

The fact that values exist in relation to each other does not mean that anything goes. Some values are better than others.

We have talked about how important it is to be clear about your values. This does not mean that because you are clear about what you value you are making the right choice.

Also, a value may be important even though many people do not accept or follow it. Some values are worthwhile whether or not they are recognized and practiced.

What if you don't believe that caring, family, trust, respect, and responsibility are important values? You do not have to choose anyone else's values or agree with them. You are free to choose. But these values are good whether you choose them or not. They form the basis of our democratic society. And they have been important values for centuries. Values are not like money, for example. You can make money and use it up. The more you spend the less you have. Permanent values grow with use.

When you choose to be a caring person, to value family, to be trustworthy, to have respect for yourself and others, to be responsible, you are choosing to be a better person. You are choosing to make the world around you a better place for everyone. As you grow, you will develop additional values. Some of these values will be held by most people you know. Others, such as religious values, will be shared by particular groups of people.

As you read and study this book, you will see that it is about relationships between people. It is about becoming more than you are, imagining possibilities beyond what you can see right now. It is about growing up caring!

SECTION REVIEW
STOP AND REFLECT

1. What is the difference between positive values and negative values? Give two examples of each. Explain why your examples are positive or negative.

2. Describe the kind of person you want to be. This is a goal.

3. Look back at the list of values you made on page 11. How will your values help you reach your goal to become the person you have described?

4. Tell how you can develop new values or change old values to become the kind of person you want to be.

5. Tell what positive values the goal you set (in question #2 above) will build.

6. Describe how achieving your goal will be good for you and for others.

CHAPTER 1 REVIEW

Putting Your Values To Work

STRENGTHENING YOUR VALUES

Reread *A Teen Speaks* **on page 3. Then answer the following questions.**

1. What does Leslie's mom think he values?
2. What does Leslie really value?
3. Why do you think Leslie's mom doesn't know what he really values?
4. How could Leslie help his mom understand what he values most?
5. Describe a time when you or someone you know misjudged another person's values.
6. When might valuing how you look be positive? Negative?
7. If you could give Leslie some advice and knew he'd listen, what would you tell him? What would you tell his mom?
8. If you were Leslie, what would you say about your values?
9. Do you think that other people know what you value most? Explain your answer.
10. What could you do to make other people more aware of what you value?

INTERPRETING KNOWLEDGE

1. Write a child's story about the values that are most important for children. Then, choose one of the following:
   - Make your story into a book. Illustrate it with drawings or pictures cut from magazines. Or get a younger friend to help you with the illustrations.
   - Make an audiotape or a videotape for children instead of a book.
   - Write music for your story and turn it into a ballad or folksong for children.

SHARPENING YOUR THINKING SKILLS

1. What do you think is the most important way people get their values? Tell why you think so.
2. Give an example of a value that you've been taught. Tell who taught the value to you.
3. Give an example of a value you have learned by trial and error. Tell about the experience or experiences that helped you learn it.
4. Give an example of a value you have picked up from someone you admire a lot. Tell who the person is. Tell why you admire that person.
5. Give an example of how someone can gain a value by growing and maturing.
6. Why do you think that the people who signed the Declaration of Independence were willing to give their lives for the values in which they believed?
7. In your opinion, what are the most important values for people in our country to have today? Tell why you think so.

APPLYING KNOWLEDGE

1. Think of some ways you can share the story you have written about important values for children.
2. Interview an elderly person in your family, extended family, or neighborhood. Ask that person to answer these questions:
   - What are the values that have guided you in your life?
   - How did you learn those values?
   - What would you do differently if you had life to live over?

CHAPTER 1 REVIEW

Putting Your Values To Work

PRACTICING DECISION-MAKING SKILLS

Read about each situation. Then answer the questions.

Situation A: Refer to Patty's situation on page 12.

1. What did Patty's friend have to decide?
2. Why do you think the decision was a problem for her?
3. What values do you think Patty's friend has that make this a problem?
4. If Patty's friend were to decide that she doesn't like being pressured and doesn't want to drink, how should she get this across to the others? What should she say? What should she do?
5. What values should be reflected in the friend's choice?
6. Suppose someone says to Patty's friend "grow up!" Everybody drinks a little. You aren't going to become an alcoholic just because you let go and have a little fun." If you were Patty's friend, would this make any difference? How could this make it harder for her to say NO?

Situation B: Boyd's little brother, Hank, who is a fourth grader, wants to go everywhere with him. Sometimes it's okay, but at others it's really a bother. Sometimes Boyd's friends tease Hank, but most of the time they don't seem to mind if he tags along. One night the guys are all going out for burgers and then to a ball game. Boyd would really like to be alone with his friends, but he knows how much Hank wants to go along.

7. What are Boyd's possible choices?
8. What positive results could each choice have?
9. What negative results could each choice have?
10. What are the values reflected by each choice?
11. If you were Boyd, what would you do? What would you tell Hank?
12. Suppose Boyd's friends say, "Hey Boyd, are you going to let that dumb little brother of yours come along? Do you think Boyd's action would be any different? Explain your answer. What values are reflected in your answer?
13. Suppose Boyd's friends say, "Hey Boyd, why don't you bring your little brother along?" Do you think Boyd's action would be any different? Explain your answer. What values are reflected in your answer?

Situation C: Marietta has been going out with an older guy for over a month. Lately he has been asking her if she is ready to have sex with him. He has told her that he loves her more than anybody he's ever been with and that he really needs her. Marietta has said she isn't ready, but she wonders if she will hurt the relationship if she keeps saying NO. Besides, she doesn't think anything bad will happen to her—she can't imagine getting pregnant. Suppose Marietta has confided to you as a friend.

14. What needs to be done?
15. What are Marietta's choices?
16. List the positive results of each choice. What negative results could each choice have?
17. What are the values reflected in each choice?

CHAPTER 2

Who Cares?

A TEEN SPEAKS

Trudy had been looking everywhere for this one kind of jacket. So when I saw one at the mall, I called her right away. Then she told me she'd had a fight with Norma. I was glad because Norma was always putting me down.

I told Trudy I thought she was being used. Norma was always getting Trudy to do stuff like cheating on homework.

After I got home, I got a call from Norma. She wanted to know if I had talked to Trudy and what Trudy had said. I told her what she said was in confidence—but I did say the gist was that Trudy was upset about what happened between them. Norma started saying stuff about Trudy. So I told Norma I thought Trudy was insecure and she shouldn't let herself be used. But I said Trudy already knew what I thought because I'd told her myself—and I had.

Right after I hung up, the phone rang again. It was Trudy, and she said, "You've had it!" Then Norma jumped in and said, "You think you're so hot."

They kept calling me all afternoon. They'd call me and I'd hang up and they'd call back again, so I finally unplugged the phone.

It really hurt me. Trudy was somebody I trusted. I really cared about her. I thought she cared about me, too.

I realized that Trudy had only been friends with me because Norma and some of the other girls wouldn't accept her. She was just using me until she could get in with them.

Metty

SECTION 1

CARING IS A VALUE

Why do some people act in uncaring ways? And, when you give yourself to others, why do you sometimes end up being hurt by them? These are tough questions that don't have any easy answers. But one thing is certain: it is impossible to care without the risk of being hurt. Even so, caring may be the most basic human value.

What Is Caring?

Life on earth could not exist without care. Human life could not exist if we did not have the ability to care for others, especially in times of need. Animals care for their young and for each other. And, in a way, the earth has cared for life by providing an environment that supports life. The terms *Mother Earth* and *Mother Nature* suggest that all human, animal, and plant life depends on the earth. Our ability to exist in the future will depend upon our understanding of the caring relationship between all living things and the earth as home for all living things.

Caring is necessary for life. The only reason you are alive today is because someone has cared for you. Before you were born, you needed your mother to care for you by taking care of her own body. In the first weeks and months after your birth, you needed somebody to feed you. You needed someone to change your diapers, to see that you were dressed, and to reassure you. Even at this young age you needed someone to talk to you. You were completely dependent on others for everything. A baby left alone will not survive.

Everybody needs to be cared for. We all need food, shelter, and clothing. We need to be cared for when we are sick. When these physical needs are met, we know that we are cared for. But we need care in other ways, too. Sometimes we need someone to understand us and say, "It's going to be okay." At other times, we need to know that we matter to someone. We know we are cared for

Showing that you care often results in others showing care for you.

when we feel that we matter to someone. We know we are cared for when we are valued by others.

> Timmy was walking home from school when three guys jumped him. They demanded his money. He gave them everything he had. Still, they slapped him around, knocking him to the ground. As they were leaving, one of the guys kicked him in the face. "Hey, you could kill him that way," yelled the others.
> "So who cares?" the guy yelled back. Several kids who walked home that way saw Timmy lying on the sidewalk, but they didn't go up to him. They were afraid to get involved.
> Mary came along later. At first she didn't recognize Timmy. As she got closer, she knew he was from school. She could see he needed help. She put her coat over him and ran to the telephone to call 911. Then she waited with Timmy until an ambulance came.

Mary did not have to know or like Timmy to care. Sometimes we act in a caring way just because we know it is right. That's the way it is with caring.

Caring is the way we show the value that we see in ourselves, in others, in life,

Caring Comes in Many Forms

Think for a minute about how people you know show they care. Do they hug each other or pat one another on the back? Do parents suggest how things should be done or set a strict curfew? These are all ways people show they care.

Caring is a value practiced and demonstrated among people around the world. But caring is shown in very different ways, depending on the culture and its beliefs.

In many European and Latin American countries, caring between friends may be openly demonstrated through hugging and holding hands. It makes no difference if the friends are male or female.

In American culture people compliment one another to show they care. Many Asian cultures believe that Americans compliment one another too much. They doubt the sincerity of our expressions.

Find out how your classmates show they care. You will probably find that caring comes in many forms.

and in the world. It is a basic human experience. Caring moves us to
- love,
- have friends,
- plan for the future,
- do good for others,
- be kind to animals,
- marry,
- have children,
- respect the environment.

It is caring that causes us to work for the security, happiness, and protection of our family and others. It is caring that causes us to be concerned about the future of the earth. Caring forms the foundation for many of our values.

Caring involves feelings and desires as well as thoughts and actions. Sometimes people act in a caring way even if they don't feel like it. For example, if their infant daughter cries at 3 o'clock in the morning, parents will get up and care for her, even though they're sleepy and don't feel like getting out of bed.

Learning To Care

Your feelings of being worthwhile come from early experiences of being cared for. Because you are loved you are able to love others. If others love you, you know that you must be worth loving.

> When I have a boyfriend I think I matter to somebody. That's how I got started having sex. It makes me feel loved. But then it's over and you're all alone again.

If you could look inside this girl, you might find someone who is afraid she isn't worth loving.

Caring For Yourself — "I'm worth it!"

You can't care about others until you care about yourself. In fact, caring even comes before self-respect. That is, self-respect comes from a caring attitude toward yourself. But, how do you respect yourself if you aren't sure who you really are?

> When the family gets together for holidays with all the relatives, I always have to go through hearing everybody say that I am just like my mother. "Just look at Celia," they tell her. "She is you all over again."
> And my mom will say, "Well, maybe she does favor me,

Caring people show care even when it's inconvenient.

> but she has her dad's eyes and Uncle Bob's hair—just look at the way it curls."
> Sometimes I'd just like to start screaming at all of them, "I'm me—I'm not my mother. I'm not my dad—I have my own eyes. I'm not Uncle Bob— this is my curly hair!"

Your sense of self—who you really are—is the sum total of your ideas and attitudes about who and what you are. For most teens, **self-identity,** your sense of who you are and what you want to become, seems to be changing all the time. That is, **adolescence,** or the teen years, is a time of finding out who you are. You know that a big part of you comes from heredity. You may have eyes like your dad's or hair like your Uncle Bob's. Another part comes from your environment. You have all your experiences with family and friends. These make up a big part of who you are, too. But your sense of self also requires reflection on your values—who you love, what you believe, what you know, what you hope for. When you ask, "Who am I?" you are also asking, "Who will I be?"

Accepting Yourself

To know who you are and who you will be, you can ask yourself questions such as these:
- How do others see me?
- What are my strengths, and how can I use them?
- What do I like about myself?
- What are my limits?
- What am I ashamed to admit, even to myself?
- What and whom do I care about?
- What is worth my best effort?
- What am I willing to give up to help someone else?

These are tough questions. Once you dare to admit that you have such questions, you can begin to look at them honestly. When you care enough to listen to yourself, you are beginning to accept yourself. You realize that strengths and talents are a gift. You can be glad for them without feeling superior to those who do not have the same gifts.

Accepting Your Limits

When you accept yourself, you can also look at your limits without a lot of self-blame. For example, if you are short, and your parents are short, there is not a lot you can do about it. Being short isn't your fault. Being short probably means that you will not be a professional basketball player. In the same way, being tall probably rules out being a jockey.

> Okay, my mother is an alcoholic. It's not my fault. I can still become the person I want to become.

Take time to get to know yourself. Think about the person you want to become.

"Okay, I'm short. So who says it's bad to be short?"

It may be easier to look at strengths, but you can learn to look at your limits, too. Everybody has done things or thought about things they regret. When you let yourself begin to think about the things you are afraid to admit, you are ready to tackle some new questions:
- What can I learn from this?
- Why does this keep bothering me?
- Do I need someone to help me work through it?

Sometimes it is helpful to look for care from others. Some good places to look for care outside your family are
- a trusted adult friend,
- a teacher,
- a school counselor,
- a rabbi, priest, or minister,
- a social worker,
- a support group.

When you accept yourself — with all your strengths and limits — you're more likely to show care to others.

When you ask what and whom you care about, you begin to learn about your values. When you listen to yourself and really think about what you hope to be, you are going to respect yourself even more.

A support group is one place to look for care from others. Can you name other places one might find support and care?

Caring For Others — "You're worth it, too!"

You learn caring not only from being cared for but also from caring for others. At first, you may not feel like caring. Still, you can act in a caring way because it is the right thing to do. Then as you begin to act in a caring way, you will find that caring feels good. Caring grows when you practice it. When you fail to care, caring disappears.

Many people say that teenagers are self-centered. It is true that your teen years are a time of focusing on yourself and deciding who you are. But often, teenagers seem to be selfish simply because they don't have opportunities to be caring. Many teenagers find ways to care through programs such as Special Olympics, or scouting.

SECTION REVIEW
STOP AND REFLECT

1. Name someone who cares for you.
2. How do those people show they care?
3. Whom do you care about?
4. How do you show caring?
5. Reread *A Teen Speaks* on page 23. How might all three girls have been more caring?
6. Answer the questions about yourself on page 28. You don't have to share the answer to "What am I ashamed to admit, even to myself?" with anyone.

SECTION 2

CARING SHOWS WHAT YOU VALUE

When you let yourself care about others, you may be in for some pain. Caring for and respecting others sometimes requires you to give up personal wishes and share somebody else's burdens. To be a caring person, you have to be willing to become involved with other people. You have to share. You have to be willing to take the time and make the effort to be caring.

Some Don't Care

Selfishness is always a threat to caring. Overcoming selfishness may be one of the hardest challenges you will ever face. Many people find it impossible. These people spend their lives looking out for number one. They always think of

- **their lives.** "You have no right to tell me what to do. It's my life and I'll live it as I please."
- **their bodies.** "If I want to eat chocolate and fries all day, I'll eat chocolate and fries. It's my body."
- **their things.** "I bought this for myself. If you want to try it, you can go get your own."
- **themselves.** "I didn't have time to study and I have to pass the exam, so I'm copying from your paper."

Selfishness — "Let's take turns. I'll go first. You'll be next!"

Selfishness is the opposite of caring. People who are selfish are concerned only with themselves. Selfishness comes in three major forms: self-indulgence, self-protection, and self-righteousness. Selfishness causes people to lie, break promises, break the law, and act in uncaring ways.

Self-Indulgence

Self-indulgent people think that what they want is more important than anybody else or anything else. Self-indulgence is the most common kind of selfishness. It is easy to spot uncaring actions that come from this form of selfishness. Here are two examples:
- "I know I promised to go to the game with you. But I've changed my mind. I'd rather go with somebody else," Sarah told her boyfriend.
- "So you say it's against the law to do crack. Maybe I want to do crack anyway. I've just got to do my own thing," Jim told his friend.

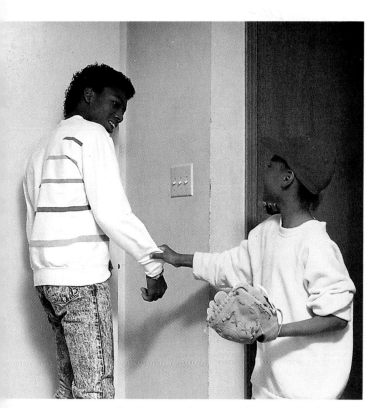

"Look, I know I promised, but something's come up." This teen is being selfish.

Self-Protection

Self-protecting people sometimes lie, try to cover-up or deceive, and blame others, all from a fear of being caught. They want to do as they please without having to accept the consequences. They don't want to be responsible for their own actions. They are interested only in protecting themselves. For example:
- "I can't go to the ball game with you because my folks have grounded me," Sarah lied. She went with somebody else.
- "I don't do crack. I'm holding this crack pipe because these guys gave it to me to keep for them. I've got to do what they say," Jim told his friend. He knew his friend did not approve of using drugs.

Self-Righteousness

Self-righteous people want laws and rules to apply to other people, but not to themselves. They decide that they are above the law and the rules of common courtesy. If someone offends them or breaks the law in a way that hurts them, they are angry and upset. For example:
- "So what if I did go to the ball game with someone else. That doesn't give you the right to go asking out other people when we're supposed to be going steady!"
- "A lot of people can't do crack without getting hooked. I can handle it," Jim explains.

Sometimes people act selfishly because of faulty reasoning. They simply aren't thinking straight.

CHAPTER 2 WHO CARES?

What might this teen be learning from the small child? What might the child be learning from the teen? What values are portrayed here?

Twyla has been known to steal other people's money. But, when Bobby steals her lunch money one day, she tells the teacher. She demands that he give it back. Her reasoning goes like this: "People aren't supposed to steal my lunch money!"

Sometimes people act in uncaring ways because they overestimate the cost of doing right. They imagine that it takes an extraordinary person to be caring.

Jesse's mother asked him if he would be willing to stay with their elderly neighbor after school one afternoon a week. Jesse likes Mrs. Carver a lot. He knows she needs someone to be with her all the time. He knows too, that it is hard for her to get care between 4:00 P.M. when her nurse leaves and 6:30 P.M. when her daughter gets home from work. He feels it is the right thing to do, but he is afraid he can't do it.

Jesse thinks, "If I say with Mrs. Carver, I'll miss basketball practice. I probably won't be able to keep up, and I'll lose a chance to be on the team. I won't get to hang out with the rest of the guys. They'll think I'm a wimp for staying with an old lady."

Jesse imagines that taking care of Mrs. Carver one afternoon a week would require an exceptional sacrifice. In reality, it may not require as much as he imagines. There may be a way he can help out without having to give up activities that are important to him. And he isn't giving his friends much credit for caring. Because he thinks it would require an unusual effort to be caring, he isn't making a thoughtful decision. In the next chapter you will learn how to make difficult decisions.

Other people are not caring because they underestimate the cost of failing to do right. These people aren't thinking about the long-term results of their actions.

Cheating, in any form, is bad for your self-esteem.

Belinda takes a great deal of pride in her grades. Even so, she wishes she didn't have to study so hard.

One day Belinda was dumping all the junk out of the bottom of her purse and accidentally threw away her pen. As she was digging through the wastebasket to get her pen, she found a copy of Mr. Kerr's history quiz. After that she found a way to raid the trash can before every test. Belinda was pleased with herself because she could make good grades without spending so much time studying.

At the end of the semester, Belinda's history grade went up. Mr. Kerr even sent a letter home to her parents. He suggested that Belinda be moved to an advanced history class next semester. Belinda's parents were thrilled and agreed. Belinda is really upset. She has gotten out of the habit of studying for history.

Belinda wanted both good grades and time free from studying. The caring person must often sacrifice short-term benefits in order to achieve long-term advantages, knowing history or self-respect. It may seem easier to lie, deceive, hide, or disregard the rights of others than to confront a problem head on and accept the cost of being a caring, honest person.

Why Care?

Life is better when you care about others. You have more self-respect. You learn about relationships as you show care to others. You are able to develop compassion.

Compassion is a tough emotion. When you have **compassion** for another person, you are able to feel with that person. Compassion does not come easily. Only those who practice caring and accept the consequences of caring can be compassionate. To become compassionate you must draw upon your own feelings of joy, sorrow, pride, hope, and pain.

We have these feelings as the result of terrible, tragic, or wonderful things that happen to us. When someone we love loves us back, when we are injured, when we can't have something we desperately want, when a close friend dies, when we work hard for something and get it, we feel deeply. These feelings help us to develop compassion. Through remembering them, we share what someone else is going through. The compassionate person says, "I share your joy. I will be with you in your pain."

It would be a better world if everyone showed compassion toward
- others,
- all living creatures,
- the earth itself.

We come to care only through having been cared for and through caring for others. We have to be directly, personally involved in caring. It isn't always easy to care.

Only those who practice caring can be compassionate.

SECTION REVIEW
STOP AND REFLECT

1. In what ways do you sometimes act selfishly?
2. Which form of selfishness best describes each item on your list?
3. Give an example that shows compassion.
4. In your journal write a list of ways in which you care for yourself and for others.
5. List at least five other things you could start doing to show that you care.

Putting Your Values To Work

STRENGTHENING YOUR VALUES

Reread *A Teen Speaks* **on page 23. Then answer the following questions.**

1. How did Metty get hurt? Why?

2. Why did Norma and Trudy "team up" the way they did? What were they trying to accomplish?

3. How does Metty view her phone conversations? Does she think she did anything wrong? Why?

4. How do Trudy and Norma view the phone conversations? Do they think they did anything wrong?

5. What could Metty have learned about herself from this situation?

6. What might Metty have learned about caring from this situation?

7. Suppose you know all three girls. What would you tell or ask Metty if she told you her side of the story? What would you tell or ask Trudy and Norma if they told you their side of the story?

INTERPRETING KNOWLEDGE

1. Using only colors (no pictures or drawings), create a color collage of how you feel when you care about someone else and that person does something to show he or she cares about you.

2. Think of a song about the pain that can come with caring. Write down the line or lines that you think best capture that feeling.

SHARPENING YOUR THINKING SKILLS

1. List five ways you feel cared for.

2. List five ways you show you care about somebody else.

3. Briefly describe a specific incident that taught you (or someone close to you) that you were cared for and that you could care for others.

4. Do you think you have to care for yourself to care about others? Explain your answer.

5. Is it necessary to know and like someone before you can care about him or her?

6. Why do some adults feel that teens are self-centered? How might teens move away from such a strong focus on self?

7. Describe a time when you or someone close to you was selfish. Why do you think the person acted selfishly? Was the selfish action an example of self-indulgence, self-protection, or self-righteousness? What were the effects of the selfishness on the person who acted selfishly?

APPLYING KNOWLEDGE

1. The next time you catch yourself (or someone else catches you) being selfish, ask yourself why you are acting that way. Describe the situation in your journal.

2. Choose someone younger than you. Every day for a week, do something with or for that person that you think will teach him or her how to care.

CHAPTER 2 REVIEW

Putting Your Values To Work

PRACTICING DECISION-MAKING SKILLS

Read about each situation. Then answer the questions.

Situation A: Neighbors of yours have an 8-year-old mentally retarded daughter named Angela. Whenever Angela sees you, she comes running up to give you a big hug. You like feeling that she is so happy to see you. Her parents, though, have shared with you how frightened they are that she hugs everyone—even strangers. They are afraid that one day she could be hurt by being so trusting and giving.

You are asked to babysit Angela one afternoon and evening when her parents go out. At first it's fun to be with Angela—she really looks up to you. But you get tired of all the constant attention. Then a friend calls and wants to stop by. You've had a crush on this friend for a while now, and you're really excited. When the friend arrives, Angela hugs both of you and says "I love you." Your friend thinks this is funny and starts mocking Angela.

1. List three actions you might take. For each of these actions, identify the person (yourself, Angela, or your friend) who would benefit from the most caring.
2. Briefly describe what might happen after each of these actions.
3. How would you feel after each of these actions? How would Angela feel? How would your friend feel?
4. What will your parents think when they find out?

Situation B: Same as Situation A. Suppose that your friend, in an effort to be funny, imitates Angela's slow speech.

5. Why might your friend do this?
6. Why is this behavior wrong?
7. Once again, list three possible actions. This time, make one choice that clearly puts your caring for Angela first and another choice that puts caring for your friend first. Make the third choice the best one you can think of. The new actions may be similar to your first three (question #2) or they may be totally different.
8. Briefly describe what might happen after each of these actions.
9. How would you feel after each of these actions? How would Angela feel? How would your friend feel?
10. What values are present in each action?
11. What values are lacking in each action?

Situation C: Same as Situation A and B. As the teasing continues, Angela's parents come home unexpectedly. They angrily scream at your friend to go home. Then they tell you that you "are worse off than Angela."

12. What do they mean?
13. How is your looking for attention the same as Angela's? How is it different?
14. What values did Angela's parents assume you had? Why would they make this assumption?
15. Do you think it will be possible to "fix" your relationship with Angela and her parents? How might you go about this?
16. How could you have been more caring? How could you help your friend to be more caring in the future?

CHAPTER 3

Your Decisions Show You Care

A TEEN SPEAKS

I like going out with JoNell. She is one great person. She doesn't look that bad either! You should be so lucky! But I can't stand taking her out to eat. It's like she makes a federal case over what to order. The woman can't make up her mind. The waitress gives us the menu, and I say, "Sure, I'll have the deluxe bacon burger and fries." It's no big deal. Maybe it's the best thing they make at this place. Or maybe I'll get something else. The point is, either I know what I want, or I look at the menu and I choose something. If it's a place I haven't been, I think about it, but I make up my mind.

Not JoNell. JoNell says, "Do I want the tuna melt? Or no, maybe I'd rather have the chili dog." She goes on and on over what to order. Then—and this is the one that really drives me up the wall—"What are you having?" What difference does it make what I'm having? I say, "Have what you want, JoNell." Or I say, "Well, I've got ten bucks so that means neither of us goes over five." Then the waitress is tapping her foot and rolling her eyes. "Maybe I'll come back later," she says. And JoNell chimes in, I'll have the same thing he's having."

Lately I've been thinking about going steady. We're pretty much going steady now. Then I think about finishing school and later—like if we want to get married or something. And I think "Do I want to spend the rest of my life waiting for JoNell to decide if she wants the tuna melt or the deluxe burger?" If she can't make up her mind what to eat, how is she going to decide about really important things? And how come I'm so impatient with her when she needs a little time?

Anthony

SECTION 1

MAKING YOUR OWN DECISIONS

At times all of us have trouble deciding what to eat, what to wear, and what to do. Just listen. How often do you hear

- What'll I do?
- I can't decide.

Some people don't want to choose. They would rather let other people make their decisions for them. Then, if things don't turn out right, they try to blame somebody else.

Other people seem to have few decisions to make. Everything seems to be taken care of—where they live, where they go to school, who their friends are, what they should do in the future. They wish they had an opportunity to decide things for themselves.

What's To Decide Now?

Every day you face situations that test your ability to make decisions. Most of the time you rely on **habits,** what you usually do. Or you rely on **custom,** what others usually do, in these situations. You make hundreds of decisions like this every day, so many that you hardly notice. Here are some examples:

- Shall I start my homework now or wait until next week when it's due? (What do you usually do?)
- Shall I wear my jeans jacket or a sweater? (What do others do?)

Even decisions like what to order to eat, what to wear, or where to go with friends can be tough. But these seem like simple choices compared to other decisions you have to make, such as

- how you want to look,
- what kind of person you want to be,
- whether you should take a part-time job,
- where you will go to high school or college,
- what you will do with your life,
- what courses you should take,
- whether or not you will have sex with your steady boyfriend or girlfriend,
- whether or not you will stay with someone who is into drugs,
- whom you will marry.

Dropping out of school may affect others more than you think.

Decisions Affect Others — "This is your concern too!"

Every day you face situations that test your ability to make decisions. All of your decisions, big and small, show who and what you care about and value.

Sheila tells her parents over and over how important her grades are. But, she always decides to do things other than study. How much does she really value good grades?

The decisions you make are important. They establish and define your character. Through your decisions, you have a hand in creating who you will be.

- You can seize or run away from possibilities.
- You can be adventuresome or play it safe.
- You can stand up for yourself or drift along with the crowd.
- You can put off decision making or you can decide.

Dave decides to start putting litter in the trash basket rather than dumping it on the ground. "It's no big deal," he says. But Dave's little brother Sid and his friends look up to Dave. They start putting litter in the trash basket too.

The caring person is interested in how his or her decisions will affect others. Decisions that reflect positive values such as trust, respect, and responsibility make it possible to live in harmony with other people.

Factors That Influence Decisions — "It's not just me!"

We seldom decide anything alone. There are many factors that influence our decisions. These factors range from how we feel about ourselves to random things over which we have little or no control.

Self-Worth

The way you see yourself has an impact on the decisions you make. It can keep you from even considering many decisions.

> I wouldn't dream of entering the talent show. I'd be too embarrassed.

Ann decided not to try. She was afraid she would look unattractive and awkward.

> I didn't show up to babysit because I didn't think it mattered.

Reggie doesn't see himself as a responsible person.

Peer Pressure

Sometimes friends and those we would like for friends tell us what to decide. Even when they are indirect, we can feel the pressure they put on us to do things their way. Through facial expression, gossip, and actions, they tell us what they think we should decide to do, to wear, and even to think.

> I didn't want to enter the talent show, but all the others girls are doing it.

Ann decided to risk looking unattractive and awkward rather than be left out.

> I didn't babysit because the guys said I had to go with them.

Reggie is letting other people decide for him.

Resources

Resources are money, materials, and time. They all can influence decisions—either positively or negatively.

> I wish I could enter the talent show, but I work after school. I don't have the time to rehearse.

Ann is saving money for college. She doesn't have time to do everything she would like.

Decisions Other People Make

Suppose you have to decide whether or not to go to a baseball game tonight. You are faced with two alternatives:

A lot of factors are involved in decision making. Can you explain how each of these factors might affect your decisions?

- going to the game,
- staying at home.

You probably won't be anxious about the decision unless you feel strongly about the alternatives. If you don't care much, staying home may be as interesting as going to the game. On the other hand, you might really want to go because

- your friends will be there,
- it is the key game of the season.

But you'd also like to be at home because

- your all-time favorite movie is on television and you don't have a VCR,
- the girl you've been wanting to ask out said she might call.

Your choices become even more complicated when other people make decisions that affect yours. For example, your mother might decide that this is the

evening she'd like to go shopping for a new compact disc player. She would like to have you and your sister help in deciding which one to buy. Now you have another possibility and the **consequences** or results are different:
- staying at home and missing both the game and shopping for the CD player,
- going to the game and missing the movie, the phone call, and shopping for a CD player,
- going shopping for the CD player and missing the movie, the phone call, and the game.

Your older brother might decide to collect the twenty dollars you owe him. If he does, you won't have the money to go to the game. Now the consequences also include going to the game and having your brother refuse to lend you money again.

Both your mother and your brother made decisions that affect your decision about whether to go to the game. But different consequences are involved. This adds to your confusion about the best thing to do.

Random Factors

Decisions may also be influenced by random factors like the weather. If you are trying to decide whether to go to the game and it looks like rain, you still have the same alternatives — going and not going — but the consequences now include the risk of getting rained on.

Values

The most important factor influencing your decisions is your values. If you make a decision because of peer pressure, you are showing that you value what your friends think. If you make a decision because of self-worth, you are showing that you value yourself. It is true — your values are always showing, even in the decisions you make.

SECTION REVIEW
STOP AND REFLECT

1. Give two examples of little decisions that you make every day on the basis of habit or custom.

2. Give two examples of big decisions that you have made in the past.

3. What factors do you think influenced your two big decisions most?

4. How did your big decisions affect other people?

5. List two difficult decisions that you will need to make in the future.

6. Tell why each of these decisions is worth worrying about.

7. Describe how each of these decisions is likely to affect other people.

SECTION 2

HOW PEOPLE MAKE DECISIONS

One of the reasons decision making can be hard is that people are afraid they will make the wrong choice. We all want things to turn out right. If your decisions don't seem to turn out right, maybe you aren't making good decisions. You may be more worried about avoiding unpleasant consequences than about making a wise choice.

Risks In Decisions

Sometimes people become so concerned about consequences that they try to see into the future in order to avoid anything unpleasant. They don't want to take any risks. They end up spending time trying to second-guess what will happen so they can avoid the worst.

Focusing On Consequences — "Where's my crystal ball?"

People who make decisions on the basis of consequences often want to avoid discomfort or frustration. Some people "go for the sure thing" or those things that seem guaranteed. Unfortunately, there are few sure things in life! Often we don't know all the facts.

> Your dad asks if you'd rather clean the kitchen and wash dishes after dinner or polish the car. You decide to polish the car. You think that polishing the car is more fun and that cleaning the kitchen takes longer. So you go for the sure thing. Later you discover the car is caked with mud.
> Not only that, but the family decides to send out for pizza for dinner and there is almost no mess in the kitchen.

In this case, what looked like a sure thing turned out to be the less pleasant choice.

Other people try to think about all of the bad things that could happen if they make a decision. Then they try to choose the least-bad consequence. They are "avoiding the worst." They also may be avoiding interesting possibilities.

> Benny has an opportunity to go on a blind date. The girl is visiting one of his mother's friends. Right away that puts him off. He thinks his mother's friend is ugly so her niece will be ugly, too. Later, when he meets the niece, he finds out that she is really something. Most guys could only wish to go out with a girl like her, and he passed up the chance!

People also try "going for the least regret." They want to decide on the basis of what will cause them the least discomfort.

> Eva wants to go to the school prom. Allie, the one guy she most wants to go with, hasn't asked her out. Eva can't decide if she should go with another guy for fear of missing out on the prom or if she would always regret going to the prom with somebody besides Allie.

Of course we should think about consequences when we make decisions. But trying to go for the sure thing, focusing on the worst possible outcome, or avoiding regrets are negative ways of dealing with decisions They may keep you from looking at the positive side of alternatives.

Focusing On The Possibilities — "Why so many choices?"

Decision making would be easier if we had few choices. When we have freedom to think about and choose from many alternatives, we have more to think about. You want to make the right decision. But you can't be sure it is right until after it has been made!

When you want things to turn out just right, you probably do a fair amount of worrying. Sometimes you "weigh the options:"

- Zade found a sweater he really likes. "Should I buy the sweater or save my money? Or could I get Mom to buy it for my birthday? Or would I rather have her get me something else?"
- Gayle sees girls selling crack in the restroom at school. "Do I pretend I didn't see, or do I report them?"

Sometimes you debate with yourself:
- "If I buy the sweater, I can get at least two years wear out of it. But if I save my money, I may be able to buy it on sale and have some left."
- "If I report them, they'll know who told and I'll be in for it. But if I don't tell, I'll feel responsible."

Or you ask other people their opinion:
- "What do you mean, I have too many sweaters already?" Zade is sorry he asked a friend for advice.
- "You mean I might talk to somebody besides the assistant principal?" Gayle learns about a new option.

Seeking information from various sources prior to making a big decision is helpful.

Then you decide:
- "I'm going to save my money so I can buy a whole outfit later."
- "I'm going to the assistant principal. I have to stand up for what is right and I trust her."

But then you change your mind:
- "No, I'd better buy the sweater. It may not be on sale later or, I may not be able to find it in my size."
- "Maybe I shouldn't do that. The assistant principal is okay, but it could get those girls in a lot of trouble."

So you feel guilty:
- "I should have saved my money."
- "I'm copping out."

Maybe you decide on something else:
- "I'm returning the sweater and getting new slacks instead."
- "I'm leaving this one alone."

Finally, you go back to your original decision or settle on a compromise:
- "I'm saving my money," Zade decides.
- "I'm going to the girls' counselor. I know I can trust her." Gayle decides.

Sometimes we end up frustrated and unsatisfied with the decision we have made. It is impossible to remove anxiety from decision making. But the more we learn about decision making, the stronger and less anxious we become.

This teen did not make an ethical decision when he decided to talk the others into trying some new drugs after school.

Ethical Decisions

Sometimes it is impossible to know if you've made an ethical decision. An **ethical** decision is one that is moral or right. You do your best. But some decisions are more right than others. That is, some choices will lead to unethical outcomes—those that reveal bad character and conduct. They go against such values as caring, trust, respect, and responsibility. However, in most situations, there are many ethical possibilities.

Jamie has just been assigned a social studies term paper. It is due in three weeks. She is also in the school play and has an afterschool job. She must decide what to do about her term paper.

Here are some of the possibilities:
- quit the part-time job,
- establish a schedule for time to spend on the play, other school assignments, the job, and the paper,

- borrow a term paper and change it enough so it won't look the same,
- drop out of the school play,
- not turn in the assignment and tell the teacher she lost it,
- talk to the social studies teacher about her options.

If Jamie wants to make an ethical decision, she needs to be able to decide which possibilities are ethical. It would be wrong to borrow someone else's term paper. By borrowing someone else's paper, she would not be taking responsibility for her own learning. And she would be lying to say that she had done the work.

It would also be a lie to tell the teacher she had lost the paper on the way to school. These are obviously unethical choices. They are wrong. They are bad. They do not show respect for herself or others, and they do not show that she is responsible.

Suppose that when Jamie interviewed for the job, she promised she would stick to it for the year. In that case, quitting because she had something else to do would also be an unethical choice. She would be breaking the trust placed in her by her boss.

To make an ethical decision, Jamie needs to weed out the unethical alternatives. After doing this, Jamie has fewer choices:

- quit the part-time job (assuming she hasn't made any promises about how long she will work),
- establish a schedule for time to spend on the play, other school assignments, the job, and the paper,
- drop out of the school play,
- talk with her social studies teacher about her options.

Jamie still faces a tough decision. Making ethical choices is not always easy. But the person who is concerned about values will make ethical choices, even when it is hard. In the long run, the hard choice may help you to take charge of your life and become a responsible, caring person!

SECTION REVIEW
STOP AND REFLECT

1. Describe a decision that you think would be really hard for a teen to make.
2. What risks are involved in the decision?
3. What would be the danger in going for the sure thing?
4. What possibilities does a teen have to choose from in making the decision?
5. How are values related to the choices?
6. Describe the unethical possibilities.
7. What would the ethical choices be?
8. How would you advise a friend to make this decision?

SECTION 3

HOW PEOPLE MAKE WISE DECISIONS

Making certain decisions will never be easy. But you can take some of the stress away by having a plan for decision making. No plan will work all of the time, but a good plan should

- take into account what is required to make ethical decisions,
- allow you to choose in light of your values,
- be practical enough to work,
- help you think about the consequences of your choices.

How Can You Decide?

To make a caring decision, you need to follow a decision-making plan. One way to do this is to ask yourself a series of key questions.

- **What is the situation?**
- **What is the challenge?**
- **What are the possible choices?**
- **What are the consequences for myself and others?**
- **What will I decide, given all that I know?**

What Is The Situation?

The situation is the reason you have a decision to make. The situation may include persons, things, or events that are planned or unplanned. To more fully understand the situation, ask yourself questions such as these:

- Why is there a decision to make?
- Why am I in this situation?
- What is involved here?
- Who is involved?
- How much time do I have to decide?

Imagine that your parents have decided to get a divorce. You are one of two children. Your parents have agreed that you and your younger sister will each decide which parent to live with.

The situation includes all the factors involved.

- **Why is there a decision to make?** Because your parents are getting a divorce, you can no longer live with both of them.

One way to test the impact a decision will have on your feelings of self-worth is to imagine a picture being taken of you implementing your decision and that picture being shown to your parents.

- **Why are you in this situation?** Because of an earlier decision made by your parents—to get divorced. You didn't create this situation. Your parents did.
- **What is involved?** Aside from where you will live, you need to consider how often you'll get to see the parent you don't live with. You should also consider what school you will attend.
- **Who is involved?** You, your sister, and both parents.
- **How much time do you have to make your decision?** This will be determined by your parents. It may be as soon as your dad finds another place to live.

What Is The Challenge?

Making decisions can be a challenge. A decision can challenge.
- **independence.** Which parent gives you more freedom? How much freedom are you ready for?
- **social relationships.** If you move in with your dad, you'll be moving away from friends. Would you like the opportunity to make new friends?
- **self-worth.** How will you feel about yourself if you choose to live with one parent instead of the other when you love them both?
- **health.** Will you take better care of yourself or get better care with one parent than with the other?

Around the World

Freedom to Choose

Few cultures offer as much freedom of choice as America does. People are encouraged to make their own decisions from childhood on. In this way, our culture teaches self-reliance.

In some other cultures, people don't make all their own decisions. For example, arranged marriages are still common in many parts of Asia and Africa.

Egyptian parents often take responsibility for their teens' decisions, including choosing an occupation, and a spouse.

Restrictive societies or religious practices may limit the need to make some decisions. Because of its large population, China has a "one-family, one-child" policy. Drinking alcohol is against Islamic law. Because liquor isn't readily available, the alcohol dilemma is not a decision most Moslem teens must face.

Americans have the freedom to choose. But with choice comes the responsibility to choose with wisdom and caring.

- **future opportunities.** How will your choice influence who and what you will become?
- **goals.** Are you more likely to get support for meeting your personal goals from your mom or your dad? Or are both supportive of you?
- **values.** What will happen to your values when you choose one parent over the other?

There aren't any right answers to these questions. You ask them in order to understand what you are really facing in making a decision. Thinking about them will help you decide what you are willing to give up in order to gain other things.

What Are The Choices?

How many choices do you have? Who is affected by each? If you look at the choices in deciding which parent to live with, there seem to be only two:
- living with Mom,
- living with Dad.

But there may be more. At this stage of decision making, it is important to brainstorm as many choices as you can. You may even want to ask others to make suggestions about choices. There may be more options:
- living with your grandparents (you often spend part of the summer with them and you like it there),
- living with both parents—one during school days and one on weekends,
- splitting the time with each parent in half and living with each one an equal amount of time (commuting to school from your Dad's apartment across town when you are with him).

What Are The Consequences For Others And For Me?

If you make a particular choice, what is likely to happen? Who will be involved? Do you have the resources? Is the choice legal? Is it healthy? Would you make the same choice if others knew it? Do the consequences allow you to be the person you want to be?

Possible Consequences

If I live with Mom:
- Dad will be angry.
- I'll have more money.
- Mom is a better cook.
- My friends are comfortable with my mom.
- Mom doesn't have a VCR.
- The dog has to stay with Mom.
- I'll have my own room.
- I'll stay in the same school.
- I'll miss Dad's funny stories.
- I'll have to help out more around home.

If I live with Dad:
- Dad will be lonely if no one lives with him.
- Mom will be angry.
- I won't have as much money.
- We'll eat out more.
- My friends won't be able to get to Dad's house.
- Dad has a VCR.
- Dad can't have a dog at his place.
- I'll have to share a room at Dad's.
- I'll be in a new school if I go with Dad.
- I'll miss having Mom help me with my homework.
- I'll have more time for myself.

Make A Decision

When you've thought through all your options and their possible consequences, you need to decide. You may decide you're going to live with your mom so you can stay in school without too much hassle. Perhaps you'll spend the summers and every other holiday with your dad.

What Happens After You Decide?

Once the decision is made, the process doesn't end. You need to
- rehearse the decision,
- take action,
- evaluate the decision.

As you think through a decision be sure to ask yourself "Is it legal? Is it healthy?"

Rehearse The Decision

It helps to rehearse in your mind the way you will act on your decision. Play out in your head the responses you may get from your mom, your dad, and your sister. Then you can plan how you'll react. You'll feel a lot more comfortable if you think through the process before acting.

Take Action

After rehearsing your decision, it's time to tell others what you have decided and to get on with it. In the case of deciding which parent to live with, you tell your family what you've decided. You tell your friends you'll be living with your dad in the summers and give them your summer address as soon as you know it. You plan to spend time with each of your parents as they work out their new living arrangements.

Evaluate

After you have made a decision and the process is complete, it is a good idea to evaluate it.
- How do you feel about your decision?
- Did it work out?
- Are you satisfied with the consequences?
- Would you make the same decision again in a similar situation?

By thinking back over your decisions, you can learn how to strengthen the decision-making process. This will lead to better decisions in the future.

The fact is, in real life you do not always have time to think through each step in a decision-making plan. But by learning about a plan and practicing it, you will be able to make more thoughtful decisions even when you are under pressure.

Mature persons see decision making as a challenge. They learn how to decide for themselves. They use the help of other people, and they follow a decision-making plan. They learn to make sound decisions that are good for themselves and for those they love and care about.

SECTION REVIEW
STOP AND REFLECT

1. Describe the situation for an important decision you have made recently.
2. Describe how you thought about the decision.
3. Tell what you chose to do.
4. Tell how you planned to put your decision into action.
5. Describe the actions you were actually able to take.
6. Describe the consequences of your decision. How did it turn out? Who was affected?

CHAPTER 3 REVIEW

Putting Your Values To Work

STRENGTHENING YOUR VALUES

Reread *A Teen Speaks* **on page 39. Then answer the following questions.**

1. Why do you think JoNell has problems making decisions?
2. Why might Anthony find it easier to make decisions?
3. Do you think Anthony's impatience with JoNell is fair? Why or why not?
4. How would you deal with JoNell's inability to make decisions?
5. What words would you use to describe JoNell? Explain your answer.
6. What words would you use to describe Anthony? Explain your answer.
7. How might Anthony help JoNell learn to be a better decision maker?
8. Are you more like JoNell or Anthony in the way you make decisions? Give some examples to support your answer.

INTERPRETING KNOWLEDGE

1. Write a short story in which two people make different decisions about the same question. Show how their decisions affect the rest of their lives.
2. Create a mobile, chart, or poster illustrating points to consider when making your own decisions.

SHARPENING YOUR THINKING SKILLS

1. What are some reasons why people let others make their decisions for them?
2. Name several little decisions that you make every day that you would like to change. Why would you like to change them? What would you need to do to change them?
3. The way a person sees himself or herself can have an impact on the decisions that person makes. Give an example of this from your own life.
4. How have some of your past decisions been influenced by your resources?

APPLYING KNOWLEDGE

1. Use newspapers and/or magazines to find examples of ethical and unethical decisions made by others. Discuss your selections with the class.
2. Keep a record of your activities during the next 24 hours. Total the time you spend in each type of activity, such as sleeping, eating, attending class, studying, and so on. Examine the results. Write a paper discussing the following:
   - Are you getting all you want from your 24 hours?
   - Where are the weak spots?
   - What role do you play in deciding how your time is spent?
   - In what ways do peers influence your decisions about how to spend your time?

CHAPTER 3 REVIEW

Putting Your Values To Work

PRACTICING DECISION-MAKING SKILLS

Read about each situation. Then answer the questions.

Situation A: Refer to *A Teen Speaks* on page 39.

1. What are some reasons why Anthony decided to date JoNell in the first place?
2. What values do you think they share?
3. What changes might Anthony make if the relationship is really important to him?
4. What changes might JoNell make?
5. Name some of the possible choices each of them has in continuing their relationship.
6. What are some of the consequences of each choice?

Situation B: Sheila gets her grades for the semester and learns that she has failed two courses.

7. What advice would you give her parents, who believed Sheila in the past when she told them, "Good grades are really important to me"?
8. Name some of the reasons why Sheila may often have decided to do things other than study?
9. In what ways do you think Sheila's decisions have affected others?

Situation C: Harry has decided to drop out of school at age 16. His parents have told him he can no longer live at home unless he returns to school or gets a full-time job.

10. What are some of the possible long-term consequences of Harry's decision?
11. What are some of the possible long-term consequences of his parents' decision?
12. Do you agree with Harry's decision? Explain your answer.
13. Do you agree with his parents' decision? Explain your answer.
14. What would you do if you were Harry?
15. What might cause you to change your decision if you were one of Harry's parents?
16. Do you know someone in a situation similar to that of Harry and his parents? What decisions were made? What were the consequences?

Situation D: Someone offers you quick cash for selling drugs in school. You really need the money to buy clothes and to pay for a class trip.

17. What would you do? Make a decision following the decision-making process.
18. Would your decision change if your parents were sure to find out?
19. Would your decision be different if your friends were to see you making the arrangements?

Situation E: Suppose a doctor has just told you that you have only one year left to live.

20. Would you make any changes in your life? Explain your answer.
21. Would you choose to be alone while rehearsing your decision to change or would you prefer the presence of another person? Explain.

UNIT 2
Caring For Yourself

CHAPTER	4	EATING SMART, STAYING HEALTHY
CHAPTER	5	EXERCISING AND STAYING FIT
CHAPTER	6	CARING FOR YOUR MENTAL HEALTH
CHAPTER	7	SAYING *NO* AND FEELING GOOD ABOUT IT
CHAPTER	8	STAYING DRUG FREE
CHAPTER	9	RECOVERING: COMING TO TERMS WITH DRUGS
CHAPTER	10	PLANNING FOR TOMORROW

CHAPTER 4

Eating Smart, Staying Healthy

A TEEN SPEAKS

I have this friend Paulette. I really like her a lot. But there's this one thing that bothers me about Paulette. She is really overweight. She says she has a thyroid problem. I don't even know what a thyroid problem is! If she can't help that she is fat, then she just can't help it.

That's the point, though. I think she can help it. She eats junk all the time. I know it is none of my business what she eats, but I have said something a couple of times when she's started eating potato chips on the way to school. She'll say she skipped breakfast or she really needs the energy or something like that.

When she does eat a meal, like at school, she wants to have stuff like a double cheeseburger, fries, and a shake. If I have a salad and soup or something like that, she'll say, "I don't see how you can get by on that. I have to have some real food!"

To make it worse, all she wants to do is to sit and watch movies or talk. Will she go for a walk and talk with me? No way. Paulette needs to start thinking about what she is doing to herself.

Carolanne

SECTION 1

FOOD CHOICES AND YOU

What do you usually eat for lunch? Does it really matter what you eat? Everybody knows you get fatter by eating too much and thinner by eating less—assuming the amount of exercise stays the same. But who stops to think how else food affects the body? What you eat affects more than your body's weight and shape! It affects your energy level, your alertness, even your moods. In fact, *good nutrition,* which is the habit of eating a variety of healthy foods, helps build your body and keeps it running for a lifetime!

Food And Values

Food decisions reflect your values. Practicing good nutrition is one way you care for yourself. A **balanced diet,** is one that gives you the variety of foods your body needs. It is essential for your growth, energy, and emotional and physical health. A balanced diet provides your body with all the nutrients it needs. **Nutrients** are tiny substances that are found in food and that are necessary for the body to function, to grow, to provide energy, and to repair itself.

Providing food for others shows that you care, too. When you help with family meals, have a friend over for dinner, or go out to eat together, it shows you care. Giving food to others says you care, too.

Making healthful food choices is a responsibility you have to yourself. At first, other people make decisions about what you will eat. Before you were born, what your mother ate directly affected her health and yours, for example. When you were a child, somebody provided all your meals.

Now you have more responsibility for food decisions. Even if your family decides what's for dinner, only you can decide whether or not to "pig out" or overeat.

> You get me started on chocolate chip cookies and I can't stop!

Good nutrition takes self-discipline. It takes conscious effort to eat a balanced diet. To maintain your best weight, you may need to resist second helpings and

With so many food choices available to you, you must learn the facts of nutrition before you can make healthy choices.

snacks. Saying *NO* to some foods and *YES* to others is part of self-discipline.

To make caring, responsible, and healthful food choices, you need to know the basics of good nutrition. Your challenge is learning the facts, not the myths. Many books, magazines, and television and radio shows talk about nutrition. Don't trust everything you read or hear. For example, you may have heard a friend say, "If you eat grapefruit after every meal, it burns off the fat!" Or you might read, "If you go on the banana diet you can lose ten pounds in a week."

These comments aren't based on nutrition facts. Food fads such as these can be harmful to your health. As much as you like them, you can't count on your friends to know what your body needs. And as much as you'd like to believe what you hear on television or read, you can't always count on its being right for you. Look for a reliable source—your doctor, a registered dietitian, your health teacher or your home economics teacher.

Food Choices

Before you can make healthful food choices, you have to recognize and understand the factors that influence you. Your family and friends influence your food choices, and you influence theirs. Your **culture**—that is, the beliefs and social behavior of your racial, religious, or social group—influences your food choices. Other influences come from advertising and from limits on your time and money.

Don't Believe It!

Myth: Eating chocolate causes pimples.

Fact: Neither chocolate, fries, nor soft drinks cause pimples. A balanced diet promotes healthy skin, but poor food choices don't cause acne. Hormone changes during adolescence and stress may trigger acne.

Myth: Pizza is a junk food.

Fact: Pizza is actually nutritious. It contains a bread crust, tomato sauce, and cheese—all nutrient-rich foods. Sometimes it contains vegetables such as green peppers or meats such as sausage—other sources of nutrients. One problem with pizza is that eating too much provides a lot of calories. If they can't stop at one or two slices, pizza can be a poor choice for people who are trying to lose weight.

Myth: Vitamin pills offer extra energy.

Fact: Vitamins don't have calories so they can't provide an energy boost.

Family Ties

Family has the biggest influence on what people eat. From the time you were born, your family influenced your food experiences. They taught you to eat with your fingers, then with a spoon and fork. They encouraged you to follow certain rules. Even now, your family influences what, when, how, and even where you eat.

Peer Pressure

The older you are, the more your friends influence your food decisions. Perhaps your friends have introduced you to foods you've never tried before. Or maybe their families cook the same foods that your family does but in different ways.

Peer pressure can affect food choices in other ways, too. For example, some teens may go on weight-loss diets because of peer pressure to be thin. And they may follow a fad diet just because a best friend thinks it works. Others may "go with the crowd" and order a milk shake after the movies just because everyone else does.

How can you overcome peer pressure? You might order first—let others follow you. Be clear about your decision. Know why you're making a choice that's different. For example, know why you won't follow a diet that's unhealthy. Then choose what's best for you. Doing what you think is best shows self-control and self-esteem.

Peer pressure affects what you eat. How might this scene represent both positive and negative peer pressure?

Culture

Why do Americans enjoy pizza, tacos, and stir-fry? The cultures of Italians, Mexicans, and Chinese, among others, have influenced our food choices.

Food customs include the way food is grown, processed, purchased, cooked, prepared, served, and eaten. Religious beliefs, celebrations, and superstitions are reflected in our food customs.

Because of cultural differences, people eat a variety of foods prepared in different ways. For example, flour is made into bread: pita in the Middle East, crusty loaves in Italy, croissants in France, tortillas in Mexico, chapatis in India, and popovers in England. Even within the United States, each region has its food culture: sourdough bread in the West, spoonbread in the South, and brown bread in the Northeast.

A caring person respects the cultural food preferences of others. An open-minded person will try new foods when offered and often likes them.

You'll be healthier if you limit your intake of empty-calorie foods.

Advertising

In an average year, people see or hear many food ads on radio, television, and billboards and in magazines and newspapers. These ads are created to influence food decisions and to make people aware of choices.

Powerful persuasion techniques are used in food advertising. They appeal to your values. Some suggest you'll have status and sex appeal if you buy the product. Others tell you part but not all of the truth; for example, a food may be low in fat but not low in calories. Or an ad may suggest that you'll look or feel like the model if you buy the product.

Use ads to learn about foods in the marketplace. But listen and judge the messages carefully. You, not advertisers, should control what you buy and eat.

Time And Money

Time and money may limit food choices. Both affect food decisions and fitness more than you may realize.

In the past, people took more time for food. They still do in many parts of the world. In America today, many people eat on the run. Families take less time to enjoy meals together. Work and school schedules often get in the way of family mealtime.

People resolve the time issue in many ways. Some skip meals and snack throughout the day. Some rely heavily on microwave cooking. For others, convenience foods at home and fast foods in restaurants often substitute for home-prepared foods. With good planning, many people choose nutritious foods in spite of time pressures. Others make poor food decisions.

Money affects food decisions, too. People who wish to spend less money may choose ground beef instead of steak, or tuna instead of shrimp. They may cook from scratch instead of buying convenience foods, which often cost more. They may eat out less often. Even on a limited budget, most people can choose foods that promote health.

Unfortunately, many Americans are undernourished because of poor food choices. They may have enough money to buy healthy food, but they spend their money on foods with calories but few nutrients.

Some people don't have enough money for food. Hunger is a growing problem in America. Without adequate food, people can't remain healthy. And

Hunger—A Worldwide Dilemma

Most Americans don't go to bed hungry. Our nation is blessed with an abundant food supply. In fact, many Americans waste food, or eat too much. Some Americans along with many people in other parts of the world are starving, however. In Africa a serious drought killed about two million people during 1984 and 1985. In 1988, six million of the 46 million people living in Ethiopia faced starvation and suffering. Many in Bangladesh couldn't get food because floods covered most of their nation.

Why do we face a global food problem? The reasons are complex. Droughts and floods ruin crops and kill livestock. In some places the population is growing faster than food can be produced. Farmland has been taken for housing. Poor farming practices make land unproductive. And war and politics interfere with feeding people.

Through international relief, people address problems of hunger. Organizations such as CARE and International Red Cross send food. Others, such as the Peace Corps, provide agricultural assistance. How might you care for the hungry at home or abroad?

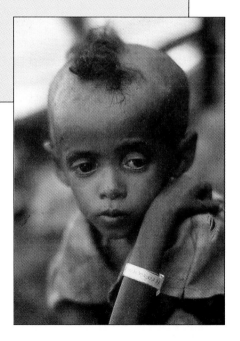

they lack the energy and drive to control their lives. If you aren't getting enough to eat, for whatever reason, you can get help. Talk to a teacher, the school counselor, or a member of the clergy about what to do. Nobody should have to go hungry.

Providing for the hungry is one way society cares for others. Government and private agencies provide services for them, such as food stamps, commodity foods, food pantries, and soup kitchens. But more help is needed.

Food And Emotions

Eating well and sharing food can be ways to show caring. Preparing and eating food can bring pleasure and a sense of emotional well-being. Can you think of a time when emotions were an obstacle to good nutrition?

> I was so stressed out I just kept eating! Then my mom wanted to know what happened to the chocolate cake. I'd eaten the whole thing!

You may realize that you are rewarding yourself with food or eating to resolve boredom or **stress,** the strain or tension that can be caused by changes in your life. This kind of eating usually involves too much of the wrong foods. Or, if you lose your appetite under stress, you may miss out on the nutrients you need. Overeating and not eating enough are inappropriate solutions to emotional problems.

Inappropriate food behavior may be a symptom of deep emotional problems. Anorexia nervosa and bulimia—two eating disorders associated with teenagers and young adults—are caused by personal conflicts.

Anorexia is a behavior involving the extreme fear of becoming overweight, which results in severe weight loss from self-starvation. **Bulimia** involves a rapid consumption of a large quantity of food in a short period of time followed by vomiting to get rid of it. This disorder, like anorexia, involves a very serious desire to be thin—too thin.

These are not problems you can solve alone. If you think you may have either of these problems, get help. People with these problems need help from trained professionals. A teacher, the school nurse, or a guidance counselor can refer you to the proper professional.

SECTION REVIEW
STOP AND REFLECT

1. If you could choose to eat with somebody else's family, whose would it be? Why?

2. Tell about a time when you were pressured to eat food you didn't need. Did you give in? Why or why not?

3. Describe a television commercial that has gotten you to eat something that isn't good for you. Describe one that caused you to eat a nutritious food.

4. List all the foods you ate yesterday. For each food, note the approximate time you ate it, describe the situation (lunch, watching TV, etc.), and describe how you were feeling at the time (happy, bored, etc.). What determined which foods you ate at a particular time?

SECTION 2

EATING FOR FITNESS

In many ways your body is like a machine. If you want it to go, you have to give it fuel. If you want it to do more than just chug and sputter along, you have to give it the right fuel. If you really care about yourself, you'll eat for fitness —with health as your goal.

Dietary Guidelines

When people start talking to me about food groups, my reaction is "BORING!" I've had this stuff since third grade. So who cares? All I really care about is what tastes good.

Like this teen, you may be tempted to skip out when it comes to talk about dietary guidelines. But the seven Dietary Guidelines for Americans, listed below, were developed to help you get fit and stay fit.

- **Eat a variety of foods.**
- **Maintain a desirable weight.**
- **Avoid too much fat, and cholesterol.**
- **Eat foods with adequate starch and fiber.**
- **Avoid too much sugar.**
- **Avoid too much sodium.**
- **Avoid alcoholic beverages.**

By following these guidelines, you can eat enough foods with the nutrients you need and avoid the excesses you don't need. Giving your body what it needs is another way of taking charge of your life.

Eat A Variety Of Foods

Have you ever heard the phrase, "Variety is the spice of life?" Variety is also the key to a healthful diet. By eating many kinds of foods your body gets the nutrients in the amounts you need.

You need the essential nutrients in adequate amounts so your body works right. But no food has all these nutrients in the amounts you need. That's why it's good to include several types of food in each meal and in your snack choices.

Food group guidelines were devised to help you get enough variety and, as a result, enough nutrients. Four of these groups have recommended servings and serving sizes.

The fifth group—fats, sweets, and alcohol—has no recommended servings. These foods aren't necessary. The nutrients provided by foods in this group can be obtained easily from other foods. Actually, consuming alcoholic beverages and eating too many high-fat and sweet foods may harm your health.

Eat To Keep Your Best Weight

> Judy is so tiny. Everything she puts on looks great. When I stand next to her I feel like a cow. Why do I have to be so tall and have such a big rear? It isn't fair!

Your best weight is what's healthy for you, not for your best friend and not for the models in magazines. Because people have different body types, there's no simple weight guideline. Some people are tall. Others are short. Some have small frames. Others have large frames. Whatever your body type, keep it fit and lean, without excess fat.

Nutrients

For health, growth, and energy, your body needs over 40 different and essential nutrients. You may know their six categories and purposes:

- Carbohydrates provide energy.

- Protein helps build and repair your body cells, and it provides energy.

- Fat provides energy and carries some vitamins into your body.

- Vitamins, such as thiamin, riboflavin, and niacin, help regulate body processes.

- Minerals, such as iron and calcium, help regulate body processes, too. Calcium is necessary for growing bones and teeth.

- Water is another regulator, making up about two-thirds of your body.

Why Keep In Shape?

Being your own best weight looks and feels good. Your body works better when you're in shape. A lean body—with good muscle tone and some, but not too much, body fat—is healthy.

Being too fat isn't healthy—physically, emotionally, or socially. **Obesity,** which involves being 20 percent or more above your best weight, puts a strain on the body. Moving around isn't as comfortable. So heavy people may not feel like being active. Obesity increases the chance of health problems—heart disease, diabetes, cancer, and other conditions—later in life. It may even shorten a person's life span.

Being obese can also affect **self-image** or how you see yourself. Depression, feelings of low self-worth, isolation, and rejection often go along with being too heavy. The problem can become a cycle. Some people overeat because they feel rejected or depressed. Overeating makes them obese which, in turn, lowers their self-image even more. Positive thinking, on the other hand, often leads to healthy eating, weight loss, and an improved self image. The cycle can be reversed.

Statistics show that obesity is a problem for today's teens. One in five adolescents is obese. And today's teens have significantly more body fat than teens did 20 years ago. There is a family tendency to overweight, meaning that if your parents are obese, your chances of being overweight are greater. It's not clear yet how large a role inherited characteristics play in this situation. But we do know that the wrong food decisions and a lack of exercise contribute to obesity.

By eating a variety of foods, you'll get the nutrients your body needs.

Trim Up!

Maintaining your best weight is simple arithmetic. **Calories** are the measure of the energy food gives you and the energy your body burns. You need to balance the number of calories you eat with the number of calories your body burns. If you eat more, you gain weight. If you eat less, you lose. The more calories you burn through exercise, the more pounds you lose.

Watch Out – Not Too Much Fat

Too much fat on your body isn't healthy. Neither is too much fat in your food. Over time, a high-fat diet may increase your chance of an early heart attack or stroke. High-fat foods are also high-calorie foods!

MEAT-POULTRY-FISH-BEANS-EGGS GROUP

Foods
All cuts of beef, pork, veal, and lamb; poultry such as chicken, turkey, duck, goose, fish and shellfish; eggs; dried beans or peas; nuts; peanuts and peanut butter

Nutrients
Protein, iron, B vitamins

Servings per Day
2 servings

Sample serving sizes
2 to 3 oz. (60 to 85 g) lean cooked meat, poultry, or fish
2 thin slices of meat
1 hamburger patty
1 small chicken leg or thigh
4 fish sticks
1/3 cup (75 mL) canned tuna or salmon
2 medium eggs
1 cup (250 mL) cooked dried peas or beans
4 Tbsp. (60 mL) peanut butter

FRUITS AND VEGETABLES GROUP

Foods
Citrus fruits, apples, bananas, peaches, salad greens, tomatoes, carrots, pepper, pumpkin, potatoes, and other fruits and vegetables

Nutrients
Vitamins A and C, calcium, phosphorus, iron, B vitamins, fiber

Servings per Day
4 servings
- At least 1 serving a day should be a citrus fruit or other source of vitamin C, such as cantaloupe, strawberries, tomatoes, cabbage, and broccoli.
- At least 1 serving should be a dark green or deep yellow fruit or vegetable as a source of vitamin A, such as cantaloupe, spinach, sweet potatoes, carrots, or pumpkin.

Sample serving sizes
1 orange
2 plums
1/2 banana
1 cup (250 mL) fresh berries or raw leafy vegetables
1/2 cup (125 mL) fruit juice
1/2 cup (125 mL) cooked vegetables

BREADS AND CEREALS GROUP

Foods
Whole-grain or enriched breads, cereals, rice, cornmeal, oatmeal, noodles, pasta, and other grain products.

Nutrients
Carbohydrates, iron, B Vitamins

Servings per Day
4 servings

Sample serving sizes
1 cup (250 mL) ready-to-eat cereal
½ to ¾ cup (125 mL to 175 mL) cooked cereal, rice, or pasta
1 slice whole-grain or enriched bread
1 bagel
2 tortillas

MILK AND MILK PRODUCTS

Foods
Milk, buttermilk, cheese, yogurt, cottage cheese, ice cream

Nutrients
Calcium, phosphorus, protein, riboflavin, vitamins A and D

Servings per Day
4 or more servings

Sample serving sizes
1 cup (250 mL) milk, buttermilk, or skim milk
1 cup (250 mL) plain yogurt
2 slices American or Swiss cheese

OTHERS CATEGORY

Foods
Butter, margarine, oil, salad dressing, sugar, honey, candy, jam, jelly, soft drinks

Nutrients
Fats

Serving per Day
Foods in this group are not needed for good health and may even harm health. No recommendations are made for a minimum daily amount.

Why all the Attention on Cholesterol?

Cholesterol is found in every cell of your body. It is a fatty, waxlike substance that helps your body produce substances it needs. It also helps protect nerve fibers.

Your body makes its own cholesterol, and you get even more from the foods you eat. Though your body needs cholesterol, too much in your bloodstream can be unhealthy. Large amounts of cholesterol can build up on the walls of blood vessels. This makes it harder for blood to pass through and can lead to heart disease. To help keep your cholesterol levels low, avoid eating too many eggs and too much animal fat. Be aware. A lack of exercise, as well as genetics, can also raise cholesterol levels in blood.

Understanding calories is one of the first steps to weight reduction.

For good health, everyone needs to eat some fat, however. Fat transports vitamins A, D, E, and K into your body. Fat helps keep your skin healthy and promotes growth. And fat is a highly concentrated energy source.

These tips will help you cut down on the amount of fat you eat:

- Eat fewer fried foods. Enjoy baked, broiled, roasted, simmered, or raw foods.
- Use less dressing on salads. Or use low calorie dressing instead. Use less gravy and less butter. Try just a little margarine, butter, or mayonnaise on sandwiches, and use less butter or sour cream on baked potatoes.
- Cut off the fat you see on meat. Remove the skin from poultry before cooking; the fat layer is under the skin.

- Spoon off the fat layer that forms on soups or stews when they're chilled.
- Substitute fish or seafood for beef or pork.
- Fill up on low-fat foods, such as vegetables (without butter or cream sauce), fruit, and rice and pasta (without cream sauces).

Eat Enough Starch And Fiber

As a teenager, you need plenty of calories to fuel your growing body. Assuming that your exercise level remains about the same, you are likely to need more calories now than later in life. When you exercise a lot, your calorie needs go even higher. Foods high in starch are a terrific source of healthy calories.

Starch, a carbohydrate, has less than half the calories that fat has. Starchy foods, such as rice, pasta, potatoes, and bread, usually have many essential nutrients and fiber, too. At the same time, most are low in fat.

Contrary to popular belief, starchy foods aren't fattening. The excess calories come from high-fat or high-sugar foods people often add, such as butter or margarine, sour cream, mayonnaise, gravy, jam, or jelly.

> Just about every time I turn on the television there is some commercial for high-fiber cereal. So what's the big deal about fiber?

Vegetables can increase the amount of fiber in your diet.

Fiber isn't a nutrient, but it is an important part of many fruits, vegetables, and cereal foods. Your body can't digest fiber. That's the benefit. Instead, it helps move waste through your digestive system to help prevent constipation and other intestinal problems. Eating enough fiber throughout your life may help protect you from cancer, heart disease, and diabetes later!

Like most Americans, you may benefit by eating more high-fiber foods. Try vegetables and fruits, especially with edible skins, whole-grain breads and cereals, corn tortillas, brown rice, whole-wheat pasta, and popcorn without salt and butter.

Control Your Sweet Tooth

Harold eats candy all the time. Even in class he sneaks candy. If it's sweet, he likes it. At lunch, anybody who doesn't want dessert automatically sends it down to Harold. I've never seen anybody with such a sweet tooth!

Look for hidden sugar in your favorite foods. Read the labels.

Do you want cavity-free teeth? Do you want the most nutrients for your calories? Are you watching your weight? Then watch the added sugar!

Added sugar provides almost no nutrients. In fact, foods with a lot of added sugar are called **empty-calorie foods** because they contribute a lot of calories and few, if any, nutrients. Sweet snack foods—sugared soft drink, cake, and candy—are empty-calorie foods.

Eating too much added sugar has risks:

- High-sugar foods may replace more nutritious foods. A soft drink, for example, may take the place of milk. As a result, you might lack enough calcium for your bones.
- The risk of tooth decay goes up when you frequently eat sugary foods. Sticky, chewy foods that stay on your teeth increase the risk.
- Too much added sugar can add extra calories, which can lead to overweight and obesity.
- Obesity, in turn, may result in inactivity, poor self-image, and cause some disease.

You can make wise choices about sugar. Cut back on foods with added sugars—soft drinks, candy, cakes, fruit in heavy syrup, and sugar-coated cereals. Read food labels so you can choose foods with less added sugar. As a hint, ingredients that end in "-ose" are sugars—for example, sucrose, glucose, and dextrose. If, like Harold, you have a sweet tooth, cutting back on sugar will be hard. But you can learn to enjoy things without sugar if you decide to do it.

Do you add salt before you taste? Can you shake the habit?

Shake The Sodium Habit

Salt is made from two chemical elements—sodium and chloride. Your body needs a little of both to keep you healthy.

Most people use too much salt. For some, high-sodium diets increase the risk of high blood pressure. High blood pressure, in turn, increases the chance of health problems, such as heart attacks and strokes, later in life.

No one can predict who will get high blood pressure, so many health professionals recommend that people of all ages eat only moderate amounts of sodium.

Like sugar, sodium is added to many foods you eat.

- Do you nibble on fries, chips, and nuts? Chances are, these snack foods are highly salted. Eat only moderate amounts.
- Because they preserve and flavor, sodium-containing ingredients are in many processed foods. You can find these ingredients and the sodium content on food labels. Buy lower-sodium foods when you can.
- Using a salt shaker without tasting isn't a good health habit—so shake the habit!

Avoid Alcoholic Beverages

Keep a healthy body healthy—avoid alcoholic beverages! Besides being illegal for teenagers, alcoholic beverages have calories and few nutrients. An ounce of alcohol has almost twice the calories of an ounce of sugar. You'll learn about health risks related to alcohol in Chapters 8 and 9.

It does matter what you eat. Learning to make wise choices isn't always easy. But it is a way of being responsible and caring for yourself.

SECTION REVIEW
STOP AND REFLECT

1. Make a list of everything you ate yesterday, including snacks.

2. Did you meet the requirements of the food groups? If not, how might you improve your diet?

3. Describe someone you know who makes unwise food choices for snacks and at mealtimes. What do you think it would take to convince that person to eat right?

CHAPTER 4 REVIEW

Putting Your Values To Work

STRENGTHENING YOUR VALUES

Reread *A Teen Speaks* **on page 61. Then answer the following questions.**

1. What is Carolanne concerned about?
2. What does Carolanne want to do for Paulette?
3. What does Carolanne want Paulette to do for herself?
4. If Carolanne came to you and said, "I'm concerned about Paulette and I don't know what to do," how would you respond?
5. Are Paulette's eating habits Carolanne's business? When is a friend's business your business too?
6. What do you think Paulette is really feeling when she says "I don't see how you can get by on that (soup and salad). I have to have some real food"?
7. Have you ever been in either Carolanne's or Paulette's situation? If so, describe how you felt and why. If not, describe what you might say to Carolanne and Paulette if they were your friends.

INTERPRETING KNOWLEDGE

1. Develop a play or skit about good nutrition to present to an elementary school class.
2. Create a mural that expresses your feelings about world hunger.

SHARPENING YOUR THINKING SKILLS

1. How is good nutrition a way of caring about yourself?
2. What do your food choices say about your values?
3. Why is it important for you to make your own food choices?
4. Think about some of the food you eat that is not good for you. Who or what has influenced you to eat it?
5. Do you think food decisions reflect good eating habits? Explain your answer.
6. Describe the food choices you make when you are angry, sad, or frustrated. Are they different from the choices you make when you're happy, excited, or content? If so, how are they different, and why do you think this is so?
7. Describe how you help prepare meals for yourself or your family. What values are reflected by how you handle this activity?

APPLYING KNOWLEDGE

1. Draw a time line that shows who and what influences our eating habits over our lifetime.
2. Think of a food commercial or advertisement that you believe is misleading. Describe the ways in which it is misleading.

CHAPTER 4 REVIEW

Putting Your Values To Work

PRACTICING DECISION-MAKING SKILLS

Read about each situation. Then answer the questions.

Situation A: Sam sits at the same table in the cafeteria each day for lunch. On most days the same group of students eats with him. Not long ago another student started sitting at this table, too. This student is new to the school. He and his family have recently come to the United States from Vietnam.

Sam has begun to notice that the other students at the table make fun of the new student. They tease him about the food he brings for lunch and the manner in which he eats it. The student doesn't respond, and the kids who sit with Sam say that he doesn't understand English. Because the new student is in some of Sam's classes, Sam knows that he understands and speaks English.

1. What choices does Sam have in this situation?
2. What are the possible consequences of Sam's choices?
3. Which decision should Sam make? Why?

Situation B: Some of the kids who tease the new student are Sam's good friends. Sam also finds himself working with the Vietnamese student on a special project for a class. Sam's friends have told him to make a choice between them and the new student.

4. Does this change Sam's choices? Explain your answer.
5. What values are represented in this decision?

Situation C: It is the beginning of the school year. You notice that your friend Pat has lost weight over the vacation. You are happy for her because she was overweight and she was often teased by other students. You compliment her and ask how she lost the weight. She replies that she completely changed her eating habits. Now she very carefully watches everything she eats, doesn't eat empty-calorie foods anymore, and is getting lots of exercise. She tells you she does aerobics in the morning at home, gets to school early to work out on the track, and is swimming every day after school so she can try out for the swim team in the spring. You begin to notice, however, that she hardly eats at all. By Thanksgiving she is starting to look underweight, and by winter vacation she looks very thin. You finally tell Pat you are concerned that she has lost so much weight and that she doesn't seem to eat. She says you're just jealous because you are not as trim as she is. She adds that because everyone in her family is overweight, she needs to stay underweight for protection against being fat again.

6. What is the situation here?
7. What is your challenge?
8. What are the possible choices you have?
9. What are the consequences of each choice?
10. What decision would you make? Why?
11. How might you rehearse this decision?
12. How might you take action?
13. What values are represented in this decision?

CHAPTER 5

Exercising And Staying Fit

A TEEN SPEAKS

My school has a terrific basketball team. But only a handful of kids can make the team. My friends and I love to play, but we aren't good enough to make the team. For a while, we did nothing. Then we realized that we could form our own team and play after school.

Everyone has such a good time when we play. I enjoy bringing the ball upcourt, and my friend Mark likes to shoot. Bill is a big, muscular guy who likes to rebound, and he does it well.

My sister, Amy, doesn't care much for team sports, so she swims and runs. She keeps a log so she knows when she improves on her time. She says she wants to do her personal best.

Sometimes on weekends a bunch of us get together and go for a day hike in the mountains. It's great to get out in the fresh air with nature all around. Boy, do we feel tired when we get home! But it's a good kind of tired and we always feel really good after a good night's sleep.

It's so much fun to share something you enjoy with your friends. We may never be good enough to make the school team, but we sure have a good time—and it's great exercise, too.

Aldo

SECTION 1

VALUING EXERCISE

Playing basketball or other sports is one way to exercise. But you don't have to play a team sport to keep fit. Everybody needs exercise, and there are plenty of ways to get it. It can be fun. And exercise makes a difference in how you look, how you feel, and how you act. Regular exercise in everyday activities is part of your responsibility for your own overall good health.

Exercise

At every stage of your life, exercise promotes wellness. **Wellness** is the process of becoming and staying physically, mentally, emotionally, and socially healthy. The types of exercise you do may change, but the benefits of exercise don't. This is why many teens have already begun to make exercise a lifelong habit.

Exercise And Health — "I feel really good!"

I had such a good time in the fun run.

People who care about themselves make regular exercise a priority for good physical health. Lack of exercise has serious consequences, now and in the long run. For example, heart disease, obesity, and bone disease later in life are linked to inactivity.

Regular exercise has a number of important benefits. You feel well physically, you look and act better, and you learn to care about yourself and others.

Through regular, vigorous exercise, your body becomes stronger and healthier, you develop better coordination, strength, and a healthy heart. You also breathe easier and improve your endurance and flexibility. Your body and mind depend on your overall health, and regular exercise can improve your overall health.

Improve Your Coordination

Like many teens, you might feel uncoordinated at times—and you probably are! While some teens do seem to be natural athletes, most go through periods when they are uncoordinated. "Watch me swing the bat and miss." "My feet don't move where they're supposed to when I dance." These comments are not "klutz" talk. They are about some of the awkward moments of growing up.

Coordination is being able to move your muscles in a harmonious way. That is, your arms and legs do what you want them to do. Girls usually develop good coordination before boys do. You can give yourself a head start. Take up swimming, table tennis, softball, or another similar sport. These activities develop coordination.

Increase Your Strength

> I get so tired of being called a shrimp.

> It just doesn't make sense. My younger brother can do more pushups than I—every time.

You don't develop strength by complaining. For young men and women, regular exercise is the only way to build muscle strength. The stronger you are, the easier it is to move and lift objects, to withstand physical stress, and perhaps to get out of some dangerous situations. Some activities that can improve your strength are cycling, swimming, hiking, and running.

Walking briskly with friends is good exercise and fun.

Keep Your Heart Healthy

Your heart is a muscle. Like the muscles in your arms and legs, it needs exercise. A healthy heart is more efficient than a weak one. It can circulate your blood without working too hard. The only way to give your heart a workout is to get it pumping, with rigorous activity, for 20 to 30 minutes about three times a week. Playing basketball or soccer, going cross-country skiing, or taking long bike rides are ways to give your heart a workout.

Breathe More Easily

Why does exercise make you breathe faster? That's your body's way of telling you that you need to take in more oxygen for the amount of work that you are currently doing.

Benefits of Exercises

Different exercises have different payoffs. A balanced exercise program has a variety of exercises that promote different aspects of physical health—flexibility, strength, coordination, and endurance. These physical characteristics can be improved by taking up some of the following activities:

- **for flexibility**—certain calisthenics (arm circles, hamstring stretches, trunk twists), dancing, gymnastics, tennis, soccer, basketball.

- **for strength**—certain calisthenics (push-ups, bent-knee sit-ups, pull-ups), running upstairs, soccer, tennis, weight-lifting, mountain climbing, bicycling, hiking, swimming, skating, basketball.

- **for coordination**—tennis, swimming, softball, dancing, soccer, skating, basketball, badminton, mountain climbing.

- **for endurance**—running, jogging, bicycling, walking, swimming, jumping rope, aerobic dancing, basketball, handball, racquetball, cross-country skiing, badminton, hiking, rowing, tennis, soccer, skating.

When you're in shape, your lungs don't need to work as hard. An in-shape body needs less oxygen than an out-of-shape body for the same amount of work.

Improve Your Endurance

When I start to run, I get this surge of strength and energy. But it doesn't last.

Endurance is the ability to handle prolonged physical stress. Have you ever felt tired and begun to breathe heavily after walking up a flight of stairs? People who are in shape can play sports or do physical activities for a longer period of time before getting tired. Long-distance runners and basketball players have excellent endurance. Endurance requires strong muscles. It also requires a healthy heart and healthy lungs to deliver nutrients and oxygen to body tissues and remove wastes so you can keep going. You build endurance by pushing yourself a little more each time you exercise.

Keep Your Body Flexible

I feel so stiff when I exercise. But Jerry moves so easily.

Flexibility is the ability to move your muscles—stretch, bend, and twist—to their fullest extent. Poise and graceful movement come from flexibility. Being flexible feels good! Everyone can become more flexible by doing stretching exercises and staying fit. Gymnasts and dancers have excellent flexibility. But you have to stay in shape to be flexible.

Exercise And Appearance — "I like what I see in the mirror!"

Regular exercise can help you look your best. When you know you have good muscle tone, you feel better about yourself. A firm body looks trimmer than a flabby body, even when the weight is the same.

Maintain Your Best Weight

While it's important to watch the calories you eat, exercise is the key to weight loss or weight maintenance. Exercise, after all, burns calories.

Some activities burn more calories than others. Three factors determine how many calories exercise uses:
- **the type of activity,**
- **how long you exercise,**
- **how intensely you work out.**

You may bicycle with your friends, but whoever bikes longest and pedals fastest burns the most calories.

Rather than increasing your appetite so you eat more and perhaps gain weight, exercise actually may decrease your appetite! You feel hungry when your blood sugar level drops. When you exercise regularly, your blood sugar level is more stable.

You think you're too thin? A regular workout can help develop muscle so you have a fuller body shape. Think you're too fat? Increase your level of exercise.

Improve Your Posture

You've probably heard people say, "Stand up tall" or "Sit straight." Strong torso muscles help you stand and sit erect. People with good posture stand and sit erect. Strong torso muscles help to achieve that good posture.

Good posture enhances your figure or body build and helps your clothes look good on you. At the same time, it

Exercise of any kind can give you a natural high.

helps your organs function normally, allows you to move and sit comfortably, and helps your bones and muscles develop properly.

Stay Alert

> After I work out, I really feel pumped up. I feel sharp.

Exercise helps keep you alert because more oxygen is pumping to your brain.

A brisk walk does wonders to revive you if you feel tired during the day. Of course, if your body does not get enough rest, it needs sleep. Proper rest is also important for staying healthy.

Relax

> When I get stressed out or really angry, I go play handball. Or I go for a run. Afterwards, I feel calmer.

An exercise workout can help you relax and relieve tensions that often cause or aggravate physical ailments. Tight muscles, stomach discomfort, shortness of breath, and a rapid heartbeat are all caused by emotional tension. When you exercise hard, you forget your problems for a while.

Most people feel depressed occasionally. Research shows that exercise can be an effective way to fight these feelings. Exercise, however, won't get rid of extreme depression. Being positive makes you feel better and makes you more fun to be with.

Exercise And Values — "Go ahead, build my character!"

Playing sports and exercising are fun and teach you a lot about yourself and others. You learn to respect people's different abilities, develop sportsmanship, and work with others to achieve common goals. Sports and exercise reinforce values you will need throughout life. Learning about cooperation, teamwork, trust, and commitment is impor-

If you're concerned about your appearance, get into the regular exercise habit. It will help you to look your best.

Exercise . . . How Do Americans Compare?

Compared with people in other nations, Americans may have a life that is "too soft." We spend too much time sitting and not enough in vigorous exercise. As a result, obesity is a greater problem here than in many other countries.

In many places of the world, getting enough exercise is a normal part of the day:

- In Japan, exercise starts the work day. Each morning workers participate as a group in calisthenics at their desks or work stations.
- Many Chinese people start the day with tai chi, very slow, fluid movements that promote muscle control and flexibility. Because this form of exercise is so slow, it's also mentally relaxing and refreshing.
- Most Europeans live in cities rather than suburbs. Instead of riding by car or bus from place to place, many walk or bicycle to work or school.

tant to you now and later. While working out or playing sports may be fun, these activities may involve hard work. Sometimes this isn't fun, but you realize that it's worth doing because you want to become stronger or healthier. Sports and physical activity also build your confidence and give you the chance to share enjoyable activities with family and friends.

Sports build confidence, self-esteem, and friendships.

Learn Cooperation and Trust

Playing on a team is a great experience because you feel that you are part of something special. Almost everyone has played on a team, whether it is basketball, softball, or volleyball. Teammates work toward common goals and help each other to do their best through cooperation. Players learn to trust and respect each individual's abilities, and everyone works together for the best interests of the team. Teams that are successful usually win because they work well together.

Being a member of a team teaches you how to play fairly and how to be a gracious winner or loser. It's nice to have teammates around to share the joy of winning and to be supportive when you lose. Through playing together, team members become great friends!

Cooperation and trust are also developed and reinforced in individual sports. Friends can train together and help each other. For example, you share responsibility for safety when you swim with a friend or when you spot someone in gymnastics.

CHAPTER 5   EXERCISING AND STAYING FIT

Exercise is good for everyone.

Build Self-Esteem and Confidence

Exercise can build self-esteem and confidence whether you're playing a team sport or just working toward an individual goal. People who find other things difficult often excel and develop self-esteem and confidence in sports.

Lack of sportsmanship and respect for others, however, can lessen a person's feelings of self-esteem and confidence. Consider how destructive these comments are: "Don't pick her for our team; she can't hit the ball;" "You missed the basket—again!"

Exercise and Care

When you care about other people, you find ways to show them. These are ways you might link physical activity to showing how much you care:

- Encourage others to do their best in the activities they choose. You might say, "You can do it; I'm pulling for you."
- Support others' dedication to exercise. Don't force them to give up their workout for your own selfish reasons. Better yet, join them.
- Teach younger children to swim, or take time to play catch, jump rope, or go for a walk with them.
- Take a walk with an elderly relative or with a blind person. Or play basketball with a wheelchair-bound friend. Be sensitive to their limitations, but encourage their abilities.

Exercise doesn't have to be an organized sport. It can be just for fun.

Instead, these comments build self-esteem: "Let's give her a chance on our team;" "We'll help you learn how to play."

Develop Commitment and Self-Control

Consider the champion skier with only one leg, the Olympic athlete, or the young person preparing to compete in Special Olympics. Athletic success doesn't just happen. It results from commitment. Excellence in sports requires hours of practice. It takes a careful training routine with good nutrition and rest. Sometimes commitment and discipline are needed to overcome personal obstacles. That self-control might mean saying *NO* when friends urge that training rules be broken.

You can learn commitment and self-control from regular exercise, even if you don't plan to be a star athlete. By applying these values to other aspects of your life, you may be pleased with what you can achieve!

Exercise And Friends — "I'll go with you!"

A game of volleyball, a canoe trip, or one of many other sporting activities can help build relationships among friends and within families. Exercise doesn't have to be boring. Find sports you enjoy. You may not be the world's greatest, but neither is everybody else. Find friends who don't take themselves too seriously and can play for fun. Exercise should be so satisfying that it becomes a lifetime habit.

CHAPTER 5 EXERCISING AND STAYING FIT

What kinds of exercise are these kids getting?

Mom and I go for a walk sometimes. It's funny, but when we are walking I can talk to her about stuff and she seems to take it better.

Certain types of exercises, such as walking, hiking, or biking, offer good opportunities for communication with a family member or friend.

SECTION REVIEW
STOP AND REFLECT

1. Think about the benefits of exercise. Which ones would make a difference in your health or appearance?

2. Do you prefer individual or team sports? Why?

3. Refer to A TEEN SPEAKS on page 81. What sport do you play on a school team? If not, how can you organize a group of friends to play your favorite sport?

4. Imagine that you are in charge of a family vacation for fitness. Where would you go and what recreation, exercise, or sports would you plan for the family?

SECTION 2

GETTING ENOUGH EXERCISE

I'm not good at sports—I just don't have the time to exercise. People give many reasons for not exercising. Some are very real; others may be excuses that show people don't value exercise as part of keeping in shape.

Obstacles

Obstacles to exercise can be overcome by understanding the reasons for the obstacle, setting exercise goals, using self-control, and applying decision-making skills.

Wheels — "Why walk when I can ride!"

Instead of walking to school, many students ride in cars or school buses. Some take public transportation. They get rides to the store and to friends' homes. Some even get their own cars during high school.

You probably wouldn't give up wheels for foot power. Why not take advantage of both?

- Ride the school bus in the morning, but walk home with a friend when you have time.
- Give up after-school television one day a week. Instead, take a walk.
- Join an after-school intramural team, then walk home—if appropriate.
- Going to a shopping mall video arcade? Walk a few laps around the mall, too.

Lack Of Interest — "I'm no jock!"

You can enjoy sports in a variety of ways. Join school teams or teams organized by your church, synagogue, or recreation department. You don't have to be a great athlete. There are sports programs for people of all ages and abilities. You can enjoy participating in sports with people who have the same interests or ability levels as you. You may find participation in sports activities a lot of fun if you simply have a group of friends who get together several times a week to walk, skate, play tennis, or take part in another activity. It has been shown that when young people play on sports teams their grades improve, their self-esteem improves, and certainly their fitness levels improve. Playing sports can be an excellent way to stay in shape, meet new friends, and learn new sports skills.

You may not know that there are sports programs for people with handicaps as well as people without handicaps. Young athletes in these programs play basketball, run, swim, play tennis, ski, and skate even though they are blind, have cerebral palsy, are amputees, are in wheelchairs, or are persons with mental retardation.

There are few excuses for not being able to play sports. Consider the fact that many of our nation's top athletes have a handicapping condition. Imagine a professional baseball player who was born with only half of his left arm. Or imagine a professional golfer who is legally blind or an Olympic gold medalist

Even though these athletes have no use of their legs, they're still able to participate in sports.

Special Olympics offers competition to athletes with mental retardation.

who was born with little use of his left arm and hand. There are many more athletes who are able to play sports and to excel despite their disabilities.

Special Olympics is a program of sports training and competition for people with mental retardation. Some Special Olympics athletes can high jump 6 feet and run the 400 meters in under 50 seconds. Others play basketball, soccer, or other sports. They go to international games and compete against athletes representing almost 100 different countries, including Brazil, Poland, India, Australia, and Kenya. Winners receive medals to recognize their accomplishments, but in Special Olympics achieving one's personal best is more important than winning. Many teens participate in Special Olympics as volunteers or, if they have mental retardation, as athletes.

Training is the key to success for any athlete in any sports program. It is a lot of hard work. The fun part comes from seeing your skills improve, feeling your muscles, lungs, and heart getting stronger, and—most important—just looking and feeling good!

In most schools, young people are required to do some sports training in physical education classes or to participate in organized sports. If you train for your favorite sport, like softball, and do not make the school team, don't give up. Instead, join a team at the local YMCA or YMHA, Boys or Girls Club, or recreation department. Or ask your coach to start a unified sports team through Special Olympics; these teams combine young people with and without mental retardation on the same team. Athletes of the same age play on the same team with other athletes of similar ability levels. You don't have to be a jock to play, you just have to enjoy playing sports.

Some people think that playing organized sports is too serious or too competitive, and that takes the fun out of sports for them. Some sports leagues are very serious. The athletes in these leagues generally like a high-level, very competitive atmosphere. Other leagues are for less skilled players. Find one that is best suited for you. Choose one that will give you the most enjoyment.

No Motivation — "Give me one good reason!"

My mom is always after me to exercise. What's the big deal? I'm not overweight. I'm healthy. I've got better things to do!

Regardless of a person's age or their ability level, there is a sports league that's just right.

Sometimes it's hard to see the payoffs for exercising. Some are years down the road. Exercising to get into the habit for later may not be enough to get you going. But your motivation for exercise is best if it comes from within yourself. Try these tips for self-motivation:

- Find out which benefits of exercise seem right for you. Do you want to feel better or look better?
- Set realistic exercise goals to work toward . . . 10 push-ups this week, 20 next week, and so on. It can be satisfying to keep track of how well you're doing.
- Give yourself rewards for success. The feeling of fitness is a reward in itself! Or buy the new pants you couldn't quite button before.
- Exercise with a friend.

Fear Of Failure — "I can't stand losing!"

Exercise doesn't have to be competitive. Many individual activities — such as skiing, weight lifting, and archery — are noncompetitive. There are plenty of noncompetitive games, too. You can find books about noncompetitive games that will give you lots of choices. No one needs to fail. Anyone who takes time to exercise is a winner!

Competition in sports and games isn't all bad. Competition can teach you how to be a good loser as well as a gracious winner. After all, life is filled with both success and disappointment.

No Time — "I've got other things to do!"

Sure I should exercise, but who's got the time?

People commit the time for what they really value. If time is limited, try these exercise suggestions:
- Go to bed earlier, so you have a half hour in the morning for jogging or calisthenics.
- Do sit-ups while you watch television.
- Make exercise part of your social time with family and friends.
- Take full advantage of your physical education class at school. Avoid standing and talking with classmates when you could be working out.

Too Expensive — "I can't afford the equipment!"

I'd sure like to get involved in sports, but I don't have the money. Have you seen the sneakers those guys wear on the basketball court? If I go out there in these old shoes, nobody will take me seriously!

Training and equipment for some athletic activities can cost a lot of money. But many sports cost little or nothing. For example, running requires only the cost of a good pair of shoes. You may not be able to buy high tech shoes. The fact is, you don't need them as long as you have shoes that give you good support. You may want them to be like everybody else's. But that is another problem. It takes courage to be different. If you have a bicycle, use it. It doesn't have to be expensive to give you all the exercise you need. Use the school or community tennis court, basketball court, or track at no charge. Buy a used tennis racket. Enjoy swimming at a public or school pool. You may not have the best of everything you could buy for sports. But you can be proud of what you have. If your friends don't understand, then your friends have a problem.

A pick-up game of football is good fun and good exercise. What safety precautions should be taken here?

Live Life In Action

One, two, three—go! Creating an exercise plan that's right for you takes knowledge of exercise, precautions against accidents and injury, and a planned exercise routine. It also takes commitment.

A smart plan includes daily activity, regular physical workouts, and exercises that promote flexibility, strength, coordination, and endurance. A better plan includes friends to exercise with you.

Daily Activity— "I've got my backfield in motion!"

My Aunt Milly loves sports. She sits to read the sports page of the newspaper. She sits in her favorite chair and watches games on television. And she sits in the stands to watch my brother play soccer.

Watching other people play sports doesn't offer the benefits of exercise!

Keeping track of your exercise program in a log book helps you to see your progress. It also helps you to set realistic goals.

Being sedentary is, in fact, a health hazard. **Sedentary** means having an inactive way of life that is mostly spent sitting down. Think of all the times and places you sit during a single week.

Sometimes you have no choice. Sitting, for example, is appropriate for the classroom. Sitting burns only two calories per minute. Walking burns 5 calories per minute. Running may burn 18 to 20.

Consider the ways you could add activity to your daily schedule. Then add up the health benefits. You might:
- use stairs rather than an elevator.
- walk or bike rather than ride in the car or on public transportation—when appropriate.
- play pick-up basketball, volleyball, or some other sport with friends after school rather than watch television.
- walk and talk with a friend rather than sit and talk.

Working Out— "I'm into this!"

A smart exercise workout has three parts: warm-up, workout, and cool down. By doing all three, you get the most out of the workout without risking injury.

Aerobic activity is rhythmic, nonstop, vigorous activity that exercises the heart. Examples are jogging, long-distance cycling, or aerobic dancing. These activities raise your heartbeat

Warm up gradually and stretch before doing vigorous exercise.

from a resting rate to a training rate. For teenagers, the resting heart rate is about 70 to 80 beats per minute.

In the warm-up phase of a workout, you gradually increase your heart rate to a training rate. A low training heart rate for teenagers is 140 beats per minute; a moderate rate is about 155; a high rate is 170. Then you work out at the training rate. In the cool-down phase of a workout, you slowly bring down your heart rate again.

Take your heart rate before starting your warm up. Place the first two fingers of one hand in the hollow of your neck below your jaw. Count the heartbeats for 15 seconds, and then multiply by four. This gives the heart rate per minute.

Exercise vigorously at least 20 minutes and at least 3 times a week.

Don't forget to cool down after exercising — perhaps by walking or stretching.

Later, while you're exercising strenuously, count your heartbeats again. Did you reach a training heart rate? After the cooling down, your heart rate should return to a resting rate.

Step 1: Warm-Up

Warm-up prepares your body for exercise. Start by loosening up for 5 to 10 minutes. Slowly your body starts to heat up. Your heart gradually beats faster so exercise doesn't put as much strain on your heart and blood vessels. Your body uses oxygen more efficiently so you tire less quickly. Your muscles and tendons stretch. In this way, you help prevent muscle strains and pulls.

How do you warm up? Start with light exercise for about 5 minutes — walking, for example. Slowly increase your heart rate. Then you can do moderate exercises — without bouncing so you don't strain a muscle. Calisthenics, such as knee-bends, arm circles, and toe-touching exercises, are good. Then do light exercise for about 5 minutes. You might walk to slowly increase your heart rate. Gradually, make these activities more rigorous.

Step 2: Workout

Now give yourself a vigorous aerobic workout of 20 to 30 minutes. Work up gradually so you don't strain yourself.

When you're in shape, you'll be able to work out longer and harder.

A good workout should be tiring but not overly exhausting. If you're out of breath or feel chest pain, slow down. You should be able to talk during your workout.

Step 3: Cool Down

Cool down as gradually as you warmed up, for about ten minutes. This helps your body slow to its less active state. If you stop exercising suddenly, you may feel dizzy, nauseated, or faint. And you'll be more likely to get sore muscles.

How do you cool down? Gradually slow down your workout activity, and then end up with stretching exercises.

Safety— "I'm not into pain!"

Have you ever sprained your ankle or pulled a muscle when exercising? Like many other things, exercise has risks along with its benefits, especially when people don't take necessary precautions.

Exercising to avoid injury and accidents is one way of caring for yourself. That includes warm-up and cool-down activities. It also includes other safety precautions.

Wear proper shoes for your workout. For example, for running or walking, you'll need well-cushioned soles and wide, built-up heels to avoid foot injury.

On long exercise workouts be sure to drink plenty of liquids.

For mountain hiking, you'll need thick-soled shoes with ankle support.

If you run, walk, or bicycle in the dark, wear reflective strips on your clothes, this way motorists can see you. Better yet, work out during daylight.

Avoid exercising alone in deserted places. You can't get help if you injure yourself. And you risk personal assault.

If you're exercising outside in cold weather, wear enough clothing—layers of clothes, mittens, and a hat—so you don't get hypothermia or frostbite. With hypothermia, your body cools below its natural temperature. With frostbite, the skin tissue freezes.

Avoid prolonged exposure to the sun when you're active outdoors. Use an appropriate sunscreen. Wear lightweight clothes that breathe. Besides getting sunburn, your body may not be able to cool down effectively. This can be life threatening.

Drink plenty of water before and during exercise to avoid dehydration. **Dehydration** is excessive water loss; it can cause nausea, dizziness, headache, and other symptoms.

For water sports, learn to swim well enough to survive an emergency, but don't overestimate your skill. Always swim with someone else. Swim only in supervised areas, and follow the rules posted. Swim safely away from diving boards. Dive only where you know the water is deep. Always have life preservers with you when you're boating. And stay out of the water when you're chilled, overtired, or overheated. Muscle cramps are especially dangerous while swimming.

Always know the safety rules for any sport you participate in. And be prepared to administer first aid when necessary.

Staying Fit

Get fit and stay fit to stay well and feel good. It's worth the effort. It's also worth the effort to help those you care about to stay fit, too. Being concerned about fitness is a way of caring about yourself and others.

Diving with a friend shows self-respect and respect for others. Diving alone is not safe.

SECTION REVIEW
STOP AND REFLECT

1. Do you know someone who could use some exercise? What makes you think so?

2. What excuses does this person give for not exercising?

3. How might you help this person develop a habit of regular exercise?

4. Describe the type and amount of exercise you get on a regular basis. Based on your weight and your overall health, do you think you should increase your level of exercise? If so, how might you do it?

CHAPTER 5 REVIEW

Putting Your Values To Work

STRENGTHENING YOUR VALUES

Reread *A Teen Speaks* **on page 81. Then answer the following questions.**

1. How do you think Aldo feels?
2. What do you think Aldo cares about?
3. Can you identify with Aldo? Why or why not?
4. What values are present in this situation?
5. When have you misjudged others by how well they play a sport?

INTERPRETING KNOWLEDGE

1. Write a two-page biography about your favorite athlete.
2. Write a story about the relationship between team sports and the group projects you work on in other classes.
3. Build a model of your ideal sports club or gymnasium. Be prepared to give tours to your classmates encouraging them to join.
4. Draw a time line that shows the history of your favorite sport or exercise.
5. Write a play about the importance of exercise. Use your favorite athletes as the characters.
6. Design a world map showing typical exercises and sports in different countries.

SHARPENING YOUR THINKING SKILLS

1. Why is it important for you to make regular exercise a lifelong habit?
2. How does regular exercise reflect that you care about yourself?
3. Who or what has influenced your exercise habits? Is this influence positive or negative?
4. What do you do when you feel anxious, depressed, or tired? Why can exercise change the way you feel?
5. Why is dieting alone not enough to lose weight or maintain your ideal weight?
6. In terms of exercise and working out, some peoiple say, "No pain, no gain." Discuss whether you agree or disagree with this statement and why.
7. Why is exercise particularly important for teenagers?
8. Describe the relationship between self-worth and regular exercise.

APPLYING KNOWLEDGE

1. Participate in a walk-a-thon for the March of Dimes or other charitable organization.
2. Design a series of afterschool activities that involves only noncompetitive games.
3. Volunteer to work with an organization that helps keep public playgrounds clean and safe.

CHAPTER 5 REVIEW

Putting Your Values To Work

PRACTICING DECISION-MAKING SKILLS

Read about each situation.
Then answer the questions.

Situation A: Your best friend Stuart is tall and lanky. He's a fairly good swimmer and an excellent tennis player. The two of you have great fun working out, and he has helped you improve your tennis game. But lately he's developed one strong desire in life: He wants to lead the school football team to the state championships, just like his brother did three years ago.

When tryouts for the football team are announced, Stuart is the first to sign up. After tryouts, the coach tells Stuart that he is not built for football and he cannot play on the team. He encourages Stuart to try out for the tennis team or the swim team. He thinks that Stuart has a good chance of making either team. Stuart pleads with the coach. The coach tells Stuart that the team needs someone to sit on the bench to hand out towels and water. Stuart decides to accept this position.

Stuart confides in you that he plans to say that he will be playing on the team whenever he is needed. Stuart makes you promise to keep this secret. Stuart also tells you he will no longer be swimming or playing tennis with you, since all his free time must be spent with the football team.

1. What is your challenge?
2. What are the possible choices you have?
3. List the positive and negative consequences for each choice.
4. What values are reflected by each choice?
5. What decision do you think you would make? Why?
6. How might you rehearse your decision?

Situation B: One afternoon Stuart's brother stops by to watch the football team practice. After practice, he tells you he thinks Stuart is not really on the team.

7. He wants to know what you think. What do you tell him?
8. Would your decision be different if it was one of your classmates who made this comment? Explain your answer.

Situation C: Lee is an excellent musician who plays first violin in the school orchestra. She has studied music for several years and has won awards at recitals. She is also captain of the girls' varsity basketball team, of which you are a member. The team has not lost a game this season. Lee comes to you one day after basketball practice and is very upset. She tells you she cannot do both anymore. There just is not enough time to study violin, play basketball, and do her homework and chores. She is ready to give up both music and basketball and just take it easy.

9. To determine your response, make a decision following the decision-making model.
10. Would your answer be different if you were in the orchestra? Explain. Would it be different if you were not on the basketball team? Explain.

CHAPTER 6

Caring For Your Mental Health

A TEEN SPEAKS

All day long my friends have been asking me if something's wrong. Even my best friend said, "Hey! What's up with you? You're dragging your feet. And I haven't seen you smile all day. So what's wrong?"

What's not wrong with me? That's really the question. I feel like a blob of nothing. I don't get along with my folks. My sister and I argue all the time. My girlfriend dumped me. My grades are the pits. Coach asked me if I'm trying out for baseball this spring, but I just don't know if I have it anymore. I feel like slime. Sometimes I wish I hadn't been born.

I've been working on a career notebook for my social studies class. The trouble with that is I can't figure out what I want out of life, much less what I want to do and be. Sure, it would be nice to be rich and famous—or both.

And I'll admit it, I'd like to get married someday. But what good would any of it be if it doesn't mean anything?

Sometimes I get really down on myself. So how am I supposed to figure out what I'm going to be when I can't even get it together now?

SECTION 1

SPIRITUALITY AND SELF-WORTH

have you ever felt like you couldn't get it together? Or have you ever stopped to ask why things are the way they are—if there is any meaning to life? Thinking about the meaning of life is part of growing up. So is getting discouraged from time to time.

The more you begin to think about the future, the more likely you are to wonder about life and your place in it. Some of the questions you have may not even have answers! But, learning how to deal with the questions and the feelings of uncertainty that go with them is part of caring for your mental health.

Mental Health

Mental health is very different from physical health because you can't actually see and touch what goes on in your mind. If you cut your finger, you feel it and see the blood. You know you need to take care of the cut in order to stop the bleeding. When your mental health is damaged, you don't always see the result. For example, you may find that you are constantly discouraged, but you don't know why. It isn't so easy to see what needs to be taken care of. Being constantly discouraged is a mental health problem.

Mental health is the state of being comfortable with yourself, with others, and with your surroundings. It has to do with memory, thinking, perceiving, feeling, and wishing. These are spiritual rather than physical qualities. Spirituality is an important part of your mental health. It may or may not involve religious feelings. Many religious people talk about spirituality. But the idea of spirituality is not religious in itself. It is an inner quality of being. **Spirituality** is the part of yourself that allows you to know you are alive and to feel the wonder of being alive. It is the part of yourself that nudges you to ask the big questions such as, "Why am I here?" and "So what's the use?"

This is why spirituality and self-worth are closely related. **Self-worth,** sometimes called **self-esteem,** is the value or importance you place on yourself. When you value yourself, you are likely to have good mental health. So you are able to

- like and accept yourself even when you question and doubt yourself,
- express emotions in a positive way instead of running away from them or feeling guilty about them,
- face the problems and stresses of daily living and keep on asking "Why things are the way they are."

You Are Special

If someone asks who you are, what do you say? "I'm Chris Nelson." "I'm a Texan." "I'm Catholic." Each of these is an important part of who a person is, but none of them taken alone truly defines a person.

Who you are includes the image you have of yourself now. This image or how you see yourself is called **self-image** or **self-concept**. It includes the person you think you can and will become in the future.

Being Unique — "Who am I?"

Liking what you see in the mirror is good for your mental health.

Biologists would classify you as an animal. Whether or not you agree with biologists, you would probably realize that you are unique among all the animals in the world. You have the ability to think. You can make choices. You experience emotions such as love, fear, and anger. But these are not things that make you unique. Some animals seem to be capable of thinking and experiencing emotion.

What makes you unique is your spiritual nature. Your spiritual nature enables you to see the world around you and reflect on it and your relationship to it.

Religion helps people understand life.

Making Sense Of Things — "Who am I?"

It is our spiritual quality that urges us to make sense of things. People want to discover meaning. We look for it in many directions.

The great world religions are ways through which people try to make sense of life and find meaning. In studying and comparing religions, you would find that from the beginning of time, people have recognized their spiritual nature and looked for ways of expressing it. For many people, religious beliefs and practices are important to their spiritual wellbeing and mental health.

While many people express their spiritual nature through religion, others believe that life itself has meaning totally independent of anything or anyone outside ourselves. Life simply is. And people must give meaning to their own lives.

People with this viewpoint may not use the term "spiritual" to describe the capacity to search for meaning. They tend to believe that meaning is discovered as we seek to develop all of our abilities and interests, thereby becoming the best we can be.

This belief, like religious beliefs, is closely connected to mental health. Persons who believe that life has meaning independent of religious faith and practice, like those who are religious, connect value to that meaning.

Early Experiences — "I was just a kid!"

Why do some people feel their worth when others don't? How you feel about yourself comes from experiences you've had since birth. When you were just a toddler, you were already trying to figure out how you fit into things. You wanted to feel that you had a place in the world. Your sense of self-worth was reinforced by the way other people saw you. When they treated you as a person who had an important place in the family, you felt good about yourself.

For example, when a baby is learning to talk, he or she does a lot of babbling. Sometimes when family members are trying to talk with each other, the baby babbles louder and louder. The baby is trying to imitate what others are doing.

Family members may react in a gentle and positive way. They might say things like, "Isn't that sweet?" Or they may even correct the baby by saying, "You want to talk, too! But we can't hear each other when you are so loud!" When this happens, the baby feels as if he or she belongs. Even when corrected, the baby doesn't feel put down. Many such experiences of belonging and being valued lead to later feelings of worth.

Family members might also ignore the babbling. They could make fun of it. And they might react in a mean and angry way by saying, "Shut up! We can't hear what people are saying." Then the baby is put down and excluded. Many experiences of being excluded lead the growing baby to feel that he or she doesn't belong. Such experiences lead to feelings of worthlessness.

Your feelings of self-worth are affected by childhood experiences.

Values And Self-Worth — "I have value!"

Values help us to express our spiritual nature. What you are able to get out of life depends on you. We are a person of unlimited worth.

The wonderful thing about your worth is that you are born with it. You don't have to work for it. You can't buy it. And, because of your unique, spiritual nature, you are able to reflect on your own value as a person. You can ask what it means to be you and why that is so important. You have self-worth, or feel good about yourself when you recognize the value that you have as a person. Understanding self-worth is the key to understanding your mental health. And it is the key to valuing others.

When you were younger, you may not have thought about taking charge of your mental health. You knew you felt up or down. You felt good about yourself sometimes and rotten about yourself at other times. Now you can analyze your moods. You can think about how you feel about yourself and why. Self-worth affects every aspect of your life:

- your performance in school,
- your creative talents,
- your energy level,
- your relationships with others,
- your overall satisfaction with life.

UNIT 2 CARING FOR YOURSELF

Good relationships require nurturing and caring.

Your Self-Worth

Some people think their value as people depends on things like the ability to
- make good grades,
- be elected to the student council,
- be a good athlete,
- look good,
- please their parents.

This is a big mistake. These things may well add to your feelings of self-worth. But they are not what makes you of value. Remember, you are born with worth. It comes with the territory of being human. Some people seem to accept and feel their own value. Others don't. What you believe about your own value is very important.

Building Relationships

Often you hear people talking about relationships. **Relationships** are special bonds formed with other people. A good relationship is a good friendship. How do you build good relationships? You build good relationships by **nurturing,** or caring for, them over time. How well you get along with other people is determined by how well you
- care,
- show empathy,
- accept others,
- respect others,
- trust others and are trusted,
- assume responsibility,
- communicate.

True friendship comes when people give and receive affection. Affection is more than a romantic feeling. It happens when people care about and emotionally support one another. When you "feel with" others, trying to understand and support them, you are being **empathetic.** It is a nice feeling for them as well as for you. When you care, with empathy, you discover that you are somebody capable of such feelings. And you realize that you are of value. When you accept yourself, you are more able to accept others. This builds a circle of caring, empathy, and affection that you give and that comes back to you.

Trust and responsibility are built over time, together. Honesty, respecting rules, commitment to others, offering consistent support, and sharing are responsible actions that show you are worthy of trust. You can be counted on. Lying, betraying or rejecting someone,

Everyone experiences rejection from time to time. It is important to learn to cope with it.

breaking promises, and acting in an uncaring way all destroy trust and relationships.

All the values and ways of behaving that promote good relationships require honest, open communication. You will learn about communication in Chapter 11.

Repairing Relationships

When relationships with family and friends aren't so good, where do you go for help? A school counselor, social worker, member of the clergy, or mental health professional can help. The individual you turn to and what you do will depend on the problem. Some of the more common problems we experience are conflict, shyness, rejection, alienation and loneliness, envy and jealousy. Conflict is such a powerful force in relationships that all of Chapter 12 is about how to deal with conflict.

Shyness

Shyness is a defense mechanism that people learn. In American culture, many people try to overcome feelings of shyness. But in some cultures, shyness is encouraged.

People who want to avoid being with others are often shy. Being shy can go with lack of self-confidence or insecurity. Shyness tends to keep people from developing relationships.

It's possible to overcome shyness if you really want to.

> I really don't want to go to the birthday party at Kiko's house.
> I don't know any of her friends.
> I wouldn't know what to say.

You can do some things to overcome shyness:
- **Become a good listener.** People enjoy being with a good listener.
- **Find out about the other person.** Get that person to talk by asking questions about hobbies, what the person likes and doesn't like at school, a movie.
- **Remember your own value.** You are somebody worth knowing.

Everyone feels shy sometimes. There is nothing wrong with being shy. However, shyness can become negative. Shyness is negative when it leads to feelings of loneliness and rejection.

Rejection

Rejection can send your feelings of self-worth on a downhill slide. But being turned down or excluded can be hard to take.

> I've asked three people to the dance and nobody will go with me.

When you feel rejected, there are some questions you can ask yourself to work through your feelings:
- **Am I being rejected or am I just having a bad day?** Don't jump to conclusions. Sometimes you feel rejected when you haven't been. You may just be feeling down, or you may have misunderstood what somebody said or did.
- **Why am I being rejected?** Sometimes people are rejected because they have different values.
- **What am I willing to give up in order to be part of their group?** Others may want you to do things that go against your values. Then it is time to ask if acceptance is worth it.
- **Do I have the time and energy to be friends with everybody?** You don't have to be friends with everybody. You'd have to be superhuman to keep up with so many different people wanting to do a lot of different things. And you'd have to be pretty superficial not to have some people dislike you some of the time!
- **Is there an outside chance that my personality isn't completely perfect?** Nobody is going to be completely perfect! But, take a look at yourself. Are you conceited? Are you too crit-

ical of others? Prejudiced? Insincere? Thoughtless? If so, can you honestly blame others for not wanting to include you? Maybe you need to make some changes.
- **Is this the worst thing that could happen?** Don't let rejection get to you. Think about the friends you do have, or think about how you can work on some of the problems you have that turn people off.

Alienation and Loneliness

When people feel **alienated** they feel cut off from others—or lonely. You are likely to hear comments like these: "My parents don't understand me. They never have." "Nobody ever calls me just to talk."

Sometimes feelings of rejection can lead to alienation and loneliness. All of us feel lonely sometimes like when a friend lets you down. But if you experience loneliness most of the time, it can be hard on your self-worth. It can cause you to withdraw even more and become alienated from others.

If you handle rejection in a positive way, you are likely to deal with loneliness and alienation in the process. You can also nurture friendships and look for opportunities to be caring. Volunteering to work with others, for example as a tutor for a child, can be a way of directly dealing with loneliness.

Envy and Jealousy

"Sure, I'd have a lot of friends too, if I had a car like that!" That comment reflects envy and jealousy. Envy and jealousy are powerful emotions. **Envy** is

Believing in yourself and your abilities makes it possible for you to try new challenges.

wanting something someone else has that you can't or don't have. **Jealousy** is feeling hostile or suspicious of another's possessions or good fortune. Sometimes envy and jealousy are unproductive feelings. But sometimes they can help us grow.

For example, you may envy a classmate who seems to get all the solo parts in the school chorus. This could motivate you to practice longer and more seriously.

Everyone can and does experience social problems. Problems with relationships are a reminder that we have real differences in values, beliefs, and ways we like to do things. Learning how to repair or deal with problems in relationships enables us to have an even stronger sense of self-worth.

Believing In Yourself — "I'm okay!"

Childhood experiences where your feelings, goals, thoughts, and dreams were valued helped you to have feelings of self-worth. Without enough of these experiences, it is hard to feel your own worth. But it isn't impossible.

What you believe about yourself is important to your mental health. Your beliefs about yourself will help you deal realistically and honestly with your emotional feelings. With all the changes that are taking place in your body and your life, you probably feel that there are several different versions of you. Sometimes you feel unpredictable. You may find yourself doing things that confuse you or even make you feel ashamed. Knowing and believing that you are a person of worth will enable you to like and understand yourself. No one can give you a magic formula that will guarantee you personal self-worth. You have to work it out for yourself. But you can begin by learning how to understand and deal with your relationships.

SECTION REVIEW
STOP AND REFLECT

1. Describe some of the experiences you have had that make you unique.

2. Tell what is most likely to destroy your feelings of self-worth. Tell what builds your feelings of self-worth.

3. Describe how your family makes people feel that they belong.

4. Give an example of how you might help others in your family feel that they have a special place in it.

5. Describe a memory you have of the way you were treated as a child.

6. Tell how the way you were treated affected your feelings of self-worth.

SECTION 2

EMOTIONS AND SELF-WORTH

Strangely enough, the way you feel may not have much to do with the way things really are. You can't help how you feel about things. That is what is tricky about emotions. For example, have you ever felt like this?

> My little sister is such a brat. Sometimes I just can't stand her. She's always getting into my stuff without asking. I get so mad at her, I could just smack her with my hairbrush. Then I get so ashamed of myself for getting mad at her.

Being angry with someone is not bad. Hitting somebody with your hairbrush is!

Emotions

Human beings are feeling creatures. Anger is just one of the emotions we feel. Emotions are the feelings we have of joy, sorrow, love, hate. Fear, bad moods, depression, frustration, guilt, grief, pride, happiness, good moods—all of these are emotions. We have the capacity to think about our emotions and to act in appropriate ways in expressing those emotions.

Emotions come from your past experiences of pleasure and pain. Sometimes your emotions may surprise you. You may not know why you feel angry or frustrated or jealous or sad. It just happens.

Feelings are neither good nor bad. They're neutral. The actions, or how you deal with your feelings, are good or bad. The best way to deal with emotions is by acting on them in a positive way. For example, when you are angry, put your anger into words that describe how you feel.

The experiences that cause you to feel an emotion are as varied as life itself. Emotion is involved in everything in which you are involved. You experience emotion
- when your desires are fulfilled,
- when you are prevented from doing something,
- when you are threatened or harmed,
- whenever something happens that affects (negatively or positively) the way you see yourself,
- whenever you discover what you can and cannot do.

It's okay to show emotion. Major athletes do.

The way you handle your emotions affects your sense of self-worth. If you don't express emotions or if you dump them on others, you can feel guilty or down on yourself.

Some of our strongest, most emotional feelings come from self-doubts. For example, suppose you've been asked to represent the students at your school by speaking on a local television talk show. You may be thrilled to be asked to be on television, but you are secretly afraid you can't do it. You don't want to make a fool of yourself. As a result, you have extreme feelings of anxiety.

Anything that appears to threaten your sense of independence, your self-confidence, or your feelings of self-worth is bound to create emotion! It doesn't matter whether the attack comes from within yourself or from someone else.

Naming Your Emotions — "Boy, am I mad!"

It's important that you experience emotions and that you name them for what they are. When you're angry and then deny that you're angry, you're doing something that can become unhealthy.

Instead of admitting to yourself that you are angry, you put the anger into a little imaginary box inside you where you keep it hidden. Something else happens that makes you angry. You put it in the same little box. After you've done that a few times, the box is full.

Now, it only takes one unimportant comment or event to make that box burst open. Your anger erupts like a volcano. When that happens, you might say, "I don't know what got into me. I overreacted." What really happened was that you let out all of the anger that had been piling up.

Perhaps you have heard that you shouldn't show your emotions. Your parents may have said, "Don't cry." A teacher may have told you, "There's nothing to be afraid of." However, your good mental health depends on your ability to recognize your emotions and to express them appropriately.

Emotions that feel unpleasant or stressful are usually harder to handle than those that help you feel good. Sometimes the best decision is to wait and act when you are calmer. Then you can act objectively. For example, if your best friend gossips about you, you may need to wait until you can have a good cry. Later you can tell your friend how you feel.

Even though emotions are a part of us, they don't have to rule us. We have the capacity to think about our emotions and act in appropriate ways to express them.

Love

Love is an emotion that is very much like caring. Tenderness, feeling empathy with another person, and the desire to cherish and protect are all part of loving. True love is given freely without asking what it will get in return. We all need to give and receive love from family and friends.

But when we talk about love, most of us think of romantic love. "Will I ever fall in love?" "How can I get someone to love me?" These are common concerns for teens. The fact is, most teens fall in love several times. The experience of falling in love can help you to grow in your understanding of yourself and of the opposite sex. It also helps you learn to relate to others.

Being in love can involve all of your emotions, from joy to anger. It helps you to grow and to mature. But, when you are in love you can be so overwhelmed with feelings that you ignore weaknesses in the person you love. Or you may ignore conflicts and differences in values that could lead to later unhappiness. In fact, being in love may involve attitudes and desires that are not healthy. You can see these unhealthy attitudes and desires in the person who

- feels unloved at home and seeks love from a boyfriend or girlfriend to make up for it,
- uses others to make himself or herself feel more important,
- makes sexual conquests in order to drop someone and create pain,
- creates dependence in order to control someone else,
- uses love as an excuse to avoid growing up and making his or her own decisions.

Unhealthy love says, "I want you for what you can give me." Healthy love says, "I love you. What can we give to each other?"

True love is friendship at its best. It is caring. It is responsible and shows respect for the attitudes, beliefs, and goals of the other. True love does not include taking advantage of another person sexually. If it's true love, there is enough caring and respect to wait. True love never says, "If you love me, prove it." True love is honest.

Anger

As you know, being angry isn't bad. Rather than hiding your anger in an imaginary box, you can find ways to work through it. Yelling, arguing, or fighting usually doesn't solve the problem. Positive anger is the anger you put into words: "I get furious when you get into my stuff without asking. I feel like smacking you." Negative anger is the anger that simmers. It explodes in violent actions like hitting and acting out. Try these approaches to avoid negative anger:

- **Give yourself time to cool off.** Take a walk or count to 10. This will give you time to put your feelings into perspective.
- **Express your anger in a nonviolent way.**
- **Discuss the problem rather than argue.** A verbal attack can break off communication.
- **Vent your anger.** Write down your feelings or talk to a friend.
- **Think about why you feel so angry.** Are your feelings reasonable?
- **Work with your family and friends.** Learn how to accept feelings and show concern.

Anxiety and Fear

Have you ever felt tense or nervous before taking a test, giving a report, or walking down a dark, deserted street? Your heart beats faster. You breathe faster. Perhaps you can't eat. These are signs of anxiety or fear.

When you feel threatened, fear is a normal reaction. Some threats are real, such as the fear of injury if you're outside in a bad storm. Other fears are imagined and may be destructive. Overcoming unrealistic fears may require some personal risks, but you can become stronger as a result.

Anxiety can be positive or negative. A little nervousness can help prepare your body for action. Your anxiety can help you give a better report or do better on a test by releasing the energy to work harder. This is positive anxiety. Negative anxiety can cause panic, worry, sleeplessness, physical ailments, even an inability to function. You can cope with mild anxiety by

- learning to relax,
- talking out your anxieties,
- laughing at yourself,
- making wise decisions,
- confronting your problems.

Bad Moods and Depression

Moody and depressed people can make the world unpleasant for everybody around them. "Just leave me alone—I don't want to talk to you!" "Do what you want—I don't care!" Comments like these are harmless in themselves. But when they are all you seem to feel like saying or the only things you hear from someone, they can be hard to deal with.

People get moody and depressed for many reasons. These feelings may come from a poor sense of self-worth, lack of self-control, or feelings of anxiety, rejection, or guilt. They often develop when events cause major changes in your life. For example, a family move or breaking up with a girlfriend or boyfriend can cause you to feel depressed.

Having peaks and valleys in your feelings is normal. Your ability to experience joy means that you can also feel sorrow. The goal is not to live on a great high or down in the dumps all the time, but to find some middle ground. To live without expressing emotion would be to have others see you as an unfeeling blob. But to live at extremes isn't real either. Usually, confronting feelings, sharing them with family or friends, and being active can help you deal with extreme feelings. If you have severe depression—a feeling of being very low or sad for a long time—or if one of your parents is depressed, you need professional help. To find someone to help you cope, talk with your school counselor, a trusted teacher, a member of the clergy, or a health professional.

Guilt

> Some of the guys told my brother Warren that they'd seen us girls cheating on a spelling test. Warren nearly got into a fight with those guys. He said he knew his sister would never cheat on a test. He made them take it back. Now I feel rotten—because I did cheat.

Guilt comes from feeling responsible when something goes wrong or when you know you've done something that isn't right. Guilt may be a message from your conscience saying, "You shouldn't cheat on tests," or "You really let your brother

Learning to relax is one way to cope with mild anxiety.

down when he trusted you." You can act on these feelings to make things better.

Sometimes people feel guilty for things they can't control, though. For example, some teens blame themselves when parents divorce. They keep thinking, "If I'd just been better it wouldn't have put so much pressure on Mom and Dad."

Drawing conclusions without knowing all the facts can set you up for needless guilt. So can accepting responsibility for things over which you have no control. Talking and listening can

Around the World

Emotions: An Expression of Culture

How do you express your emotions? Are you reserved? Or do your feelings show in your every action or word? Your culture influences these expressions.

Mexicans, Brazilians, and other Latin Americans often display their feelings very openly. For example, the Spanish language is more emotionally expressive than English is, with many words to show different kinds of love. Latins use superlatives—"fantastic," "stupendous," "magnificent"—more frequently than most Americans do.

Asians tend to be reserved and polite. In Indonesia, praise offered directly is considered crude; instead, it's given through a third party. Soft speech is a sign of politeness, not shyness. Americans express gratitude by saying "thank you." But in India "thank you" is avoided because it signals finality or ending a relationship.

To the expressive Latin Americans, people from the United States are often described as restrained, even unfeeling. But the reserved Asians may view us as aggressive, hot-tempered, or excitable.

help you avoid or relieve guilt. In cases where feelings of guilt are extreme, it is best to get some help from a trustworthy adult.

Frustration

Frustration comes from feeling that things are out of control. Being frustrated may cause you to feel ineffective, discouraged, tense, or anxious. You can deal with frustration by accepting things you can't change. And you can also avoid situations that create unnecessary frustration. For example, if you are frustrated by always missing your favorite television show, make a point of doing your homework right after school. Then you'll have free time to watch the show. You can also deal with frustration by talking about it with a friend or your family.

Grief

Donna moved to the Midwest last week. Every time I walk past her house I start to cry. We were best friends. I'll never have another friend like her. Why did she have to move?

Grief is the emotion you feel when you've experienced a shattering loss, as when your best friend moves away. Death, divorce, and disappointment can lead to grief.

Grief is a normal part of accepting loss. It happens because you care. Usually, people go through several stages in their grieving:

- **Denial.** "I know Donna's parents are talking about making a move, but it'll never happen."
- **Anger.** "How can they think of moving? It isn't fair to take her out of school!"
- **Bargaining.** "Why don't you let Donna finish the school year here? She can live with me."
- **Depression.** "I really don't care about anything since Donna left."
- **Acceptance.** "Donna is gone. I miss her so much. We'll always be friends, but I need to make some new friends, too."

It is normal for grief to last. When someone we love dies, it can take months—or years—to get over it. Time and support from others will help grief to pass. That's why it is so important to be able to share feelings with other people.

People who lack self-confidence are often hesitant to try new experiences.

Emotional Support — "I need somebody!"

It is especially important to all of us to feel valued by our parents and family. After all, your family can be your best support system. They can help you get through even the worst of times. When people don't feel valued at home, it becomes easy for them to fall into a pit of self-doubt and depression.

Playing sports is a way to improve your athletic skills. It's also a way to build strong friendships and self-confidence.

You are valuable whether or not you feel that way. You are valuable whether or not anybody treats you that way. Realizing that you have value is an important step in mental health. If you don't have friends and family to support you, this will be hard—but not impossible.

Getting Help— "Who can I turn to?"

It's smart to get help when you have emotional or social problems. People often can't handle all their troubles alone. You need to be cared for. Most of the time you can turn to people at home or friends for support.

But, like physical problems, some emotional problems need professional care. Counselors, psychologists, and clergy, among others, are trained to give specialized help. A professional will be sensitive to your feelings and help you find support you can feel good about. There are people who care. They will care about you if they know you need them.

Support groups can also give personal insight and help. Support groups are people with similar problems or concerns who join together to help each other. For example, Alcoholics Anonymous has support groups for teens who are alcoholic or whose parents are alcoholic.

Some schools have peer counseling or support programs. Teens can help each other through rough times.

Self-Worth— Your Best Support System

When you are able to affirm your own value, you are able to feel good about yourself most of the time. You'll still have times of self-doubt. But these pass.

Usually, persons with a positive sense of self-worth are also able to have good relationships with other people. When you care about yourself, you are in a better position to care for others. Furthermore, you are likely to take good care of yourself.

On the other hand, people who don't have a positive sense of self-worth often withdraw from relationships with others. They show a lack of confidence and self-respect. Frequently, they also fail to take good care of themselves.

You aren't through growing up. In fact, neither are the adults you know. We keep learning new things about ourselves everyday. This means that even trusted adults can make mistakes and need to be cared about. Acceptance, empathy, forgiveness, and caring are a two-way street. Our parents, teachers, and friends need them as much as we do. You are more than just a body controlled by a mind. You are a whole person, one with body, mind, and feelings. You can reflect on events and their meaning. You can experience life and shape it instead of just letting it happen to you!

How can you improve your own self-worth?

SECTION REVIEW
STOP AND REFLECT

1. Describe a time when you knew you wouldn't trust your feelings.
2. Tell about a time when you were really angry. How did you deal with it? Describe what happened.
3. Tell about a time when you were really happy.
4. Imagine that you need to be convinced of your worth as a person. Write a letter to yourself. Tell why you are special. Include all your good points. Think of the talents you have that could be developed. Mention your values. End the letter "Love," and sign your own name.
5. Find someone to read your letter to. You could read it to a trusted friend (of any age). Or read the letter to your reflection in the mirror or to your pet.

CHAPTER 6 REVIEW

Putting Your Values To Work

STRENGTHENING YOUR VALUES

Reread *A Teen Speaks* **on page 105. Then answer the following questions.**

1. What is Joseph's attitude toward his life now?
2. What values do you think Joseph has?
3. What values do you think Joseph is lacking?
4. Suppose you are a close friend of Joseph's. What might you say to him?
5. Think of a particularly difficult period in your own life. How did you feel? What things discouraged you? What helped to improve your attitude?
6. Think of a particularly positive time in your life. What things were going well in your life? What were you doing to cause this?
7. What do you really value in your life? What people are most important to you? Who can you really count on?

INTERPRETING KNOWLEDGE

1. Write a description of yourself as you think you will be in ten years. What are you like? What is your life like?
2. Create a three-dimensional collage of things that represent you and your values. You might include photos, pictures, and objects.

SHARPENING YOUR THINKING SKILLS

1. Define *mental health* in your own words. Think of a person who seems to have it all together, who is mentally healthy. What qualities does he or she possess? How does the individual lead his or her life?
2. Look at a positive relationship you have with someone in your life. What makes it work? How do you both contribute to it?
3. Look at a negative relationship you have with someone. Why is this relationship not successful? What steps can you take to improve the situation?
4. This chapter lists several ways to handle negative anger. Do you think any of these ways might work for you? What other suggestions would you add to this list?

APPLYING KNOWLEDGE

1. Keep a daily journal describing what went on that day, include your feelings about what happened. Find at least one positive thing to say about your day.
2. Make a list of at least five things you do well and would like to continue to pursue. Next to each item, note how you will continue to include it in your daily life. Next, make a list of at least five things that you would like to improve about yourself. Note the actions you will take to make this happen.

CHAPTER 6 REVIEW

Putting Your Values To Work

PRACTICING DECISION-MAKING SKILLS

Read about each situation. Then answer the questions.

Situation A: Refer to *A Teen Speaks* on page 105. Think about a time when you have experienced feelings similar to Joseph's.

1. What was the situation in your own life?
2. How did you feel?
3. What things seemed particularly difficult for you?
4. Despite the difficulties you were having, what were some positive things in your life?
5. What circumstances changed to make the situation better?
6. Did you take any particular actions that seemed to help? If so, what did you do and how do you think this might have helped to improve the situation?
7. Were there any people in your life who helped you to see things differently and get back on track? If so, what did they say or do to support you?

Situation B: You have a very close friend who has been very depressed and down on herself lately. She has a lot going for her but also has some problems in her life. You know that only she can make herself happy and that she must face her problems and make decisions for herself, but you want to help in whatever way you can. Other people have tried to encourage her, but she doesn't seem to respond to their efforts. Despite this, you still feel that you need to try to help.

8. What are the important issues in this situation?
9. What might you say and do to help her through this difficult time?
10. What might be the consequences of each action you could take?

Situation C: Suppose this friend has been depressed over a period of several months and has shown no signs of improvement. You are extremely worried. You think she may be so unhappy that she needs professional help.

11. Does this information change the actions that you suggested above? If so, how?
12. What might you do for your friend in this situation?
13. What might you say to your friend?
14. At what point might you consider talking to a responsible adult about your friend?

Situation D: Imagine a really good day. Things are looking up. You feel good about yourself.

15. Describe your day.
16. What specifically is going right in your life?
17. How do you feel about yourself?
18. What have you done to help make this happen? What can you actually do in your life to make this day happen? What types of actions can you take? What choices or decisions do you need to make?

CHAPTER 7

Saying *NO* And Feeling Good About It

A TEEN SPEAKS

I've been babysitting for the Levys for almost two years—since I was 13. They have two little kids that I just love.

I've been saving all year for this stereo system, and I still have $100 to go. So I was real glad when the Levys said they'd be out late on Saturday night—more money for me!

Then I told Eric, my boyfriend. He started in on me and wouldn't let up: How come I take all my Saturday nights for myself? Couldn't he come over for a while Saturday night?

It was Monday when I told him. By Thursday his friends were on me. How about if just a few of us came over? They'd bring pizza.

I knew that was a bad idea. But they kept on me. I could feel myself starting to wear down. So finally I told Eric that he could come over after the kids were in bed—just him.

So on Saturday night Eric shows up with four guys and three girls. I couldn't believe it. He said, "It's okay, we'll only stay till we eat the pizza!"

I don't even want to think about what happened. Pizza got on everything. Pretty soon it was loud and the kids were awake. While I was getting them settled Eric and the others found the hot tub and the beer. When I got back, they were all stripped down in the tub, and drinking.

That's when the Levys came home, exactly four hours early.

Holly

SECTION 1

WHY IT'S TOUGH TO SAY *NO*

Saying *NO* can be rough. Saying it to a close friend can seem downright impossible. Turning down a person you care about can feel like you're turning him or her away.

> He said because we're such good friends, I should lie about how he got his black eye. He didn't want his folks to know he'd been in a fight.

And when you turn down someone you care about, you may be afraid they'll turn you away, leaving you all alone.

> I was afraid if I didn't sleep with him I'd lose him.

Inner Pressure

Maybe you're afraid that if you say *NO* to the wrong person, you'll be labeled a "chicken," "prude," or a "wimp." Or maybe saying *NO* makes you feel so guilty or foolish you'll do anything to avoid that uncomfortable feeling. Finally, you may find it tough to say *NO* because you don't really know how you feel about an issue. The questions may seem to paralyze you. *YES* or *NO*, should you let your friend copy from your test? *YES* or *NO*, should you ride with someone who's been smoking dope or drinking? *YES* or *NO*, are you really ready to have sex? To say *NO* effectively to someone else, you first have to decide where you stand on any *YES-NO* question.

> I'm no thief. But it was like my hands weren't even attached to me. Before I could stop them, they just snatched up the bathing suit and stuffed it into my bag.

Pressure doesn't always come from the outside. Sometimes it comes from within your own body or mind. It can be hard to ignore. Your emotions may be swaying you to get involved. Your mind says it's really a low idea. Your body may be telling you to pick up the pills just this once so you can stay up all night to study for the math test. Your head knows you'll feel wrecked tomorrow. The inner pressure can get you into trouble or get you hurt if you act before you carefully consider what you really value and what's at stake.

Peer Pressure

"Come on," she said. "I've never been caught."

So I took the dare. I made the dive. I dove, all right. Right into this wheelchair.

Everyone wants to be liked, accepted, valued, looked up to. That's normal. Sometimes, when you feel unsure of yourself, you may look to other people for answers, using their behavior or habits as a guide. You may look to your **peers,** people close to your own age, as models. You watch them so you'll know how to act, what to wear, where to go, or what to say. Sometimes, when you're feeling insecure, you may even look to them to tell you who you are.

Maybe you work hard to **conform,** or make yourself like other people so you will fit in. You may give in to **peer pressure,** the control and influence people your age have over you, more than you like. Sometimes peer pressure can feel like a blessing. You may enjoy not having to make decisions for yourself. Other times peer pressure can feel like a curse.

Peer pressure can be positive, as when a whole group of students decides to hold a walkathon to raise money for a charity. Or peer pressure can be negative, as when people try to get you to use, buy, or sell drugs. Going along with the crowd can be fun if what the crowd does is in keeping with what you believe and what you value. But if what the crowd is doing goes against your values, going along can lead to big trouble.

Peer pressure can be positive or negative. How is it negative here?

How Some People Manipulate You — "Just this once!"

They said they'd cough up a hundred bucks if I rode on the handlebars down Suicide Hill.

To **manipulate** means to try to control others to get what you want. It is not done openly, but in sneaky, indirect ways.

Sometimes your peers may manipulate you. When they do this, they are trying to keep you from making your own decisions. They may coax, bargain, or pester you, trying to push you in the direction that serves their purposes. They may act as if they understand the world much better than you do.

To get you to do what they want, they may accuse you of being weak, childish, stupid, or overprotected. Worst of all, they may even hold their friendship over your head.

Everyone manipulates a little bit once in a while. But if people you're close to do it regularly, it can really get to you. You may start to feel as if you are a robot and they're at the controls.

Methods of Manipulation

If a classmate wants to borrow your new ten-speed bike for the weekend, he or she can ask you directly. But if the person is trying to manipulate you, he or she might instead use one of these tactics:

- **bribery.** "If you don't let me use it, I won't tell you what she said about you."

- **blackmail.** "If I can't have it, I guess I'll just have to tell Mr. Malone who 'borrowed' his Walkman."

- **guilt.** "Must be nice to be so rich." Or, "If I can't borrow it, I won't be able to do my paper route. Then I won't be able to help my folks with the food bill this week."

- **flattery.** "Everyone knows you're the best racer in town. I'd just like a chance at being half as good."

- **mocking or teasing.** "I didn't want one with training wheels anyway."

- **insults.** "Don't be such a total jerk."

- **giving only part of the story.** "I think I'll only need it for a few hours."

- **making deals.** "If you lend it to me, I'll get you invited to the party."

- **making threats.** "If you don't let me use it, forget our friendship."

SECTION REVIEW
STOP AND REFLECT

1. Describe someone who makes it hard for you to say *NO*. Tell how this person acts and how he or she makes you feel.

2. Imagine this person is putting pressure on you about something. Write out a conversation you have with him or her in which you say *NO*. Use "I statements" in telling the person how you feel.

3. Tell about a time when you had positive peer pressure used on you.

4. Tell about a time when you had negative peer pressure used on you.

5. Make a list of things people have said to try to manipulate you. Next to each one, write the name of the tactic they were using.

SECTION 2

DRAWING THE LINE: WHEN TO SAY NO

Knowing when to draw the line—when to say *NO*—is important to your emotional, physical, and spiritual health. Saying *YES* when it doesn't feel right can leave you depressed and feeling bad about yourself. Saying *YES* when the risks are too great can leave you physically hurt or worse. It can also hurt the people you care about and the people who care about you.

> "Roman candles and cherry bombs? No problem," he said. "We know how to handle them." Then he quickly handed the firecrackers to me. I lit this cherry bomb that didn't go off. So he picked it up and the thing blew up in his hand.

Consider The Consequences

Growing up means taking risks. Risks are necessary for growth. But there are reasonable risks and unreasonable risks. Going windsurfing may be a reasonable risk. Going windsurfing in a storm, even though it may be thrilling, is an unreasonable risk. The problem is that it may be hard to tell the difference until after the fact.

If you take the right risks with care, they can help you to develop more self-confidence, courage, and control. They can make you a more caring and responsible person. They can add to the respect you have for yourself and the respect others have for you. That is, if you know when to draw the line.

> I figured getting grounded was a possibility. But I never figured anyone would call the police.

The fact is that every action you take has consequences. A **consequence** is the outcome or result of an action. There are physical and emotional consequences. There are short-term and long-term consequences. The key is to learn to consider all of these consequences before taking action.

Taken For A Ride: Roberto's Rights

Roberto is 15. He is eight months away from getting his driver's license. He has been working as a busboy on weekends and saving all his money to buy a car. Roberto thinks that his rights include getting a driver's license and owning a car. But something happened that may change Roberto's plans.

> It was a beautiful day, and everybody was in a good mood. Rick invited me over to his place to check out his brother's new souped-up Trans Am. It was an incredible machine. Then Rick, who's only 14, said, "Why don't we take her out? You drive." I jumped at the chance. "We'll just take a spin around the block a few times," I said, and he said, "Fine. My brother won't be back till after work anyway." I stood there for just a minute, asking myself, "What could it hurt?"
>
> I was a little nervous at first, but I quickly got the hang of it. That's when Rick said, "Hey, this is great. Let's just head down Route 4 for a little while." I considered saying NO, but I was loving it. "Come on," I told myself. "This is a great chance to fly. No traffic lights. Relax. You're doing fine." And I hung a hard right onto Route 4.

Here, Roberto and Rick aren't thinking about the possible consequences of their decision.

Physical And Emotional Consequences

Before taking any major action—particularly one that causes an unsettling feeling inside—you need to consider both the physical and emotional consequences.

The problem for Roberto is that he doesn't take the time to think through the consequences of his actions. If he stopped to think about the possible after-effects of taking someone's car and driving without a license, his list might look something like this:

CHAPTER 7 SAYING *NO* AND FEELING GOOD ABOUT IT   133

If either Roberto or Rick had thought about the consequences they probably would have made a different decision.

- **possible physical consequences:** damaging the car; having to pay for the damages; hurting or killing Rick, me, or someone else; getting a ticket; getting arrested; not being able to get a license or buy a car; being yelled at by Rick's brother; dealing with the anger of Rick's parents and my parents.
- **possible emotional consequences:** feeling afraid of being found out; feeling guilty; being uneasy around Rick's brother; if someone were hurt or killed, feeling terrible for a long time; feeling ashamed that I'd let my family down; feeling bad about myself and angry that I'd been so stupid; losing Rick's friendship.

> **The Long and the Short Of It**
>
> When you're being pushed by someone else to say *YES* and you need that extra inner push to say *NO*, remember this simple phrase: *PAY NOW/PAY LATER.* This catch phrase can remind you to ask yourself: What must I *PAY NOW?* What is the cost to me today if I go ahead with this? And what must I *PAY LATER?* What might be the cost to me or others in the future? Remembering this simple message may stop you in your tracks.

Short-Term And Long-Term Consequences

You need to consider the short-term and long-term consequences of your actions, too—that is, how your actions will affect you and others both now and in the future.

Many decisions you make today can have important effects on the quality of your life tomorrow and for a long time to come. That is sometimes difficult to remember.

> **Snap Decisions**
>
> When you're faced with a situation that demands an instant *YES* or *NO*, you might consider this exercise: Imagine that you are already in the middle of doing whatever has been suggested. Imagine that a snapshot is taken of you in the middle of the activity. Then think of personally handing the developed photo to those people you most care about and value, such as parents, teachers, or close friends. Think about what you would feel. If there is embarrassment, guilt, shame, or fear of disappointing them when you hand over the snapshot, saying *YES* to the situation isn't such a good idea.

Roberto was thinking only about the present. He wanted to have a good time. If he had been thinking about the risks he was taking in an organized way, he might have made a second list that looked something like this:

- **short-term consequences:** getting caught; getting hurt; hurting or killing someone else; having to face my parents, Rick's parents, or Rick's brother.
- **long-term consequences:** not being able to get my license or buy a car; having to spend my money on damages to the Trans Am; using up my parents' money on doctors' fees, hospital care, and high insurance costs if I were hurt; lifelong injury; lifelong guilt for taking someone else's life; grief for my family and friends if I were killed.

Flashing Lights: Those Inner Warnings

I don't know, it was like a little voice inside me saying, "Watch out."

Sometimes you may have a long time to consider whether to say *YES* or *NO*. But often tough decisions have to be made on the spot. In some of these situations, flashing lights or warning sirens may seem to go off in your head. Your body may send you danger signals. Your heart may race. You may feel queasy. Your head may start to hurt. These warnings may go off when your safety is at risk or when what's being suggested to you goes against your values or the values of those you care about.

If these "flashing lights" go off when your buddy asks you to tamper with the school's computer you need to pay attention. You need to train yourself to sense these signals. Stop and ask yourself, "Does this spell trouble?" If the answer is *YES*, then it's time to say *NO*.

CHAPTER 7   SAYING *NO* AND FEELING GOOD ABOUT IT

Sometimes the unexpected happens. These boys were lucky. They could have killed a child.

In A Scrape: Roberto's Result

Sometimes, even when you weigh the consequences, something unpredictable may happen. Consider Roberto.

> I did great on Route 4. I picked up speed and burned a little rubber. Everything looked fine.
>
> Then the Malinowski kid ran out into the street to get his frisbee. I swerved. I scraped the door of the Johnson's van. No one was hurt, but everyone sitting on the stoop saw. Rick started swearing. And I knew my dad was going to be furious!

Roberto could say that the situation wasn't really his fault. But actually, if he looks carefully at what happened, Roberto will see that he was responsible. He took the first action—taking the Trans Am—and that action started the whole chain of events.

Roberto could have drawn the line when Rick first suggested that they go for a spin. He had the right to say *NO*. Later, Roberto could have said *NO* to going onto Route 4. That is, if he'd known how.

SECTION REVIEW
STOP AND REFLECT

1. Tell about a time you or someone you know should have said *NO* and didn't.

2. What were the physical consequences of saying *YES* in the situation you have described? What were the emotional consequences?

3. What were the short-term consequences? What were the long-term consequences?

4. What were the "flashing lights" that you (or the other person) felt before saying *YES*?

5. What made saying *NO* hard?

6. How could you (or that person) have done it differently?

SECTION 3

HOW TO SAY *NO* EFFECTIVELY

It's not like I was rejecting her. I just didn't want to lend her any more lunch money. Saying *NO* to your friends doesn't have to mean losing them. It doesn't even have to mean hurting their feelings. You can say *NO* to a friend's behavior without saying *NO* to the friendship. The key is in learning how to say *NO* in effective and caring ways. If you do it in a way that your friend can understand, in a way that shows you respect and care for your friend, it can actually be a growing experience for both of you.

Negative Ways To Say *No*

> I got sick of listening to her problems on the phone every night, so I finally just hung up on her.

There are many negative ways to say *NO* that can backfire on you. These include being bossy, nasty, timid, or self-righteous. Such negative methods of saying *NO* always show an imbalance between you and the person to whom you are talking. Unless the method you choose to say *NO* shows that both of you are valued and equal, the method cannot be effective.

The Aggressive Way — "Shove off!"

> He didn't seem to get the point with my words, so I thought I'd let him know with my fists.

One definite way to turn people off when you say *NO* is to be overly aggressive. Being **aggressive** means being overly forceful, pushy, hostile, or destructive. It means demanding that there is only one way—your way. It often means violating other people's territory and rights.

Suppose your friend keeps getting things out of your locker without asking.

You want her to stop. Saying "Will you just get lost?" or "You'll be sorry the next time you do that" may stop her annoying habit, but it may also put your friendship at risk. And it may spur your friend to get back at you.

You don't need to throw insults to get your point across. And if you charge aggressively into a situation, you shouldn't be surprised if the people around you keep their distance or come at you ready for a fight.

The Passive Way — "Well, I don't know..."

> I didn't really know how to tell him I didn't want to lend out my tape deck any more, so I ended up just letting him take it.

Being **passive** means giving up, giving in, or backing down without considering what's best for you. It means letting other people decide for you what you should say or do. It means more than "going with the flow." It implies surrendering your power as an individual and acting according to others' values, not your own.

Suppose the kids at the back of the bus are smoking, and they want you to smoke, too. You don't want to smoke, but you give in to their wishes anyway. When you fail to take action or fail to get your message across, you give up your rights. You lose respect for yourself.

This teen didn't know how to say NO. She let others make decisions for her.

Or you may say *NO* to the cigarette but communicate by the way you say it that you are not sure of your decision. You may look down, use a weak voice, even apologize for turning down the offer.

You may think you are communicating *NO* effectively, but your passive behavior probably communicates fear or lack of commitment. Your peers may keep bugging you till you back down. And if you're passive too often, your friends may start to think of you as someone who's easily pushed around.

Around the World

When YES Means NO

Most Americans can be very candid or blunt. If you ask a direct question, you're likely to get a straight answer.

In Japan there seems to be no difference between YES and NO in conversation. The Japanese place great value in being polite. Saying "yes," even when they mean "no," shows respect. To the Japanese, the way something is said may be more important than the words themselves.

This cultural practice gets confusing for Americans. Suppose you ask a Japanese student, "Do you understand how to check out a library book?" He may answer "yes," even though he's not sure. How might you pose the question in another way?

Positive Ways To Say *No*

I just said it straight out. I just said, "No thanks. I can't study when I drink."

For a change, I said it like I meant it. And she hasn't borrowed my clothes without asking for over a week.

There are many positive ways to say *NO*. These ways always show that you value both yourself and the other person. When you say *NO* positively, you say it openly and directly without insults, crude jokes, or apologies. You say it like you mean it, but without manipulation, without trying to force the other person to agree with you. Sometimes just the simple words "No, thanks" may do the trick.

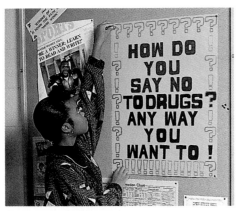

Posters can help get your message across.

The Assertive Way — "No, I don't want to."

The most effective way to say *NO* is the assertive way. Being **assertive** means standing up for yourself, for your values, and for what you believe in firm but positive ways. You don't bully, but you don't cave in either. You don't get defensive, protecting yourself from attack, and give a long list of reasons or excuses to back up your answer. You just state your decision.

You simply state with confidence what you want. Then you act.

If your friend wants you to walk through the park with her late at night and you don't think it's a good idea, an aggressive way to say *NO* would be, "No way. Do you want to be mugged?"

A passive way would be, "I don't know." But an assertive way would be, "No, thanks. There was an attack there last week."

Being honest and direct can help your self-esteem to grow. You can feel good about yourself because you are standing your ground.

SECTION REVIEW
STOP AND REFLECT

1. Tell about a time you or someone you know said *NO* in a negative way. Were you (or was that person) a bully or a doormat?

2. Rewrite the situation you described above. This time, be assertive in saying *NO*.

3. Describe someone you know who is overly aggressive in saying *NO*. What are some things this person says and does? How does he or she make other people feel?

4. Describe someone you know who is passive in saying *NO*. What are some things this person says and does? How does he or she make other people feel?

5. Are you more likely to be aggressive or passive? Explain why. What might you do to be more assertive?

6. Think of a situation you need to deal with in an assertive way. Write out a conversation in which you say *NO* in an assertive manner.

SECTION 4

WHEN A SIMPLE *NO* WON'T DO

I was really bummed out. No matter what I tried, he wouldn't take *NO* for an answer.

Sometimes a simple *NO* won't work. In such cases, this three-step plan can help you get your point across:

1. **Give information.** State that your answer is *NO*. Give the most important reasons.
2. **Let your friend know you see his or her side of it.**
3. **Restate your position.** Just keep calmly but firmly repeating your answer until your friend either gets it, accepts it, or gives up trying.

Saying *No* Can Be Difficult

Cecily is 14. Think about the problems she faced in trying to say *NO*.

I've never thought of myself as pretty or anything. Guys just have never seemed very interested. So when Carl started to show interest, it really felt good. I felt like maybe I finally had a chance.

First he started sitting with me in study hall. Then, in English, he moved to sit in the next row over. He started looking at me a lot, and he passed me notes sometimes. Finally, he asked me out. We went to a couple of movies, to the zoo, for Chinese takeout—nice stuff.

But, at the end of the school year, he started to make me nervous. He acted like he owned me or something. Even in school, he'd lock his arm around my neck and sort of pull me down the hall. At the school dance, when the slow music came on, he held me too tight. I wanted to get away, but I didn't know how. I didn't want to let him down. And I didn't want to make a scene.

Later, at the party, he asked me to go outside. I figured we could talk. But he wasn't interested in words. He started to stroke my hair and my back. That felt really nice. But then he pulled me close to him. I tried to get away, but he just kept on. Finally, I said, "Wait a minute. I want to tell you something." He said, "I've waited more than a minute. I've waited six weeks. And I'd really like to do it with you." That's when I pulled away and said, "Carl, I'm not ready right now. I'll have to think about it."

I didn't even know what I felt at the moment—except maybe like a caged animal. "I really don't know," I said. "I don't think I'm ready for sex." He told me, "Lots of girls your age do it. I know. And if you don't care about me enough to have sex, then maybe we ought to call it quits."

"I see it differently," I said. "I . . . ." And he cut me off. "Think it over. Let me know." he said. And he took off down the road.

In The *No*

Cecily was really in conflict. She didn't want to lose Carl. She didn't want to be alone again. But she didn't feel ready to have sex either. She was angry at him for pressuring her. But she cared about him a lot, too.

In a position like this, one needs to be assertive.

The first thing Cecily decided to do was talk with someone who cares. She knew if she didn't share it with someone, she would just get more confused and down. She chose her 18-year-old cousin Nicole, who's been through a lot and understands. Cecily knew she could trust her cousin. Nicole doesn't give lectures. She just listens and asks good questions. And she knows all about having sex too soon. Her baby son, Andrew, is proof of that.

The next day, after their talk, Cecily kept picturing Nicole in her tiny apartment. She kept hearing her cousin's words, "You know, in some ways I've been really stupid. I've made some of the biggest decisions of my life without really making them. I just kind of let them happen. I've had to find out the hard way that unless my heart and my head tell me it's okay to do something, I probably shouldn't do it."

Talking through a situation with someone who cares — and someone who has been through it before — can help you see all the options.

Cecily decided to take a really hard look at what she thought about having sex with Carl. She went to her room and listened to music. She talked to another friend about it on the phone. She cried.

If Cecily had thought through the situation in writing, her process might have looked something like this:

- **Identify the situation.** Carl wants to have sex. I don't. Carl wants an answer soon. I have to decide.
- **Recognize the challenge.** I have to make a decision that's in keeping with what I believe. If I can, I also want to keep Carl as a boyfriend — or at least as a friend.
- **List the possible choices.** My basic choices are *YES* or *NO*. I also have to choose which style of telling him I am going to use — the aggressive, passive, or assertive way.
- **Identify the possible consequences of each.**
  - Positive consequences if I say *YES* include keeping Carl, pleasing him, feeling grown-up, feeling wanted and attractive, not having to put up with the pressure any more, getting past the fear of sex, maybe discovering I like it.
  - Negative consequences if I say *YES* include physical consequences like getting pregnant, getting a sexually transmitted disease like herpes, gonorrhea, or even AIDS, having a bad or painful first sexual experience, getting caught in the act, getting a

bad reputation as someone who sleeps around, being expected to have sex often. Emotional consequences include feeling used, feeling shameful or guilty, going against my own and my parents' values, feeling afraid that now if I lose Carl it all will have been for nothing, feeling more jealous and possessive when he hangs around with other girls, being afraid every time we have sex till I get my next period, losing self-respect.

— Positive consequences if I say *NO* include feeling proud I stood my ground, feeling a sense of power, increased self-respect, realizing I can feel good about myself without relying on someone else to tell me I'm worthwhile, finding out how much Carl really does or doesn't care about me as a person, feeling a sense of relief, having increased confidence when I make the next major decision in my life.

— Negative consequences if I say *NO* include physical consequences like losing Carl, having to watch Carl turn to the next girl with the same request, maybe not dating for a while or having as much to do on weekends. Emotional consequences include feeling sad for a while, feeling uncomfortable about what Carl might say about me if he decides to retaliate, feeling like I let him down, feeling like maybe I passed up an opportunity.

- **Make a decision.** It's tough, but I have to tell Carl *NO*. I'm willing to take some risks with him—but not this one. I'm sure he's slept with other girls. I'm sure he will again. I don't want to catch anything. I don't want to get pregnant like Nicole did.
- **Rehearse the decision.** I'll say something like, "I really care about you and I hope we can keep going out. But I have to say *NO*. I can't have sex with you." If that doesn't work, I'll try the three-step method:
  1. I'll give information by saying, "I really care about you a lot, Carl, and I hope we can keep going out, but I'm just not ready to have sex. I'm afraid of what might happen."
  2. I'll let him know I see his side of it by saying, "I know how much you want to have sex and that it feels right to you. I understand."
  3. Then, restating my position, I'll say, "But I just have to say *NO* for now." I won't get into any argument, discussion, or debate about it either.
- **Take action.** I said *NO* the short way but Carl wouldn't accept my answer. So I just kept repeating my position as I'd rehearsed it.
- **Evaluate the decision.** Carl acted both disappointed and angry. That was tough to see. But he eventually seemed to get my point. And I learned that no one dies when I say *NO*, including me. I don't feel as bad about myself as I expected, either. In fact, I feel kind of good.

Thinking helps you make good decisions.

Cecily's Decision

In reality, Cecily didn't know how to think through a decision when she had to face Carl with her answer. But she did know that she had sensed trouble—those inner "flashing lights"—when the idea of having sex was first introduced.

Cecily also knew that in saying *NO* she didn't want to hurt Carl's feelings or make him feel in the wrong. She knew, too, once she had firmly decided *NO*, that she had to stick to her decision.

I actually did say *NO* the week before exams. It took Carl a long time to accept it, but I just kept telling him I wasn't going to sleep with him but I loved him and hoped he could hear that.

Well, he didn't call or anything for a couple of days, and then I got a note in my locker saying he was going to start dating other people but he wanted to still see me, too—sometimes.

Summer vacation started, and I mostly hung out with Rebecca and Jeannette. But then, last night, Carl called again and asked if I'd like to go to the zoo. And he says he has two tickets to a concert in July. He invited me—no strings—if I want to go with him.

Why *NO* News Is Good News

My uncle said, "So this will be our secret, right?" And I just said, "No. Not any more."

You can say *NO* and keep from getting hurt. You can say *NO* and end up feeling good about yourself. You really can say *NO* and still keep your friends.

You really can say NO to a lot of things and still have friends.

For a toddler, saying *NO* isn't tough at all. Saying *NO* means power and freedom. A small child who says *NO* is saying, "Here I am." "I count." "You'd better take notice."

For you, too, the word *NO* can be a positive and powerful tool. It can be a way of saying, "Here's who I really am." "Here's what I stand for, what I value." Or "These are my limits." It's also a way of saying, "I respect who you are, but I also respect myself."

When you're feeling unsure, remind yourself that saying *NO* can actually be a way of saying *YES*—to yourself, your values, your friends. You can say *NO* to a ride with someone who's been drinking and therefore say *YES* to life. You can say *NO* to a drink, a hit, or a fix and say *YES* to your overall health.

It's okay to be different. You are not an exact copy of your friends. It is important to listen to others, to turn to people who care when you want advice or direction in decision making. But make your own choice in the end.

Listen to those inner messages. Consider your values and beliefs. Remind yourself that this is your life, not anyone else's, and you have a right to say *NO*.

SECTION REVIEW
STOP AND REFLECT

1. Why do you think people like Carl put so much pressure on other people to do things their way? Why can't they accept a *NO*?

2. Describe a situation facing you (or someone you care about) where a simple *NO* won't work.

3. What information could you give?

4. How would you let the other person or persons involved know that you do see their side of things?

5. Restate your position. Tell what you will keep saying until the other person gets your message and gives up.

6. If you stick to saying *NO* in this situation, what good things will you be saying *YES* to?

CHAPTER 7 REVIEW

Putting Your Values To Work

STRENGTHENING YOUR VALUES

Reread *A Teen Speaks* **on page 127. Then answer the following questions.**

1. What values are present in Holly's actions on the night she babysits? What values are lacking?
2. What values are lacking in Eric's behavior when he calls Holly? When he visits her at the Levy's?
3. Do you think Eric's actions show true caring for Holly? Why or why not?
4. What do you think you would have done if you were Holly? Eric? Eric's friends? The Levys?
5. When, unlike Holly, have you paid attention to inner warnings and not given in to peer pressure? What values were you upholding by not giving in to it?

INTERPRETING KNOWLEDGE

1. Write a poem that describes how it feels to be pressured by your friends.
2. Do a creative work that captures the basic feelings you associate with being aggressive, passive, and assertive. For example, you might perform a three-part modern dance, make three masks, or create a three-paneled painting. Devote one part of your creation to each of these three methods of saying *NO*.

SHARPENING YOUR THINKING SKILLS

1. Define *peer pressure* in your own words.
2. Give some examples of outer pressures and inner pressures that teens feel.
3. List some of the possible physical and emotional consequences of drinking and driving. Then identify the consequences on your lists as short-term consequences or long-term consequences.
4. Give some examples of "flashing lights"—physical symptoms or other inner warnings—that you might feel just before taking an unreasonable risk.
5. Explain what this statement means: "Unless my heart and head tell me it's okay to do something, I probably shouldn't do it."

APPLYING KNOWLEDGE

1. Write a two-scene play with parts for three characters. In the first scene, two characters should use negative peer pressure on the third. In the second scene, they should use positive peer pressure on the same character. With two friends, present your play in class.
2. Make a videotape that demonstrates at least six methods of manipulation.
3. Imagine that someone has cut in front of you in the cafeteria line without asking. Act out ways to deal with the situation using three different approaches—aggressive, passive, and assertive.

CHAPTER 7 REVIEW

Putting Your Values To Work

PRACTICING DECISION-MAKING SKILLS

Read about each situation. Then answer the questions.

Situation A: Refer to *A Teen Speaks* on page 127. Imagine that you are Holly, and that you have not yet told Eric he can come over. Answer these questions as Holly, trying to decide what to do.

1. What is your situation? What decision do you have to make?
2. What is the challenge in this situation?
3. What are the possible choices?
4. What are the negative and positive consequences of each of these choices?
5. What is your decision?
6. What will you say to yourself in rehearsing the decision?
7. What action will you actually take? When?
8. How will you evaluate your decision?
9. Suppose Eric and his friends come over, would you make a different decision if Eric had come alone?
10. Would you make a different decision if Eric had been worried about passing geography and needed your help in the course?
11. What if one of the friends who came with Eric had just learned that his grandmother had died? If this friend needed support from the group, would you make a different decision?

Situation B: You are at the mall with a friend. You see your friend slip a watch into his pocket. You suspect that your friend has a problem with shoplifting and needs help. When he realizes you have seen the crime being committed, he begs you not to tell anyone. Just then, the security guard, a close friend of your family, confronts your friend. He denies the crime. The guard turns to you and says, "Did you see him take anything?"

12. You have to decide whether or not to turn him in. Make this decision following the decision-making process. What is the result?
13. Would your decision change if the guard were not a family friend? Explain.
14. What if you knew that your friend already had a shoplifting record and that he would be sent to a juvenile detention center for another offense? Would that change your decision?
15. If you decided to turn him in, what values would be reflected in your decision?

Situation C: Describe a recent situation in which you were faced with a *YES-NO* decision involving your peers, and they were pressuring you to say *YES*.

16. What values were reflected in your decision?
17. Do these values reflect what is important to you?

CHAPTER 8

Staying Drug-Free

A TEEN SPEAKS

Okay, so it's no excuse! I'm doing drugs because everybody else is doing drugs. But face it. Every party you go to, people smoke a little grass, do a line of coke, or pop a few pills. It's not like they're addicts or dope fiends. They're just regular kids, like me. You know, we just want to relax and have a few laughs.

But all you hear is "Drugs are bad for you. Stay away from drugs. They'll kill you. They'll eat you alive." At school they're always preaching to us at assembly programs. We have these "Drug-Free School Zone" signs around the school—what a laugh. Whoever put those signs up hasn't been to the bathrooms!

Talk about a joke. We had this essay contest about drugs and alcohol. So who won the prize for the best essay? The guy who is bombed out of his mind every weekend. He won the $50 bucks. Do I need to tell you what he did with it?

And the adults warning us about drugs? My folks are the worst. My Aunt Inez says, "Lonny, just say NO!" Give me a break. Most adults I know—except for Aunt Inez, she's okay—would never give up their beer or their cocktails. So what's the big deal—some people can't handle booze. Some people can't handle drugs. So why do the rest of us have to miss the fun?

Lonny

SECTION 1

KNOW THE FACTS

Why do so many teens think that drugs are okay? Do you know what doing drugs—any kind of drugs in any amount—is risky business? Some teens think this is an exaggeration. They imagine that adults are just trying to scare them. They are quick to say, "So what's the big deal?" But the fact is, you don't always know where drugs will take you or how fast.

> I used to say, "What's a couple of drinks?" That is, until they pulled my brother's body from the wreck. Now I really understand that drinking and driving don't mix.

The Down Escalator Of Drugs

Deciding to take any drug at all is like stepping onto the top step of a down escalator. You think you are in control and can get back off at any time. But the whole staircase is moving downward. It controls you and the direction you take. Before you know it, you may be heading straight for the bottom—at a frightening rate.

The decision to take that first step into drug use is a decision to give up control. It is a decision to stop caring about yourself and others. Picking up that first pill, joint, or drink can start a chain of events that can make you sick, ruin your health, or even kill you—or someone else.

Losing The Game

Len Bias seemed to have it all. The 6-foot 8-inch University of Maryland basketball star had just been picked as the number-one draft choice by the Boston Celtics. He had just been offered a major commercial contract by a famous athletic shoe company. That night, he went out to celebrate. The next day, he was dead.

What happened? At the celebration party, he tried cocaine for the first time. The drug was too much for his body, and his heart stopped. The cocaine in his

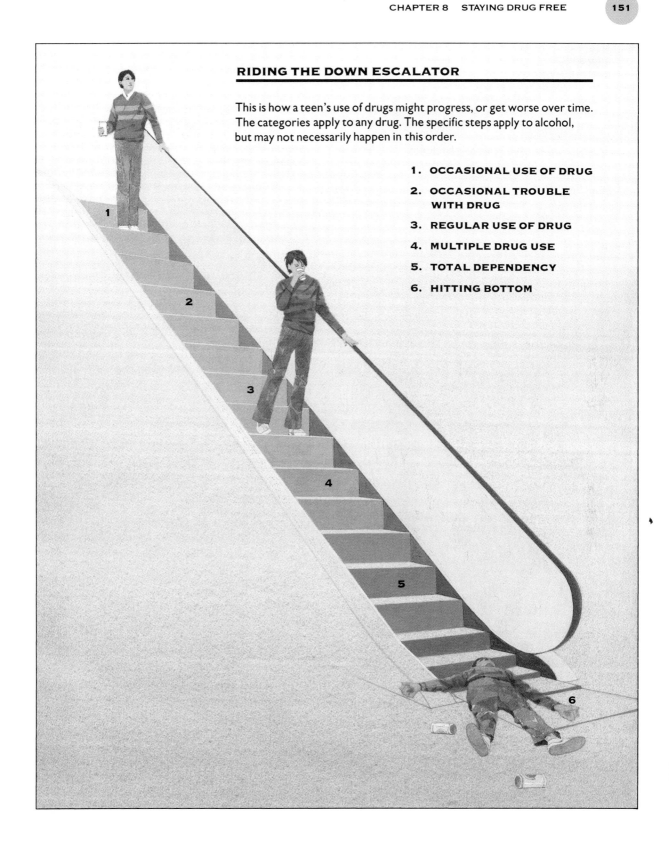

> **Tops Among Teens**
>
> In a recent survey, American teenagers were asked to name the number-one problem among today's young people. Their top answer? Drugs.

blood was at levels that don't always kill. But everyone's body reacts differently to drugs. Each person has different body chemistry and body structure. Cocaine is a strong stimulant that sometimes causes heart failure or blood vessel ruptures which could be potentially fatal. People often have no way of knowing it until it is too late how cocaine will affect them. One person's high is another person's funeral.

Maybe Bias didn't have the facts. Maybe he didn't know that drugs can act like bombs that explode, without warning, in the body and brain.

If he'd had some basic facts about drugs, wouldn't he have stopped to ask himself, "What's at stake here? Is it worth the risk? What do I value most?" Wouldn't he have made a different decision that night? If he had resisted the pressure from his buddies to "celebrate" with drugs, he might still be shooting hoops and sharing his talent with the world.

What Are Drugs?

Drugs are substances other than food that change the way your mind and body work. They can change your moods, appearance, and health. They can change how you think, act, and learn.

Drugs come in many shapes and forms. Alcohol, marijuana, cocaine, tobacco, sleeping pills, and many other substances all qualify as drugs.

Some drugs do tremendous good. The legal use of drugs as medicines helps people in many ways. Medicines cure diseases and relieve pain. When used properly, usually according to a doctor's prescription, medicines can help people to live longer and healthier lives. However, using drugs in ways that are illegal and unhealthy is called **abuse.** Drug abuse, which is often called substance abuse, also includes the misuse of over-the-counter and prescription drugs.

Most drug abusers want the "high" that comes from taking some drugs or the "mellowed out" feeling that comes from taking others. If you use drugs, you may crave the temporary feeling that everything is wonderful, the extreme lift in spirits that occurs when the abused drug reaches your brain. Depending on the drug and the amount, a "high" lasts from a few seconds to a few hours. When the high is past, you may feel let down and sometimes even more depressed than before taking the drug. This "down" often drives you to take the drug to get that "high" again—and again—and again.

Abusing drugs over time leads to bigger problems. As your body adjusts

Not Just Us

Drugs cause serious problems in many parts of the world. The use of illegal drugs is a major problem in the United States; the production of illegal drugs causes problems in other countries as well.

Colombia is the main source of marijuana and the main producer of cocaine. Laos, Thailand, and Burma are major sources of opium.

Through the United Nations' fund for drug control, countries have begun to work at preventing drug abuse. Burma encourages youth to express their views on drugs through publicly displayed cartoons and drawings. Thailand has begun a crop substitution program. Coffee and kidney beans require less human work to grow than opium poppies. This frees children from working in the fields to go to school. Pakistan has begun to offer drug rehabilitation at hospitals in its major cities. Then follow-up care is provided in local communities.

to a drug, it begins to require larger and larger amounts of the drug to get the same high. This is called developing **tolerance.** Your mind and body need more and more of the drug just to feel normal. Eventually, even the largest doses can't make you feel good. Meantime, as you increase the doses the physical damage to your body increases.

Dependence is another problem with drug abuse that occurs over time. Your body gets used to having the drug in its system. It becomes dependent on the drug just to function.

> I used to smoke crack for the rush. Now I just smoke it to keep my hands from shaking.

You can quickly develop a dependence on a drug, whether it is cocaine, nicotine from cigarettes, marijuana, or

Everybody reacts differently to drugs — legal or not. If you must take medicine, follow directions carefully.

beer. This is called **addiction.** Sometimes people refer to it as being chemically dependent or "hooked." Teens develop addictions faster than adult substance abusers because their decision-making skills as well as their bodies are still developing.

If you take the drug away from an addicted person's body, the body and mind react, sometimes violently. This reaction is called **withdrawal.** Sudden, total withdrawal, called cold turkey, can be dangerous and painful. It can even be deadly.

Sometimes severe illness or death results from a single dose of a drug. This is called an **overdose** because it's over the amount the body can stand at one time. Overdoses can happen to you if you are a first-time user or if you've been taking drugs for a long time.

SECTION REVIEW
STOP AND REFLECT

1. Tell why you think there is such a big drug problem in this country.

2. Why do you think teens want to try drugs?

3. What argument would you use to keep a younger brother or sister from trying drugs?

4. Do you think that knowing the facts about drugs will keep teens from doing drugs? Explain your answer.

5. Tell how you would try to end the drug problem if you were the principal of a school.

SECTION 2

FACE THE FACTS: WHAT DRUGS CAN DO TO YOU

So I smoke cigarettes and drink wine coolers at parties. Big deal. At least I don't do drugs.

Any time you use alcohol, tobacco, or illegal drugs, or anytime you misuse prescription or over-the-counter (OTC) drugs, you are playing with fire. You may not know you are getting burned at first. It can take time. But a drug is a drug is a drug. Whether liquid or solid, all substances that are abused harm the body. Each substance hurts your body in its own unique way.

Alcohol, The Liquid Drug

Alcohol is the most abused drug in the world. It depresses, or slows down, all of the body's systems. At first, it may feel like a stimulant, "revving up" your body's engines. But after just one or two drinks, alcohol shuts down the part of your brain that controls reason and decision making. From that point on, your emotions run the show.

Maybe you've watched people drink at a party. At first, they're all smiles and chatter. But later in the evening, angry words and fists are flying, tears are flowing. Alcohol causes people to lose physical control in other ways, too. They may stagger, slur words, fall down, or even pass out.

You may think alcohol can't hurt you because it isn't a "hard" drug like heroin or crack.

My boyfriend says beer won't get me as drunk.

Maybe you think that you can drink more than any of your friends without showing it. The fact is that alcohol kills more people than all other drugs put together. It can damage your liver, kidneys, throat, stomach, intestines, and brain.

Even if you're just a first-time drinker, taking in too much alcohol in a short period of time can cause irregular breathing, extreme shaking, even death. This can happen even if you're only drinking beer.

Early Warning Signs

These are some of the early warning signs of alcoholic drinking:
- personality changes when drinking,
- behavior that shows loss of control when drinking,
- blackouts—not remembering what was said or done while drinking,
- drinking alone,
- being overly concerned with the supply of alcohol,
- denying, trying to disprove, or overreacting to suggestions that drinking is a problem,
- being able to "drink others under the table" without showing the obvious signs of being drunk.

Alcoholism—"I just like to drink."

How can I be an alcoholic? I only drink on weekends.

People who are addicted to alcohol are called **alcoholics.** They suffer from a disease called alcoholism. **Alcoholism** is a lifelong, but treatable, disease in which a person's life becomes unmanageable to some degree due to the use of alcohol. The disease is progressive. That means it gets worse over time. Some people drink alcoholically, or without control, from the first drink. For others, it takes years for the disease to be full-blown. The alcoholic tries to control the drinking and may succeed for a while. But sooner or later, he or she fails. Relationships start to fall apart. Health declines. Life itself gets out of control. On the average, drinking shortens an alcoholic's life by ten to twelve years.

The body of an alcoholic processes alcohol differently from other people's bodies. It turns the alcohol into a poison faster and takes longer for it to get that poison out of its system. Because this chemical "poison" shuts down the judgment portion of the brain, the alcoholic really doesn't know how bad the problem is.

Alcoholism tends to run in families. In the population at large, about one in ten people are alcoholics. Children of alcoholics have a one in four chance of developing the disease. But you don't have to have the disease in your family to be at risk. Teens may become addicted to alcohol in as little as six months. And experts estimate there are at least three million teenage alcoholics.

There is no "typical" alcoholic. Alcoholism can strike teens or grandparents, women or men, rich or poor, movie stars or street people. In fact, skid row bums account for only 3 to 5 percent of alcoholics. Over 75 percent of alcoholics hold jobs, have families, and

Illegal, Designer, And Misused Legal Drugs

People abuse many different kinds of drugs. Whether these drugs are bought on the street or handed to you by a friend, they can spell trouble.

Marijuana — "It was just one joint!"

> I don't do drugs. I just smoke a little grass.

Marijuana goes by many names—grass, joint, pot, dope, reefer, weed, Mary Jane. By any name, it is a drug that is harmful when inhaled into the body. For young teens, it is especially dangerous because of its effects on growth and learning. Regular users have trouble reading, solving math problems, and concentrating. Little by little, they lose the drive to do, to be, to care.

Maybe you know people who get stoned a lot. Their eyes get glazed and glassy, and bloodshot. They may laugh a lot, or completely withdraw when they are smoking dope. They may seem happy or relaxed, but watch the heavy users closely over time. See how little they accomplish. Watch their grades dip. Notice the reactions they get from other

Teens can become alcoholics. For some, it takes only one beer.

appear, at least from the outside, to lead normal lives.

To be an alcoholic you don't have to drink "hard" liquor like whiskey. You don't need to drink in the morning. You don't even have to drink every day. You may drink only on weekends and still be an alcoholic.

You may not even get hangovers—the physical effects of withdrawal the day after drinking. However, on any given occasion when you do pick up that first drink, you can't predict what you'll say or do.

> **Pryor Commitment**
>
> Comedian Richard Pryor almost died from burns that resulted from freebasing cocaine. Freebasing is the process used to turn cocaine into its more smokable and powerful form, crack. It involves using liquid chemicals that can catch on fire and explode. And that's just what they did. They exploded in Pryor's face. He learned the hard way. But today he is grateful to be alive and he is committed to staying drug-free.

people. Watch their friends change. See them drop out of sports—even out of school. See how the drug seems to rob them of their will to make a difference.

Marijuana damages your body. It contains over 400 chemicals. And these strong chemicals stay in your body for a month or longer after one use of the drug.

Marijuana can irritate your lungs, make you more likely to catch colds, viruses, and other infections, and cause panic attacks in some people after just one use.

As a growing teen, you may be particularly interested to learn that marijuana lowers the levels of male and female hormones needed for normal growth, sexual development, and sexual functioning. Males who use the drug may even develop enlarged breasts. Females may have irregular menstrual periods.

There is one more fact that you should know: There's a good chance there will be other drugs mixed in a marijuana batch. In fact, some joints, or marijuana cigarettes, contain no marijuana at all. One common marijuana substitute is parsley, an herb, mixed with PCP, a dangerous and powerful drug. Sometimes, you may even be spending your money on lawn clippings laced with deadly drugs in lethal amounts.

Stimulants — "I wanted a little energy."

> I thought there were spiders crawling under my skin. I knew I was dying.

Stimulants, such as cocaine and amphetamines, speed up your nervous system. This gives a false feeling of extra energy. That's partly because they make your heart and lungs work harder. They shock your body into a state of emergency, and the body's alarm systems may be set off.

Sometimes stimulants cause **hallucinations,** imagined sights and sounds that seem real at the time, and other mental problems. Stimulants can also cause serious damage to your heart, lungs, and liver. And these dangers exist whether the drug is swallowed, smoked, snorted, or injected with needles into the muscles, veins, or skin.

Cocaine, often called coke, snow, and blow, can cause heart attacks and strokes even in young, healthy people. It can cause your lungs to fill with fluid. A single use can cause seizures. It can make you depressed, violent, or paranoid—imagining that people are after you. When snorted, it can eat away the lining of your nose, resulting in constant nose-wiping known as the "cocaine salute."

A newer stimulant, crack, is popular with some teens. A smokable form of cocaine, it reaches the brain within seconds. Crack creates a high for 10 to 15 minutes, then a terrible low. The low makes you feel so miserable that the only relief a user may think of is to reuse the drug. Soon you're on a roller coaster of ups and downs that is difficult to get off.

Crack is one of the strongest and deadliest of street drugs. It is also one of the most addictive. You can become hooked after just one use.

Amphetamines, also called speed and uppers, can cause "the shakes," seizures, panic, mood swings, heart attacks, high fevers, sleeplessness, coma, and death. Using speed over time, you can develop hallucinations, and you may begin to think that everyone is out to get you. You'll probably also have brittle hair, bleeding gums, weakened teeth, and nails that break easily.

A type of methamphetamine called "ice," that is sold in small crystals and smoked, is presenting serious health consequences in those who use it.

Depressants — "It was a real downer."

Depressants, sometimes called downers, include tranquilizers and barbiturates. At first they give you a relaxed feeling. But they can also cause confusion. They can make walking, talking, and even breathing difficult.

Depressants are highly addictive. They can slow your body too much, causing coma and death, especially when taken in combination with other drugs. Barbiturate overdoses play a role in about one-third of all drug-caused deaths.

Crack is one of the most addictive street drugs in existence.

Hallucinogens — "I was seeing things."

Hallucinogens such as LSD, PCP or mushrooms (sometimes called "shrooms"), cause temporary confusion of mental images. These hallucinations make you see and hear things that aren't really there. They can cause you to have flashbacks, or delayed effects, long after you have used them.

Hallucinogens can cause panicky feelings, fevers, rapid heartbeat, convulsions, coma, and death. PCP, nicknamed "bad drug" and "angel dust," can give a false sense of power and strength. In fact, more PCP users die from the results of their behavior than from the physical effects of the drug. And PCP can stay in your body for up to four years, so these physical and mental effects can erupt long after you think you are out from under the drug's spell.

All illegal drugs are dangerous. Sharing needles adds another risk — AIDS.

Narcotics — "I wanted to escape."

> I thought of shooting up as a way to escape. But I never thought about escaping into AIDS.

Narcotics, such as heroin, morphine, codeine, and opium, relieve pain and bring on sleep. They are usually injected through needles. If you share needles with other drug users, there is a high risk of diseases such as hepatitis and AIDS.

Narcotics cause you to lose your appetite, which can lead to poor nutrition. They cause lack of caring, loss of judgment and self-control, and lack of energy. Women who are addicted often have babies that are born too early, are stillborn, or are born addicted.

Inhalants — "It didn't last."

> Get real. Nobody ever died from breathing this spray stuff.

Inhalants are fumes of chemicals such as lighter fluid, paint thinner, glue, or typewriter correction fluid. They produce a brief high, or sense of extreme well-being, when inhaled. Inhalants can cause balance problems, confusion, and permanent brain damage. They can ruin your lungs, liver, kidneys, and bone marrow. They can also create problems with your vision, judgment, memory and muscles. They can lead to violent behavior. They can even cause death by suffocation.

Designer Drugs — "They said it was legal."

Designer drugs are made in a lab from substances not currently on the federal government's list of controlled substances. Often a well-known illegal drug is altered slightly so that it is no longer a controlled substance. This means it isn't illegal — yet! As the drug becomes known, it is declared illegal by the government. Such drugs can be highly addictive. Sometimes even a single dose can cause brain damage or death.

Look-Alike Drugs — "I thought it was grass."

Look-alike drugs are made from legal substances to look like illegal or street drugs. For example, look-alike "speed" might instead contain high doses of caffeine plus cold medicines. Such mixtures can cause dangerously fast heart rates, changes in blood pressure, strange behavior, nervousness, and breathing problems. Or a sugar substitute might be put in a capsule and sold as an illegal drug — at a huge profit.

Chemical Cheating: Life in the Fast Lane

In the 1988 Olympics, runner Ben Johnson won first place in the 100-meter dash, breaking his own world record. Later, after testing positive for steroids, he lost his gold medal, millions of dollars in promised contracts, and his honor.

Increasingly, athletes are being tested for the presence of steroids in their bodies. Some users have been stripped not only of their records but also of their health and their self-respect.

Steroids — "I wanted to make the team."

I was sick of my body the way it was. I wanted some muscle, some shape. I wanted to be noticed. I just wanted to look good on the beach.

Anabolic steroids are legal drugs that are often misused. **Steroids** are hormones. Doctors sometimes prescribe

Exercise is a healthy way to build muscle. Taking steroids is unhealthy and illegal.

them for cancer patients and for men lacking in certain natural hormones. These are correct uses of the drug.

Today, however, steroids are being widely used in unhealthy, unprescribed ways. The attraction is that they "bulk up" the body quickly, adding to muscle size and body weight. They create a false sense of confidence and may, for a time, boost athletic performance. Because of these effects, many bodybuilders and athletes use steroids as training shortcuts, without thinking about the many dangers.

Steroids can cause acne, sexual problems, and heart disease. They can lead to liver damage, liver cancer, kidney damage, and sterility—the inability to have children. Steroids are particularly hazardous to teens because they can slow or stop your growth. Steroids that are injected with shared needles can spread deadly infections, including the HIV virus. AIDS develops as a result of this virus.

Very often steroid abuse causes mental problems, which do not always go away. No matter how peace-loving you normally are, steroids can turn you violent, even against the people you love. They can also leave you severely depressed.

SECTION REVIEW
STOP AND REFLECT

1. Tell why alcohol is as harmful as other drugs.

2. Why do you think so many teens want to drink?

3. Do you think more people in your school are likely to drink or to do drugs? Why?

4. List four or five things you most want to do or be in the future. (These are goals.)

5. Tell how alcohol abuse could keep you from reaching each of the goals you listed above.

6. Tell how drug abuse could keep you from reaching each of the above goals.

SECTION 3

A FEW MORE FACTS ABOUT DRUGS

There is a lot to know about drugs. And the more you know, the more reasons you have to say *NO*. Still, what you know about drugs is not as important as what you choose to do. Some people know a lot of facts, but they just don't pay attention to them.

People who abuse drugs of any kind tell themselves that they can stop when they want to, but they just aren't ready to do it. That's another fact about drugs—it's not easy to stop. Sometimes it feels downright impossible.

Tobacco

I just smoke once in a while, just when I really need to relax—or when I'm with friends—or when I'm feeling a lot of stress.

"I can stop any time I want" is what a lot of people say about tobacco. It hardly seems in the same league with drugs and alcohol. But tobacco contains a powerful drug, and, it is a threat to health and life. In fact, smoking tobacco is the number–one cause of early death in the United States.

In spite of repeated warnings, many teens continue to smoke. Tobacco use is, unfortunately, rising, especially among teenage girls. The younger you start to smoke, the more likely you are to suffer illness or even death from the effects of tobacco. You begin damaging your lungs the minute you take your first puff on your first cigarette.

There is a drug in tobacco called nicotine. **Nicotine** is a stimulant that speeds the heartbeat and raises blood pressure. It affects the nerves, causing trembling and irritation. People who use tobacco regularly develop an addiction to nicotine. And it is not a mild addiction.

According to a 1988 report by the U.S. Surgeon General, nicotine is as addictive as both heroin and cocaine. That's why it is so hard for people to give up smoking once they've started. And when they do stop, they go through withdrawal.

Your health is at risk if you smoke. Your health is also at risk if you spend a lot of time around people who smoke.

Smoking cigarettes cuts down on the oxygen supply to your blood. The chemical compounds found in cigarette smoke are the same as those found in car fumes and in the air during smog alerts.

Smoking can cause heart attacks, blood clots, ulcers, wrinkled skin, damaged gums, and a variety of lung diseases. Smokers often suffer not only cancer of the lung but also cancers of the lip, mouth, pancreas, and kidneys. Of the many chemicals in tobacco, at least 43 have been found to cause cancer.

Passive smoking occurs when a nonsmoker breathes in smoke from another person's cigarette, pipe, or cigar. A passive smoker faces all the health risks that a smoker faces.

Secondhand smoke may cause up to 15,000 deaths a year. Because the smoke is not filtered by tobacco, it has twice the nicotine and five times as much carbon monoxide, one of the chemical compounds in car fumes, as the active smoker gets.

The use of smokeless tobacco—chewing tobacco or snuff—is just as harmful to your health as smoking. It may start out as casually as chewing gum, but it is quickly addictive.

If you use smokeless tobacco, you risk getting mouth sores that can develop into cancer of the lip, mouth, or throat. Smokeless tobacco causes bad breath. It cuts down on your ability to taste and smell. It discolors the teeth and damages both teeth and gums. Though it is meant for the mouth, some tobacco gets swallowed with saliva, damaging the digestive tract.

More Dangers Of Substance Abuse

I was 11 when I started with cigarettes and wine. Then I moved on to marijuana, then coke. Now it's whatever I can get my hands on.

Drugs that lead to more drugs are called **gateway drugs**. An example of a gateway drug is nicotine in cigarettes. Marijuana is another gateway drug.

CHAPTER 8 STAYING DRUG FREE 165

It is possible to become cross addicted to nicotine and alcohol — or to two other drugs.

> **The Three Rs: The Nonsmokers' Bill of Rights**
>
> As a nonsmoker, you have three basic rights:
> - The Right to Breathe Clean Air.
> - The Right to Speak Out.
> - The Right to Act.

Once a person begins using one drug, it's easy to be led into trying another drug, particularly by a pusher.

It's dangerous to combine drugs, too, even prescription medicines. This is why your doctor wants to know what other prescriptions you have before giving you a new one.

Many drugs, when used together, have a deadly effect. Even tobacco cigarettes and marijuana joints have a more harmful effect together than separately. When it comes to combining drugs, one plus one adds up to an effect many times more powerful than two. This is called the **multiplier effect**.

Often a substance abuser becomes hooked on more than one substance. This is called **cross-addiction**.

When trying to give up one addictive drug, it does no good to replace that drug with another. The brain and body of a person addicted to one drug don't know the difference between one substance and another.

An alcoholic who no longer drinks alcohol but smokes marijuana is still in active addiction. Using any addictive substance can trigger addiction to others. And often the new addiction leads right back to the old one.

Not only do you hurt yourself with drugs, but you hurt others. Almost everything a pregnant woman takes into her body also goes to her baby. That includes alcohol and other drugs.

Babies whose mothers drank large amounts of alcohol during pregnancy may be born with a condition called Fetal Alcohol Syndrome. These babies may be

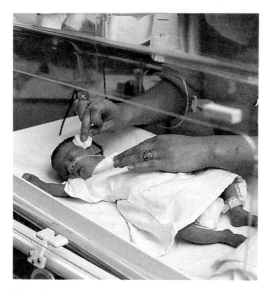

There have been cases where the mother is held legally responsible if her child is born addicted to drugs. She can be charged with giving drugs to a minor.

mentally retarded and have birth defects of the heart and face. Fetal Alcohol Syndrome is one of the three leading causes of birth defects, and the effects of the syndrome cannot be reversed.

Babies whose mothers smoked cigarettes during pregnancy are usually born with lower than average birthweight, making them more likely to suffer other diseases.

> My baby looked so tiny and helpless. She jerked and shook. The nurse told me she had to go through withdrawal because I was always getting high when I was pregnant. I didn't want to hurt my baby. I just didn't know doing crack could hurt her. Now they tell me she may be retarded.

Your parents or grandparents may not have known that tobacco, alcohol and other drugs were dangerous to their babies. Because of work scientists have done in recent years we know much more. We now understand that some infant deaths, personality disorders, learning disabilities, and physical handicaps are the direct results of how the mother used substances while she was pregnant. These birth defects are avoidable.

SECTION REVIEW
STOP AND REFLECT

1. Write a "truth in advertising" label for cigarettes. Tell what they really do.

2. What are the dangers of passive smoking? Who is at risk?

3. Explain the dangers of gateway drugs.

4. Make a list of the dangers associated with alcohol use.

5. Explain in your own words the meaning of **addiction, dependence, tolerance,** and **withdrawal.**

6. Describe a situation in which other people are hurt by a teen's use of drugs.

SECTION 4

RECOGNIZING CHOICES

It's natural to be uncertain of how you feel about things. One day you might swear you'll never do drugs. The next night, feeling shy and uncomfortable at a party, you're not so sure. Sometimes you feel like you're caught in a giant game of tug-of-war, and both competing teams reside inside your own head.

The Drug Tug

Understanding the choices involved can help to bring the tugging to a halt. So can remembering the facts you've learned about drugs and putting them to use in your decisions. Practicing in your head what you'll do or say when faced with various drug situations can make you more sure of yourself when the situations actually come up. So can looking carefully at why other teens do drugs. When you see that drugs don't really fill your deepest needs, you can be stronger in your own drug decision making.

You may find yourself struggling with questions like these:

- Should I use or not use?
- Should I use one drug instead of another?
- Should I combine drugs or not?
- Should I start smoking or not?
- Should I speak up when someone around me smokes?
- Should I hang out with a crowd that I know uses drugs?
- Should I tell my friend's parents how strung out on drugs he is and how much he needs help?
- Should I get a fake ID, borrow someone else's, or ask my older sister to buy the beer?
- Should I have a drinking party when my parents are away?
- Should I deal just a little dope for some extra cash?

It's normal for teens to wonder about such questions. You are not bad or weak because you are not always certain of your answers. Feelings are not facts. Unanswered questions are not actions.

What does matter is what you decide when you are actually faced with any of these questions. What matters is how you act on your decisions. These are "substance situations." How you respond to them may raise central questions about who you are and who you want to become.

Why Some Teens Do Drugs— "I'm hooked."

Some teens think doing drugs—whether alcohol, cigarettes, or illegal drugs—will make them more accepted by their friends and more popular.

> Anybody who's anybody drinks.

> Hey, it's everywhere. People used to drink a little beer at parties. Now we drink a little beer and snort a little coke. It's what we do.

Maybe you've used drugs to fit in, to relax, to let go. Maybe you've considered using them to cover up your fear or to make yourself seem more grown-up.

At first, drugs may seem to do all kinds of social magic. But the magic wears off. And the cost of the magic show can be high.

Some teens do drugs to escape the pain of upsetting problems. Arguments with parents, low test scores, or feeling shy around the other sex may get you

Teens with self-respect don't get involved with drugs. They don't use them. They don't sell them.

down and push you to look for chemical relief. You may even be facing major life problems like a broken relationship, a move to a new place, the divorce of your parents, the death of a friend, or sexual abuse.

At times like these, it may be really tempting to reach for a quick fix, to get high and blot out the pain. But when you come down from the high, the pain and the problems are still there. And nothing's been done to face, solve, grieve, or otherwise cope with the difficulties.

Teens sometimes say they use drugs simply because they're bored. But boredom is often a sign that something else is wrong. If you find yourself doing drugs with a "who cares?" attitude, maybe you're really saying that you don't feel valued. If you find yourself doing drugs with a "there's nothing better to do" mind-set, you could be depressed and need some help. Maybe it's time to find new ways to use your talents and get involved in school, church, synagogue, temple, mosque, or the community at large.

Other teens may do drugs because their parents do. You may reason, "If Mom and Dad can enjoy their beer after work, why shouldn't I enjoy my after-school joints?" Teens may also do drugs as a way to rebel. "I'll show my parents" or "Nobody can stop me from doing what I want to do" may be signs that you are drugging as a way to act out anger, to be noticed, even to distract your family from deeper problems. Some teens may even take overdoses of drugs on purpose as a scream to parents or friends for help.

Teens who don't have the support of a caring family or circle of friends may feel sad and lost. You may consider turning to drugs, thinking they will cure the unhappiness, loneliness, or emptiness inside. But chemicals are never a satisfactory substitute for feeling loved.

Instead of using drugs, you need to ask, "What's lacking in my life? How can I get the love, attention, and caring I need in healthy ways?" You may get a Big Brother or Big Sister, turn to a teacher, counselor, or clergy member, or work in therapy to learn ways to "parent" and love yourself. Or, join a support group.

> **Addiction Prediction**
>
> A recent study of seventh and eighth graders showed that teens' attitudes toward alcohol can help to predict who will develop a drinking problem in the near future and who will not. Which teens were most likely to problem-drink within a year or two? Those who held the beliefs that booze would help them to do better socially, think more clearly, or improve their athletic ability— beliefs that are all untrue.

Most teens who have taken drugs admit they started on drugs because friends encouraged or pressured them. Such peer pressure is probably the toughest test of your values. And teens who already have drug problems may put on the most pressure. That is because anyone who is drug-free is a threat to their drinking and drugging. They try to justify themselves by pulling you in.

Teens may give many reasons for starting to use drugs. But when teens become chemically dependent, they all use drugs for the same reason—because their bodies tell them they have to. They are hooked.

Deadly Serious

Over half of all teen suicides involve drugs. Many teens who die from drug overdoses probably never really meant to take their lives. They only meant to get attention and help, but their plans backfired. Other teens, who were never suicidal before taking drugs, became suicidal because the drugs they were taking left them deeply depressed.

Most teen suicides involve drugs — in one form or another.

Why Some Teens Deal Drugs — "I have a habit."

In an east coast city not long ago, a 10-year-old was arrested for running drugs. His pockets were full of $20 vials of crack. That same night, across town, an 11-year-old was found beaten to death, a small drug dealer done in by a bigger one. Why did these kids start dealing drugs?

> I just liked being able to do a line of coke now and then. It starts to cost a lot, especially when you want it for a party or something. And I had a lot of contacts. So this guy who always supplied me said maybe I could work with him. He was a real nice guy. We're not talking about somebody who looked like a crook.

Many young dealers don't start out by selling drugs. They start out just by trying them. Then, when they get dependent on drugs, dealers try to push more and more expensive drugs at them.

As long as the money and the drugs hold out, things seem okay. Dealers are friendly and helpful—like in any other business. But if the supply of drugs or money starts to dry up, things get ugly. And once kids are really hooked, they have to support their drug habits. Then they find themselves lying, stealing, or worse.

Some teens and even preteens are tempted to deal by what seem to be big payoffs. It's hard to look down on a 16-year-old who's driving his or her own sports car, even if the money for it came from dealing drugs.

It's hard to look down on a 12-year-old who comes to school with more money than your family can earn in a month. It's especially hard if your own parent can't pay the rent, much less buy a VCR.

Dealing drugs promises to bring in big bucks. A lot of kids have the clothes, the sports equipment, and the spending money to prove it. They can buy anything they want. And some kids never get caught.

But the quick draw of money isn't worth it for the caring person. You can't be a caring, responsible person and deal drugs, no matter how nice you act on the outside or how much good you do for your family with the money. Giving up the things money can buy isn't easy for anybody, especially for people who don't have enough money to get by. Being a caring person is a harder but more rewarding way to live.

The short-term rewards of drug money are tempting, but the real "payoffs" are too often addiction, arrests, violence, and death. You may never see these real "payoffs." Kids who have been pulled under by the dark side of dealing drugs often disappear before you ever see what it has done to them. Even when you do see the dark side, it is easy to say, "It won't happen to me." But it does happen to even the "nicest" people if they place themselves in this kind of situation.

One of the consequences of dealing drugs is dealing with the police. Others include addiction, prison, violence, even death.

Giving up money and things that you could have right now requires a strong sense of your own values. And it requires sacrifice. But dealing drugs means helping to destroy both others and yourself. A Corvette is no trade-off for a jail cell or a toe tag at the morgue.

Even if the "payoff" is less dramatic, it isn't worth letting yourself see other human beings, whose lives are as valuable as yours, as no more than possible business deals.

Stealing for Drugs

Among teens who recently called a cocaine hotline, 64 percent admitted that they had stolen from family, friends, or others to buy drugs.

Flexing His Values: Joe's Decision

Joe, age 15, has been watching his friend Owen for a few months. He says "Something's up. Last year, when we wrestled, Owen and I were in the same weight class. Now that it's fall and football season, Owen looks fantastic. Awesome biceps. A neck like a tree trunk. He seems ready to take on the world. He isn't running or working out the way he used to, either. But it doesn't seem to matter. He's got all the girls hanging all over him. Meanwhile, I can hardly get Charlene even to talk to me."

Joe knows a few guys on the team are taking steroids. He wonders if that's how Owen has built up his body so fast. Then he hears that last weekend, in a terrible rage, Owen slapped his girlfriend around. Joe knows from health class that violent outbursts are one of the side effects of steroid abuse.

The next weekend, after the game, Owen offers Joe some steroids. He tells Joe that maybe Coach Ferrara won't be on his case so much to pump iron and run laps if Joe starts taking the drug. When Joe says he isn't sure, that he has to think about it, Owen asks Joe if he really wants to be a bench warmer forever while he and the other guys are in the starting lineup. Then comes the real punch. "Come on," Owen says. "If you really want Charlene to think you're not such a wimp, wise up."

Joe knows the dangers of steroids. But he's tired of watching Owen get all the attention, tired of training hard without seeing enough results. And he's sick of having Charlene, the only girl he really likes, act like he doesn't exist.

Joe has to make a decision: Should he take the drug or not? But first he has to decide how to go about deciding.

He might use the decision-making model to examine just what is at stake.

- **Identify the situation.** Joe has been offered steroids. He can't decide whether to take them or not. He doesn't want to harm his mind or body, but he does want to start on the football team, get Charlene's attention, and get Owen off his case.
- **Recognize the challenge.** The challenge for Joe is to find ways to feel good about himself without risking his well-being. He has to ask himself, "How will taking steroids affect my health? My behavior? My relationships over time? My feelings about myself? My future goals?"

- **List the possible choices.**
  - Joe can start taking steroids.
  - Joe can avoid Owen.
  - Joe can say NO to steriods and then tell Owen why.
- **Identify the possible consequences of each choice.** Joe divides the possible outcomes of each choice into positives and negatives.
  - Positive consequences if he decides to take the drug: He may gain weight and muscle size fast. He may do better on the football field. He may get Charlene's attention. He may feel more like one of the guys.
  - Negative consequences if he decides to take the drug: He may harm his health, cause himself serious mental problems, become violent. If the coach or the principal finds out, he could be thrown off the team. He could destroy the trust his family has in him. Even if no one found out, he would lose his self-respect.
  - Positive consequences if he puts off the decision: He may buy time and avoid confrontation with Owen. He may avoid problems with Owen and Owen's friends.
  - Negative consequences if he puts off the decision: He may feel anxious each time he sees Owen. He may never have the satisfaction of settling the situation once and for all. He may feel bad about himself for not standing up for himself and what he believes.
  - Positive consequences if he decides not to take the drug: He

Violent behavior is often a result of steroid use.

  may get inspired to train harder and get in better shape. His fitness will be real and lasting. He can feel good that he has made a decision to protect his physical and mental health. He can share with Charlene who he really is, and then, if she likes him, he can know it isn't for superficial reasons. He can maintain his health and self-respect. He may even find some new friends as he stands up for what he believes.
  - Negative consequences if he decides not to take the drug: He may be mocked or excluded by Owen and the other steroid users on the team. He may lose Owen as a friend. He may spend the football season on the bench. He may lose Charlene.
- **Make a decision.** Joe decides to say *NO* to the steroids the next time Owen offers them. He decides that

Teens who have stopped using drugs sometimes become dedicated to stopping other teens from using. This is very healthy for all involved.

even though he is tempted to take the easy way out, it isn't worth the risks to his health, his behavior, his relationships, his self-esteem. He doesn't want some drug making him beat up on people. He doesn't like feeling out of control. And if Owen can't handle the decision, that's Owen's problem. He'll have to find a way to deal with it.

- **Rehearse the decision.** Joe rehearses in his mind the ways he might say that he won't take steroids. In his head, he plays out the scenes of the kinds of responses he may get from Owen. He decides not to put Owen down but not to apologize for his decision either. He decides that the next time the drug is offered, he'll say, "No thanks. It's not worth the risks." If Owen seems open to hear what the risks are, Joe will tell him. If Owen says something like, "Come on, man, grow up," Joe will just calmly repeat, "No thanks," and walk away.
- **Take action.** After a pep rally, Owen offers Joe the steroids again. Joe says, "No thanks. It's not worth the risks to me." Owen shakes his head and walks off.
- **Evaluate the decision.** Joe doesn't like the feeling that someone is mad at him, but he figures it will probably pass. Even if it doesn't, he feels that he made the right decision, that he stood up for his own well-being. And now that he has done it, he even feels some relief. He has increased his feelings of self-worth. He can feel good that he has flexed his values.

What About You?

When you step onto the down escalator called drugs, you may be in for a long ride. You can get back to the top step if you work hard enough, but the top step keeps moving down.

Some teens close up when anybody talks about drugs. They don't want to hear it. They're tired of facts, and they're tired of speeches.

Teens who have been hurt by using drugs may be the group most dedicated to stopping teen drug use. They've been on the down escalator, and they know what they are talking about.

Maybe, so far, your experiences or your friends' experiences with drugs have told you otherwise—that despite all the adult warnings, alcohol, tobacco, and other drugs aren't really all that dangerous. Maybe the way you see it, there are plenty of happy endings, plenty of teens who've partied and haven't paid the price.

As a teen, you may have seen only the good times at the top of the escalator, not the suffering at the bottom. But the facts in this chapter are not make-believe.

Sure, substance abuse may not always end up as dramatically as it did for Len Bias. Sometimes the effects of drugs just sneak up on you, eating away at you slowly like a cancer. Some people may even seem to escape drugs' effects altogether. But the facts say otherwise. The facts say that however carefully or rarely you may use drugs, drugs will take some toll on your body, your mind, your life.

Because of one drug decision, Len Bias never had a shot at life. You do. Take your best shot, drug-free.

SECTION REVIEW
STOP AND REFLECT

1. Think about where you are on the down escalator called drugs. Have you stayed off? Or how far down are you?

2. Why do you think some teens don't want to talk about themselves and drugs?

3. What pulls the teen drug users you know into drugs? Are there other attractions besides those listed on pages 168-169?

4. If you knew a teen who had been asked to be a drug runner or to deal drugs, what would you say that might help him or her to decide against it?

5. What do you think it takes to make the tough choices and sacrifices that go along with staying drug-free— instead of giving in to drugs or drug money?

6. Why do you think teens who have gotten off the down escalator of alcohol and other drugs are even more dedicated to staying drug-free than most adults?

CHAPTER 8 REVIEW

Putting Your Values To Work

STRENGTHENING YOUR VALUES

Reread *A Teen Speaks* **on page 149. Then answer the following questions.**

1. According to Lonny, what are the reasons he and his friends use drugs?
2. What are some of the messages about drugs Lonny and his friends hear all the time? Why don't they take these messages seriously?
3. What message does the behavior of most of the adults in Lonny's life give him?
4. What values seem to be present in Aunt Inez's attitude?
5. What values seem to be lacking in Lonny's parents' behavior?
6. What values are present in Lonny's attitude? What values are lacking?
7. What basic attitudes do you and your friends have about drug use? What values do these attitudes represent?

INTERPRETING KNOWLEDGE

1. Find and read a newspaper account of a teen or celebrity who died because of alcohol or other drug use. Turn the facts into a ballad, or story-song.
2. Write and perform a play in which twin villains, Coke and Crack, from the nasty Cocaine family, try to lure a teen into taking their family's deadly potion. Use actual information about the dangers of cocaine and crack use.

SHARPENING YOUR THINKING SKILLS

1. Why should alcohol be considered a dangerous drug? In your opinion, why do so many people mistakenly think alcohol isn't a drug at all?
2. What are some of the early warning signs of alcoholic drinking? Why do people who show these signs often not recognize they are in trouble?
3. Make a chart that presents some of the harmful effects of each of the following classes of drugs: stimulants, depressants, hallucinogens, narcotics, and inhalants.
4. What would you tell a friend who wanted to use steroids to increase muscle size and improve athletic performance? Why?
5. Why can't a recovering alcoholic smoke a little dope? Why can't a recovering cocaine addict drink an occasional beer?
6. The United States has one of the highest rates of teen drug use in the world. What are some reasons for this?

APPLYING KNOWLEDGE

1. Keep track of how you react each time the subject of alcohol and drugs comes up during the next month. Try to determine if there is a pattern to your reactions.
2. Imagine that you are a lawyer. You are arguing a case for having smoking banned in all public buildings in your state. What evidence will you present?

CHAPTER 8 REVIEW

Putting Your Values To Work

PRACTICING DECISION-MAKING SKILLS

Read about each situation. Then answer the questions.

Situation A: Refer to Joe's situation on pages 172-174. The next weekend, at another party, Joe hears Owen talking about buying steroids on the black market and giving them out to the team and a few weeks before the state championships. Later that evening, Joe watches Owen completely flip out—smashing a lamp and knocking over a chair. Now Joe feels that he has another decision to make. He has to decide whether or not to tell Coach Ferrara about Owen's drug use and Owen's plan to provide the team with steroids.

1. What are Joe's choices?
2. For each of the choices, list the consequences or outcomes. Put a + by those that are positive and a − by those that are negative.
3. Based on these possible consequences, what decision do you think Joe should make? Why?
4. How might Joe rehearse the decision?
5. What values would be represented in this decision?

Situation B: Freda is the youngest student in her grade. At a Friday night school dance, she feels uncomfortable and very young compared to the rest of the crowd. One of the girls she knows from gym class offers her a cigarette and says, "Come on. Don't be a baby." Freda feels like the whole group is watching to see what she decides.

6. Imagine that you are Freda in this situation. Make a decision using the decision-making process.
7. Imagine that one of the other girls said, "Don't light up. It's not worth it. My dad's dying of lung cancer." Would that change your decision?

Situation C: You are riding in a car to a football game far from home with your older brother and his friends. One of the friends opens a beer and offers it to you.

8. What would you do? Why?
9. Would your decision change if your brother weren't in the car?
10. Would your decision change if the teens in the car were younger than you?
11. What would you do if the friends were drinking beers, but didn't offer you one?
12. Would you change your decision if your brother and all of his friends were legally old enough to drink?

Situation D: Think of a recent situation in which you were faced with a decision involving alcohol or drugs.

13. Describe the situation you faced.
14. How did you go about deciding?
15. What did you decide?
16. What was the outcome?
17. How did you feel about yourself after making and acting on the decision?

CHAPTER 9

Recovering: Coming To Terms With Drugs

A TEEN SPEAKS

Being a teen mother was a lot tougher than I thought it would be. The need to have someone to love and care for was very real. But later it got a little lost in the diaper changing and all the other little nitty-gritty details you don't think about when you find out you're going to have a baby.

All she could do was cry, wet, sleep, eat, and drink. And lie there when I'd hold on to her. I was looking for something I couldn't get from a baby. So I kept getting high—alcohol or pills, whenever I could.

Then something happened that scared me so much I knew I had to get help. I was high, and I walked out of the house. I was gone for about three hours before I realized I'd left my baby there by herself. I was all the way across town in my girlfriend's kitchen. And she asked me, "Marla, how's the baby?" I totally freaked out and ran back home. She was all right, just soaking wet. She was just waking up for her feeding. It really scared me because a million things could have happened to her while I was gone. She could have hurt herself. She could have swallowed something, a toy in the playpen or anything.

I just sat there and held on to her and cried for a long time. Then I got up and fed her and everything. But that was the last time that happened. I mean, I was so scared when I realized what I had done.

Marla

SECTION 1

OFF THE ESCALATOR: GETTING WELL AGAIN

Watching someone who is chemically dependent can be frightening. Chemically dependent people are hurting themselves and others. When you are dependent on alcohol or other drugs and suddenly wake up to what they are doing to you and those you care about, you can get scared, too.

If you or someone you care for is in trouble with drugs, it is not the end of the world. It's never too late to get help. There are many ways to get off the down escalator, even if you are near the bottom—even if you're afraid you can't do it. No one is a hopeless case.

> I was afraid that if I gave up drinking I'd never have fun again.

Still Riding The Stairs: Marvin's Mess

Marvin is now 17 and a junior in high school. He's doing just fine. But at 15, Marvin was in serious trouble. He was trying and mixing all kinds of drugs. He was usually absent or late for school. He had a constant string of infections. He had dark circles under his eyes, and he stopped caring about how he dressed or looked.

> I started with cigarettes when I was 10 or 11, I guess. Somewhere around 13 I started smoking dope. Nothing major—just a toke here, toke there. Usually behind the bleachers at school or down at the park with the guys. It was just something to do.
>
> Then Leo got me to try some Black Beauties. I felt like I could fly, and I liked that feeling a lot. After a while, though, I needed more of them. And sometimes I couldn't stop the sweats or the shakes. I felt like a race car without brakes. So I started taking Blue Devils and Yellow Jackets to bring me down. After a while, I was getting into my mom's Valium. I was buying whippets and doing

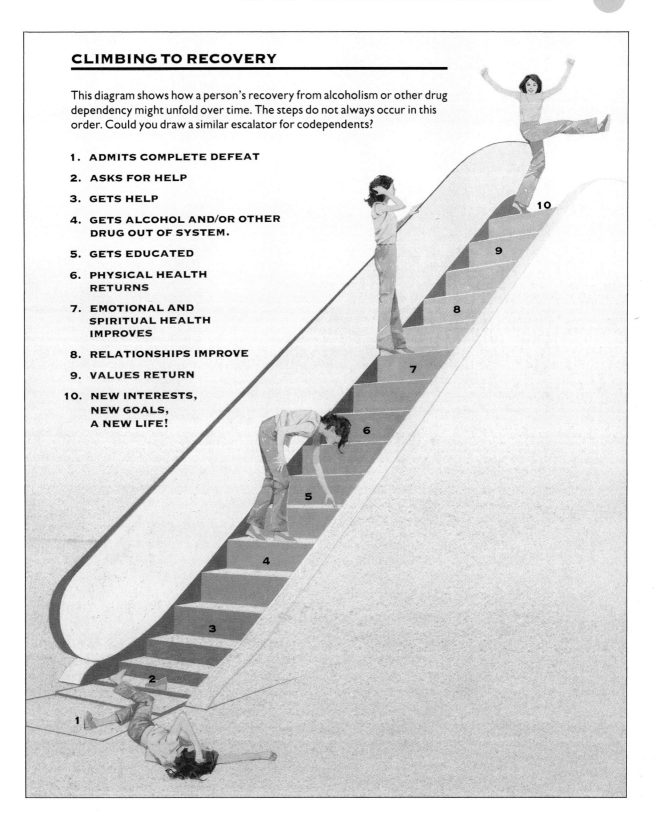

A good first step in confronting a drug problem is to talk with someone who has been there — someone who was on drugs and has given them up.

poppers, coke, whatever. I started drinking vodka in the morning before school just to stop the trembling, just to stop feeling sick. Then Leo said "Marvin, I've got this great new stuff called White Lightning. You've got to try it."

My mom was constantly after me. "Marvin, where are you going? Who are you going with?"

Then, toward the end of my freshman year, a teacher went to get some books for a kid who was absent. She opened the wrong locker — my locker! She found rolling papers, grass, a packet of pills, and a little butane torch.

I got kicked out of school. They told me that I couldn't come back at all unless I went through treatment for my drug problem.

Treatment Choices: Where To Get Help

When you or someone you care about admits there is a problem with drugs and wants to help, there are several treatment choices.

For someone who has just started experimenting with drugs, talking to a teen who has gotten in trouble with drugs and given them up may help. So might talking with a peer counselor or an understanding adult.

> I'm grateful now. But I never would have gotten help on my own.

Drawing up and signing a family contract may help, too. But if there is an ongoing problem with drug abuse, professional help may be needed.

The first thing Marvin did was to go to see a drug counselor—somebody his school counselor recommended.

> I really fought the idea inside. I thought, "Who does this guy think he is? He wouldn't know a good time if he tripped over it." But after a while, I realized the guy wasn't out to get me. He told me he used to be strung out on drugs himself, that he knew what I was going through. But then, when he said he thought I needed 28 days in a rehab, I said, "No way." It was like I'd been sentenced to prison for life.
>
> For the first few days at rehab, I had the jitters. They kept checking me over. But then I started to sleep better. I was eating, walking the grounds. Then, I don't know exactly when, but somewhere in the middle of the second week, I just kind of gave in to it all. It was like saying "Okay, I don't have to fight this any more." I looked around at the group, and the room was full of kids just like me. And all of a sudden, I didn't feel so alone anymore.

The rehab Marvin talks about is a drug and alcohol rehabilitation or treatment center. Such centers help chemically dependent people to get drug-free, and some of these rehabs specialize in treating teens with addictions. People generally stay at a rehab for four weeks or more, but they are not, as Marvin

A drug counselor knows how to help chemically dependent people, of all ages, get off drugs.

> **Back on Track: Help Through Counseling**
>
> Some counselors specialize in problems of substance abuse. Among these are certified drug and alcohol counselors. You can get the names of such counselors by contacting one or more of the following: the local chapter of the National Council on Alcoholism,
> - drug and alcohol treatment centers,
> - local counseling centers,
> - school counselors,
> - substance abuse telephone hotlines,
> - friends or family who have used such services,
> - a local place of worship or a member of the clergy,
> - offices of support groups such as Alcoholics Anonymous, Narcotics Anonymous, or Cocaine Anonymous.

feared, locked up. Rehabs are neither jails nor mental hospitals. People are free to leave if they choose to. And treatment centers are often located in relaxing settings with nice surroundings.

The typical treatment plan in a rehab includes a few days or more in "detox," the medical unit where chemically dependent people go through withdrawal under a doctor's care. This period of **detoxification,** "drying out" or becoming otherwise drug-free, is followed by a month or more of drug education, counseling, and support group meetings.

Another treatment option is outpatient rehabilitation. With this option, the chemically dependent person does not go away from home. Instead, he or she receives treatment for a few hours a day at a local treatment center. During the rest of the time, the chemically dependent person keeps up a normal schedule of school, work, and home. These programs are less expensive but more spread out over time than rehabs.

Support Groups

It's difficult to get clean and stay clean, or drug-free, on your own. But there are many groups in the community that can help if you or someone you know is trying to give up cigarettes, alcohol, or other drugs. These groups are free and anonymous—who you are and what you say are kept confidential, or secret. You don't even have to give your last name.

Around the World

Drug Abuse Links Nation to Nation

Drug abuse has historically linked one nation to another. In the 1830s, opium that was sold by Great Britain became widely abused in China. The British sold opium to make money, to damage the Chinese character, and to exert the power of their vast empire. In 1839 Chinese leaders, unable to stop either the supply or the demand for this drug, banned its import and destroyed British-owned opium stored in Canton, China. This began the Opium War between China and Great Britain.

America's drug problem today is also linked to other nations. The solution requires global cooperation. However, national leaders have different views about fighting drug abuse and illegal drug trafficking.

Some governments believe that the solution is to reduce the drug supply. If drugs aren't available, people can't buy them. That's why the United States works with Mexico and countries in Central and South America to keep drugs from coming in.

Other governments believe the United States should take more responsibility for reducing its demand. Very strict punishment of people who either sell or use drugs could decrease demand. Drug education could prevent teens from experimenting with drugs or being pressured by peers to use drugs. Probably the best way to fight drugs is to reduce both the supply and the demand.

These self-help groups meet regularly and encourage members to discuss their problems. They offer a plan of recovery—ways to give up harmful substances and return to healthy, caring, drug-free lifestyles.

> I didn't think I could have this much fun without drugs.

The people who attend these meetings get well and stay well by helping others who face the same kinds of problems. Members are available to talk day or night. They will even provide rides to meetings for those without transportation. And there are no dues or fees.

Marvin: An Update

Marvin has been drug-free now for 18 months. He regularly attends meetings of Alcoholics Anonymous and Cocaine Anonymous.

> It blew me away. At first I expected meetings full of old men who lived under bridges and slept in their clothes. But the meetings had all ages and all kinds.
>
> I felt weird being there at first. But people were really nice to me. They made me feel a part of it. We had lots of laughs. There were guys my age there, and everyone talked from the gut. And their stories—it felt just like they could read my mind. After a while, I even started to like going. Great friends. Good talks. It's keeping me sober and straight. And now it's kind of like a second home. I finally think I'm on my way.

SECTION REVIEW
STOP AND REFLECT

1. List four people you could talk to if you or a friend had a drug problem.

2. Make a list of words or phrases that tell you how you feel when you think about addiction.

3. Why do you think people get addicted?

4. What do you think someone would have to do to get off drugs or alcohol?

5. What would you want people who care about you to do if you were hooked?

6. Why do you think groups can help people get drug-free or alcohol-free?

SECTION 2

WHEN FAMILY OR FRIENDS ABUSE SUBSTANCES

growing up in a home where one or both parents are substance abusers can be very difficult, even devastating. So can having a sister, brother, or close friend who is in trouble with drugs. You may know firsthand how caring deeply about someone who is using drugs can rattle your nerves, take over your mind, and eat you up inside.

The problem is everywhere. Yet children of alcoholics, as well as children of other substance abusers, often suffer in silence, thinking there is no one else in the world like them. That's in part because they are taught never to talk about the problem outside the family.

Codependents

Children need their parents to love them, provide for their needs, and protect their safety and health. But if one or both parents are chemically dependent, they aren't always able to do those things. A child may feel deeply abandoned and hurt—at any age. But since the hurt is so great, the child learns to turn it off by going numb. He or she may even take on the responsibilities of the parents, listening to their problems, paying the bills, making the meals.

Maybe when the parents are drug-free, they are kind, caring people. But when the top pops off the first beer or the first line of cocaine is done, no one can predict what will happen. Words become weapons. Hands become abusive. Family members live in fear. They learn they can't trust. They lose respect. Worst of all, family members learn not to trust their own experience.

People who live with or are closely involved with chemically dependent people are called **codependents.** They suffer from emotional and sometimes even physical problems because they are too involved in the drug user's life. They become part of the problem, and they, too, need help.

It doesn't matter what drug the substance abuser uses. Whether the chemically dependent person drinks beer or shoots up heroin, it affects the people close to him or her in the same way.

Chances are, even if you are not directly affected by chemical dependency, or addiction to drugs, you know someone who is.

Courtney: A Case Of Shame And Blame

Courtney is 14. She has an older brother and two younger sisters. From the outside, she seems to have the perfect family. Her dad owns his own business. Her mom is a lawyer. They have a big house, two cars, and a pool. But Courtney's mom drinks—a lot. Lately it has started to get out of hand.

> Mom's sleeping a lot lately, claiming she has the flu. She gets me to call in sick for her at work. She makes me keep my sisters quiet when she has a headache. I clean up after everybody all the time. Now I've even started to do the grocery shopping.
>
> Mom used to play tennis and go to meetings, but now she spends a lot of time in her bathrobe or in her room with the door shut. I've tried everything from hiding the bottles to watering down the booze so it isn't so strong. I've begged Mom to stop, and she keeps promising she'll try, but she never does. I've even tried to talk to Dad about her drinking, but he says it's not that bad, that I'm exaggerating, and I should stay out of it. They

When Push Comes to Shove

Alcohol and other drugs play a role in up to 90 percent of all child abuse cases. If a substance-abusing parent becomes a physically abusing parent, you need to act right away. Leave home, even if only temporarily, until things cool down. Call Parents Anonymous, or Kids Anonymous, or, if you are being sexually abused, Daughters and Sons United. Ask the operator to connect you with a child abuse or crisis hotline. Tell a responsible adult. If necessary, call the police. Care enough about your health and life to protect yourself. No adult has the right to abuse a child—even if the adult is your parent.

CHAPTER 9 RECOVERING: COMING TO TERMS WITH DRUGS 189

For every chemically dependent person, there are approximately four codependents.

Drugs and Americans: A Multiplying Problem

There are now an estimated 35 million chemically dependent people in the United States. And for every chemically dependent person, there are an estimated four codependent people — people whose emotions and lives are deeply affected by this drug abuse. That means that more than half the nation's population is directly or indirectly caught up in the disease of chemical dependency.

fight all the time. He slams doors and gets silent. She screams or cries. Sometimes she even throws things.

This marking period I made the honor roll. I thought it would make Mom happy. It didn't. It's like she's there but not there. And when I do get her attention, it makes me uncomfortable. Some days she nags me, hurts me, blames me for all kinds of things. Other days she is too affectionate, kissing me and telling me how much she loves me.

The clincher was Friday night. I asked her to promise not to drink during my slumber party, but she got drunk anyway, tripped on the stairs to the family room, and cut her head — all in front of my friends. Then she started cursing at us for making so much noise. It was a real scene. I could have died. I just can't take it anymore. I've had a headache ever since. I can't sleep. I'll never have anyone over again. Never. And I'll never ever be like her!

Loaning money to a friend — when the friend has blown his money on drugs — is enabling.

Enabling: What Courtney Should Not Do

Courtney wants her mom to stop drinking or at least to drink less. She wants to stop her own pain. She's willing to do anything to try to control the situation and keep her family going. But much of her "help" just makes it easier for her mother to keep on drinking. This is called **enabling.** Enabling means protecting the drug user from suffering the consequences of his or her drug use or addiction.

Courtney enables her mother by keeping the secret that her mom's drinking is out of control She enables her by covering for her, mothering her younger sisters, doing more than her share of the housework. In short, by always trying to "fix" the situation, Courtney only makes it worse.

Enablers waste their energy trying to keep the problem from getting out of hand, instead of facing the problem.

> I didn't want to lie for my brother again. But I figured it was better than hearing Mom cry.

Non-enabling starts when you admit that there is a problem and then commit yourself not to do anything that will support and encourage the drug user's substance abuse. When the urge to "help," "fix," or make excuses for the drug user arises, ask yourself: Will my actions help or hurt the person in getting drug-free? Am I doing this just so the person will like me or not be mad at me? Am I doing this just to keep the peace?

Non-enabling does not mean turning your back on someone you love. It means valuing yourself and the person enough to say, "No, I won't become part of your sickness." In fact, often when those around the addicted person stop enabling, that person finally decides to seek help. Not enabling really means daring to care and having the courage to say *NO.*

What Courtney Can Do For Herself And Her Family

It is important for Courtney to know that her mom's drinking is not her fault. She may not be able to stop her mother's drug use, but she can stop making it the center of her life. She can stop keeping the secret that things at her house are out of control. She can get help from people who understand the effects of drugs on the family. She can begin to care for herself. That's also the first step in really caring for her mom.

Enabling: Fixing Those in a Fix

Here are some other examples of enabling.
- You loan money to your friend to go out on a date because he's blown his allowance on drugs.
- You quit dance lessons because you're afraid your mom will come to the recital and embarrass you.
- You pay a fine or bail for your older brother who was caught driving drunk.
- You get a part-time job to make up for the money your dad is spending on cocaine.
- You cover for your girlfriend by telling her parents that the hash pipe they found is yours.
- You put your mom to bed every night, wasted.
- You lie to the coach for your teammate who is too hungover to play in the game.

Non-enabling doesn't mean turning your back on someone you love.

Too Close for Comfort: Tips for Codependents

Here are some steps you can take if you are close to someone involved with drugs.

- Confront the person with his or her drug-related behavior and how it bothers, hurts, angers, or embarasses you. Do this when the person is sober and straight. A good time to do it is the morning after an "incident."
- Don't be surprised if the person becomes sarcastic or angry or denies the problem.
- Let the person know you care. Express your concern. Show respect for the person, but not for his or her drug behavior.
- Remember that addiction is not a matter of willpower or weakness. It is a disease.
- Don't put the person down or make cheap jokes. Try not to judge or punish. The person already feels bad about himself or herself.
- Don't preach or lecture. Avoid the "if you loved me" approach. Threatening, tears, or tantrums won't help. They may only give the person more excuses to use drugs.
- Don't try to control the person's life or performance. Let him or her fail or make mistakes.
- Don't provide money, transportation, or help to get drugs.
- Don't change patterns, hobbies, friends, or anything else to make life easier for the user.
- Don't promise not to tell others about the problem.
- Don't react to what the substance abuser says. Act according to what you think is right.
- Get information about chemical dependency, drug and alcohol counselors, self-help groups, and hotlines. Familarize yourself with the information first; then give it to the person. Offer to discuss it.
- Put the person in contact with a recovering substance abuser who understands. Then step out of the way.
- Offer to go to an open meeting of a recovery self-help group with the person.
- If the person relapses, or "slips" back into drug abuse, remind yourself it is not your fault.
- Get help for yourself, too.

There are many ways that people like Courtney can get help for themselves. If you live with or are closely involved with someone in trouble with drugs, you, too, can get help. You can have individual or group counseling with people who understand chemical dependency and codependency. You can get well whether or not the drug user in your life decides to get better.

> It's like my whole life is on hold till he either gets clean or dies from the stuff.

More Support Groups — "So many people cared!"

> It feels good to help the new kids who are just going through it.

If your parent or parents are chemically dependent, you can join a self-help support group such as Alateen. Free and anonymous, Alateen can help a codependent preteen or teen learn to value him-

These people all have something in common. They each live with a chemically dependent person. This support group is their life line.

self or herself and to detach from the drug user's problems. Other support groups for codependents include Al-Anon (for adults involved with alcoholics), COA (Children of Alcoholics), Nar-Anon (for families or close friends of drug addicts), and Coc-Anon (for families or close friends of cocaine users).

Intervention — "We're in this together!"

> I was afraid she'd never talk to me again. But I knew I had to do it.

Experts used to think that for a substance abuser to want to get help, the person had to "reach bottom." That meant really suffering a major problem, like causing a car wreck, being arrested, or getting kicked out of school. These days, experts believe that those who care deeply about a person in trouble with drugs can confront the person and get him or her into treatment before things get so out of hand. This is done through a process called **intervention,** a meeting of the chemically dependent person and close friends and/or relatives—people who care.

The people take part in the intervention with a person trained in the process. Over several days or weeks, without the drug user's knowledge, they meet and learn about chemical dependency. They also learn what to expect when they confront the drug user, and they talk

---

A Bias Against Drugs

Lonise Bias has a mission. The mother of Len Bias—the college basketball star who died from a cocaine overdose—wants to save other young people from what happened to her son. She has turned her tragedy into teaching, taking her anti-drug message on the road to high schools, colleges, and churches across the country.

Mrs. Bias doesn't think kids who use drugs are bad. She thinks they are hurting and can't cope. She says that to beat the deadly draw of drugs, we need to strengthen our families. And we need to learn to love one another in better, healthier ways.

A successful intervention takes careful planning and a lot of soul searching.

about ways to express how much they care. They may act out the ways the person is likely to react.

They make lists of all the drinking or drugging behaviors in which the person's drug-related behavior affected them. They list, too, how they felt.

At a meeting, they present this information to the substance abuser all at once. This is often enough to make the drug user see just how bad the problem is. The group then presents treatment options, and may take the person directly to a treatment facility.

> At first, I thought I was there for my dad. Then I realized I was there for all of us.

While in the treatment facility, the chemically dependent person may participate in group counseling. Later, the individual continues his or her own recovery through programs like Alateen and Al-Anon.

> My sister was always Miss Perfect. I was always Mr. Loser. Now we're learning ways to change all that.

Courtney: An Update

A few months after Courtney's disastrous slumber party, her mom had a car accident while drunk. A few weeks later, the family staged an intervention with the help of an alcoholism counselor. The intervention was not only a way of caring for Courtney's mom. It was also a way of caring for the family and taking a first step to save it.

Family Programs — "We're a family — we're worth it!"

Many drug rehabilitation and counseling centers have family programs for those close to a recovering drug abuser. Codependents may spend a few days, a week, or several weekends at the center learning about the disease of chemical dependency and about their part in that disease.

Courtney's mom went away to a drug and alcohol treatment center. She doesn't drink any more. The rest of the family attended a five-day program at the same rehab. Now the whole family is getting ongoing help. Courtney's mom goes to AA and her dad goes to Al-Anon. Courtney, her brother, and one sister go to Alateen. Her youngest sister goes to a group called Alatot for young children of alcoholics. The whole family is now on its way to being well. As Courtney says, "The nightmare is finally over."

Recovery: Getting High On Life

Recovery from drug addiction is a challenge, but it is possible for anyone who wants it. Recovery is an adventure. It is a lifelong process. You are never recovered but always recovering. Though staying completely drug-free is necessary for recovery, you really only have to stay away from drugs one day at a time. And living drug-free does not mean living fun-free. In fact, recovery is often both rewarding and exciting.

> I can read again, think again. I'm even running track.
>
> I've stopped being so afraid all the time.
>
> I've been sober now eight months, three weeks, and four days. And it keeps getting better.

Living drug free doesn't have to be fun-free.

The overall goal of an intervention is to get the chemically dependent person into a treatment center.

Detaching, or pulling away from someone else's addiction, can finally free you to live and enjoy your own life and care for others in deeper, non-enabling ways.

Deciding that you have a problem with tobacco, alcohol, or other drugs is a big step. So is deciding you have a problem with someone else's drug use. In fact, these are the first steps in recovering. And helping others and yourself to get well can make you feel good about yourself and hopeful about the years ahead. It can even bring you joy.

Riding the up escalator called Recovery can give you a natural lasting "high." When you get on this escalator—going up, you are valuing yourself. And you are able to recover the values of caring and respect. You will care for yourself and for others. You will have self-respect and respect others.

SECTION REVIEW
STOP AND REFLECT

1. Tell what you could do to help a friend whose mom or dad is a substance abuser.

2. Why do you think people become enablers?

3. Describe how you would feel if you realized you were enabling a parent or friend to keep up a dependency.

4. Tell what you might do to stop being an enabler to a parent or friend.

5. Write a letter to Courtney or another imaginary friend to encourage and support him or her in helping the family.

CHAPTER 9 REVIEW

Putting Your Values To Work

STRENGTHENING YOUR VALUES

Reread *A Teen Speaks* **on page 179. Then answer the following questions.**

1. How do you think Marla really feels about her baby? On the night she was high and walked out, did her actions reflect these feelings? What values were present in her actions? What values were lacking?

2. What might have happened to Marla's baby while she was away? To Marla? How might Marla have handled her frustration differently, showing caring for herself and her baby?

3. What do you think Marla learned from this scary incident? Since then, what step has Marla taken to improve her situation? What values is she reflecting in this step?

4. Think of someone you know who is in trouble with drugs and someone else who is in recovery. What values seem to be present and lacking in the actions and attitudes of the person with the active disease? The person in recovery? What can you conclude about how addiction affects a person's values?

INTERPRETING KNOWLEDGE

1. Design and paint a large two-panel mural. On one half, paint images that suggest addiction. On the other half, paint images that suggest recovery.

2. Make a sculpture or write a song about the down escalator of addiction and the up escalator of recovery.

SHARPENING YOUR THINKING SKILLS

1. Why do you think groups like Alcoholics Anonymous have been successful in helping chemically-dependent people get and stay drug-free?

2. Make two lists: one of the traits of chemically dependent people, and one of the traits of codependent people.

3. Explain the concept of enabling. Describe ways to encourage teens to stop enabling substance-abusing family members or friends.

4. List some tips for a person about to confront someone in trouble with alcohol or other drugs.

5. Why are teens often afraid to confront friends with drinking or drugging problems?

6. Describe the intervention process.

APPLYING KNOWLEDGE

1. Using the tips provided in this chapter, confront someone close to you who is in trouble with alcohol or other drugs. Rehearse, in writing, what you will say. You may not be close to anyone who is in trouble with drugs or alcohol. In that case, write what you would say if you were.

2. Set up a Recovery Resource Table in your school. Offer free information, such as pamphlets, lists of self-help groups, names of certified alcoholism counselors and rehabs, and so on.

CHAPTER 9 REVIEW

Putting Your Values To Work

PRACTICING DECISION-MAKING SKILLS

Read about each situation. Then answer the questions.

Situation A: Rob is worried about his friend Hank. Hank seems to have a serious problem with marijuana. He gets stoned every morning before school, and sometimes he has a joint or two at lunchtime. He can't seem to remember anything. He acts withdrawn, and he just doesn't seem to care about anything anymore, including his friends. Rob has a nagging feeling that he needs to do something.

1. What decision needs to be made?
2. What is the challenge for Rob?
3. What are Rob's choices?
4. What are the possible positive and negative consequences of each of these choices?
5. What decision do you think Rob should make? Why?
6. How might he rehearse the decision? What kinds of things is it important that he say to his friend?
7. How might he go about taking action?
8. What values would be represented in this decision?

Situation B: Imagine that your sister is using cocaine. She keeps asking you to bail her out of the trouble she gets into because of her coke habit. Often, you've loaned her money. To keep the peace and save your parents pain, you've lied to them about where your sister was and when she came in. You've even called the video store where she works part-time to say she had the flu, when in fact she was wasted after a night of highs. Tonight, while your parents are away, she wants you to give her the car keys your dad has left with you. She says she needs to go to the library to work on her term paper, but you suspect she is going out to party or to buy some crack.

9. Make a decision using the decision-making process.
10. Would your decision making lead to an enabling action or a non-enabling action? Explain your answer.

Situation C: Your good friend has a mom with a serious drinking problem, and your friend seems really stressed out about it. In the past month or so, your friend has stopped eating because she is so upset about her mom. You try to bring up her mom's drinking and the effect it seems to be having on your friend. But whenever you start to talk about the problem, your friend changes the subject. You think maybe some information about children of alcoholics might help her, but you don't want to pressure or embarrass her.

11. Would you give your friend the information? Why or why not? Would you bring up the subject of her mom's drinking in some other way? If so, how?
12. What values would be reflected in your decision?

CHAPTER 10

Planning For Tomorrow

A TEEN SPEAKS

Mom always worked and my oldest brother, Tony, took care of the rest of us. He was out with his friends most of the time, so we had to take care of ourselves. Life was hard for Mom. She drank a lot and hit us. By the time I was 13, I'd made up my mind to get out. So I took some of my babysitting money and caught a bus to New York City.

The bus station in the city was so big. I didn't know anyone. I hadn't thought about what I'd do when I got there. I didn't have any money.

As soon as I walked off the bus, this guy, Butch, came up to me. He asked me if I was alone. He said girls like me had to be careful in the city. He bought me some lunch. He said I could stay with him, if I'd help him out later. Now I know it was dumb for me to go. But Butch was so nice to me, and I was so scared.

The next couple of weeks were heaven! Butch bought me new clothes. And I got to dress up and go out.

Then he said he needed my help. He wanted me to have sex with some men at this hotel and give him the money. I didn't want to. But he'd been so nice to me, I just couldn't say no.

That was two years ago. Now I have a baby. What I could tell you about Butch you wouldn't want to hear. He thought he owned me. He did terrible things to me.

I did get away finally thanks to an organization that helps kids on the street. And I'm back in school. Even though I'm a single parent, I know I'm going to make it. I'm going to get my high school diploma. I thought I had no future back home. But I didn't know I was throwing away my future. Life on the street is a dead end.

SECTION 1

FACING UP TO THE FUTURE

Sometimes you can look at the future and feel trapped. If things are bad at home, you may wish you could bail out by running away. Or, things may be good at home. But if you are under a lot of pressure to live up to high expectations, you may feel like bailing out, too.

The future is always out there—waiting to happen. What your future holds may be a big question mark. But you can take steps today that will make tomorrow better for you. You've probably been thinking about tomorrow and the years to come since you were a child.

Do Dreams Ever Come True?

Remember when people used to ask, "What are you going to be when you grow up?" They may still be asking! When you were a child, you probably had an easy answer—maybe a truck driver, a lawyer, or a pro basketball player, or a doctor. But your answer may have changed every time somebody asked! Children spend a lot of time trying out their ideas about the future as they play.

> Play like we are fire fighters and I drive the truck!

> Play like I am the teacher and you are the bad kid! Now sit down and be quiet.

Planning for the future seemed easy when you were a child. You could do and be anything you could imagine. Childhood fantasy helped to prepare you to begin thinking about the future. But it's not quite so simple anymore.

The future may be as near as next year. Or it may be as far away as 10 or 20 years. If it hasn't hit you yet, you'll soon face the fact that important decisions about your future need to be made now. You may already be asking yourself questions like these:

- Who am I really?
- What do I want to do with my life?
- Will my parents ever let go?
- Do I have what it takes to be on my own?

- Do my parents really care about me?
- Do dreams ever really come true?
- When am I going to find someone to love me for myself?

These questions will be answered as you make decisions about who you are, who you will be and what you will try to do with your life. Your decisions will reflect what you value most. That doesn't make dealing with them any easier, however. This is why you may put off dealing with the questions. You may be unsure of your values, or deciding may seem too difficult. The fact is, the more you avoid questions about the future, the harder it is to deal with them. The questions don't go away. The more you avoid them, the more anxious and confused you feel.

Dreams vs. Reality

Some people go through life avoiding decisions and letting things just happen. They seem to be unable to choose and to act on their choice. But dreams rarely come true by accident. If you want something to happen, you have to be willing to think about it, to make choices, and to act. For example, you may dream of falling in love and marrying the ideal partner—interesting, fun, good-looking, and with enough money to make things comfortable. But, have you asked yourself if you are the kind of person your dream mate would really want to marry? If you have few interests, pay little attention to grooming, are selfish,

If you want dreams to come true, you have to make appropriate choices and act accordingly.

and haven't the least idea about how to manage money, you can dream away. But you probably won't see your dream come true.

Or, you may dream of creating your own successful business and then retiring at age 40 to a life of leisure. But have you asked yourself if you are becoming the kind of person who could develop a successful business? If you can't hold a part-time job, won't do your school work, and are doing drugs, you can dream away. But you aren't likely to have the positive attitudes and the skills needed to develop your own successful business.

Jim Abbott didn't let his disability stop him from dreaming — and making the dream come true.

Not all of your dreams could possibly come true, and you wouldn't want them to. Would you really want to marry the girl or boy you dreamed about in third grade? Other dreams are completely out of reach, too. Is a famous movie producer likely to see you hanging out with your friends and insist that you are the person he wants in his next movie? If you have absolutely no musical talent, are you really likely to become a famous singer? Or, if you have little natural ability for mechanical things, are you really likely to design a spacecraft?

Realistic dreams are worth working toward, however. They are within your reach if you are willing to plan and work. Yet, even these dreams will fade if all you do is sit around hoping they will come true.

How Can You Build Dreams In The Real World

By now you should have left childhood fantasies behind. You may still fantasize about your life and about other people. But you know the difference between fantasy and reality.

Sometimes the difference is hard to take. It would be wonderful if nobody had cruel and indifferent families. It would be nice if those who do have difficult families could manage to be kind and good in spite of it and end up rewarded for their goodness. Unfortunately, real life is not always like that!

Sometimes bad things happen to good people. It doesn't seem fair. Sometimes life can make you feel hopeless, that no dreams are possible. Many dreams, however, are possible if you set realistic goals along the way, and then work hard to reach them. Many people have overcome difficult and discouraging circumstances to follow their dream. And they've ended up making life better for others in the process.

The Real World— "Whom can I trust?"

As an adolescent, you start coming to terms with the fact that adults are not perfect. It may sound silly to say adults aren't perfect. Of course nobody is perfect! But, it's one thing to know that people have faults. It is another to find that people you love and trust have faults, too. When you begin to notice the difference between what some adults say they believe and how they act, you may want to write those adults off as completely phony.

> Linda Sue's dad is always complaining about inflation and taxes. "I don't know why we have to put up with such jerks in government! They don't know any more about how to run things than I do." Linda Sue wonders why her dad complains so much when he never bothers to vote.

> Jake's mom works all day. When she comes home, she has to cook dinner. She does all the housework by herself, except for the chores Jake does. When his dad comes home from work, he sits and reads the paper. He wouldn't think of doing "woman's work." Jake knows his mom resents having so much responsibility. But she never asks his dad to pitch in and do his share. When Jake questions her, she says, "That's just the way men are."

Sometimes the adults you care about seem to have few **ideals** or standards of perfection. They give in to the way things are without questioning.

When you were younger, you probably didn't notice contradictions and injustice. Now you probably have seen both, not just in adults, but also in your friends.

> Suppose a lab partner in science comes from a family that has very little money. He has had to do without a lot of the things you take for granted. In science class you can always count on him to do his share of work. Not only that, he was willing to take time to help you when you had trouble understanding the assignments. One day he tells you, "You can do anything you want as long as you work hard enough." You think he's a neat guy.

> Another classmate comes from a family that is very well off. He has practically everything he ever wants. He is a very good student, but very competitive. You can count on him to have the right answer, but don't ask him for help.

> When it comes time for an election of class officers, both boys declare that they are interested in running for class president. None of your friends will even consider voting for your lab partner. "He's nobody!" they declare.

When you seem to run straight up against something unfair, it can be tempting to give up. Why vote for a lab partner who probably won't win anyway? What is one vote? That is where your values count. Every action you take for justice or fairness is an important action — even if you take it alone. Besides, you never know who will be influenced when you take a stand.

Hanging In — "Count on me!"

Sometimes it can be hard to believe that ideals really matter. Perhaps you see little evidence that having high ideals gets you anywhere. Some teens give in to sadness or depression because they cannot deal with the difference between their ideals and reality.

> Angie was always nearly perfect. She never seemed to do anything wrong. Her grades were good. She got along with her parents. Other kids thought she was okay. She seemed to like everybody, but she wasn't best friends with anyone special. She was always trying to get people involved in some big project. One time she wanted the school to do something about nuclear war. Another time she thought the class should collect money for the homeless instead of a class trip. Nobody took her too seriously when she started talking about one of her projects.
>
> Then one day Angie didn't show up at school. Nobody paid too much attention. But she wasn't there the next day or the next. Then her body was found. Apparently Angie had committed suicide. Nobody could believe it. Later someone found her notebook. Inside the notebook were lots of little notes Angie had written to herself: "Nobody cares." "Who cares?" "Children starve and we go on class trips." "Nothing matters."

Everyone experiences feelings of being alone and helpless some of the time. And, like Angie, some teens give up. Think about the problems that face our world: poverty, racism, threat of nuclear war, drugs, crime. What can one person do?

It's true that you aren't too likely to change the world alone. One drop of water won't fill a bucket. But when you put a lot of drops together, they can fill up buckets and lakes and oceans. The same is true of actions. Every time an Angie gives up, there is that much less good in the world. When someone like Angie keeps on taking a stand for what is right, her actions — along with other people's actions — can create a lot of change for the good. Many helpful organizations and movements were started by only one person.

Caring Anyway—"Why not?"

Your values enable you to take stock of your dreams. Then you can replace old thoughts about life with more realistic dreams and goals.

You can care. You aren't alone. Other people do care. They may show you that they care in ways you don't expect. They may tell you to put on a jacket. They may tell you not to run on the stairs. They may ask you to be home by a certain time. They may ask you not to drink. You may not know how much they care unless you begin to think about what's behind their actions.

You can value your family, even if it isn't perfect. You can show respect for yourself and others, even when you and they make mistakes. You can be trustworthy and responsible, whether or not anyone else chooses to be.

You can make plans that are realistic. The questions that you keep having about the future, your doubts and feelings of sadness, all are part of the growth that is preparing you to be on your own. You can't answer all your questions about the future, about life, and about yourself. You can't always avoid mistakes, either. But, thinking about these questions and trying to answer them is important.

Thinking about your future brings you back to your dreams. By asking questions and struggling to find answers, you can make your dreams more clear. You can see what actions you need to take to make your dreams come true. And you are more likely to discover other dreams.

What Action Should You Take Now?

Many choices about the future need to be made now. These choices can begin with dreams about what you would like to do and be. But dreams need to be evaluated. How do they fit with your values? With your talents? They need to be accompanied by some realistic actions.

This teen could take a stand against graffiti—she could interest others in cleaning it off. She might even find a group of artists willing to paint a mural.

Getting a Job

There are two main steps in getting any job:

1. **Applying for a job.** You may have to write a letter of application, fill out an application form, and provide **references**. References are the names of people who can tell about your responsibility and your skill as a worker.

2. **Interviewing.** You may have a meeting with an employer who wants to ask you questions and explain the job to you. This meeting is called an **interview**. It gives you an opportunity to create a good impression and helps you find out more about the job.

Your guidance counselor may provide help in filling out job applications and preparing for interviews. Some schools provide short courses that allow students to develop and practice these skills.

The best place to begin making choices for the future is by thinking about who you want to become. What kind of person do you want to be?

You can begin to make choices about future careers by thinking about your current interests. What do you like to do? How do you spend your free time now? What hobbies do you have? Do you like working with people, or do you prefer working alone?

You can also begin to pay attention to your special abilities. What can you do well? What are your aptitudes? An **aptitude** is the tendency to learn to do something easily. For example, some people are able to learn to do auto repairs easily because they have an aptitude for mechanics. Sometimes people begin to develop a mechanical aptitude by taking apart clocks, kitchen appliances, and bicycles. You can develop your mechanical ability by volunteering to help others work on their cars or motorcycles. You can read magazines and books about mechanics. You can join a club or interest group at school or your youth center, or you can take a school course in mechanics.

The same is true for almost any ability or aptitude. Start looking for ways to develop your aptitudes.

In addition to earning money, you can learn a lot about yourself from a part-time job. What might these teens have learned already?

Part-Time Jobs — "How do I get one?"

You can look for part-time work that will allow you to start developing your aptitudes and talents. You could develop an interest in computers by getting a part-time job at a store that sells computers. There you will learn more about computers and new programs for computers. You might even develop aptitudes that you did not know you had. You could be an outstanding salesperson, or you might excel in bookkeeping. Finding part-time work that is related to your interests will allow you to develop in ways that you may not even be able to predict.

Before you get a part-time job, there are some things you need to do. Many jobs have age requirements. That means you have to be a certain age before you can be employed. Jobs like mowing lawns, running errands, and babysitting do not usually require a work permit. Agreements are worked out between you and the person who hires you. Some states require teenagers to have a work permit before they can have certain types of jobs. Those jobs also require a Social Security card.

The United States Department of Labor has guidelines for employing young people, too. For example, if you are 16 or younger, you may not be employed in a job that is considered hazardous. If you are 14 to 15 years old, you can work only outside of school hours unless you are in a school work-study program. Check with your school guidance counselor about the requirements and restrictions on working in your community.

To find a job, you can start by talking to people you know. Read newspaper ads. Ask people in local stores or businesses if they need after-school help.

You also need to take stock of yourself. Consider your personal appearance. Think about the way you present yourself. Are you the kind of person who is likely to be employed? By caring for yourself, you can develop the grooming, health, and communication skills that make you a good candidate for employment. By developing your values of caring, respect, and trust, and responsibility, you will become the kind of person who can get a job and keep it.

If your school doesn't provide any help to students who are thinking about work, don't let that stop you. Why not ask a guidance counselor or teacher to help you form an interest group? Together, you and other interested students can learn more about the application and interview process.

Part-time jobs can be used to explore your interests. If you are helping a neighbor with yard work on weekends, learn as much as you can about plants and their care. Your next part-time job might include full responsibility for the care of another neighbor's yard.

Almost any job can lead to other opportunities. Use your job to learn everything you can. Use it to develop habits that will help you to get and keep a job in the future. And, if you give your part-time job your best, you may find that it leads to full-time employment in the future.

Plan Ahead— "What's next?"

As you think about your interests and aptitudes, you can begin to examine training requirements for various jobs. If you come from a family with very little money, planning for a career as a doctor or lawyer may seem impossible. These careers require many years of expensive study. You can say, "What's the use?" and forget your dream. Or you can plan to take steps that will make your dream more likely to become reality.

For example, if you have an aptitude for working with children, you may want to think about becoming a pediatrician, a doctor who treats children. But, if your family has very little money, you will need to make realistic plans. You will want to make the best possible grades in school so that you will qualify for scholarships to college and, later, to medical school.

At the same time, you can explore other ways of working with children that could enable you to become a pediatri-

The military offers excellent educational opportunities as well as a way to travel, meet new friends, and earn money. A recruiter can explain the opportunities to you.

cian. You might prepare to teach children in school or day care so that you could work for a few years after college, before going on to medical school. This would give you valuable experience to draw upon when you become a doctor. You might even find your work as a teacher so satisfying that you decide to make it a career. Either way, you will have increased your opportunities by planning ahead and striving for a goal.

But what if you start out planning to become a doctor and your grades are not good enough? Sometimes you can't change your grades, no matter how hard you try. In this case, your aptitudes may not be in line with your dreams. This is why it is helpful to think of all your aptitudes and interests. There may be many careers that would let you develop those aptitudes and interests. Then, if circumstances like money or grades prevent you from developing along one career path, you can take another.

Deciding on a college preparation or vocational focus for high school will be important in the next years. Again, your values and your aptitudes and interests will help you decide. For example, if you have an aptitude for working with children, you may decide to take a child development course in high school.

Before selecting your high school courses, check out the possibilities. Talk with your guidance counselor. Think about what you want to do in the future.

This kind of course can help you understand children and be a responsible parent. You might prepare to be a teacher's aid until you have a family of your own. Or you might enjoy being a nanny or full-time babysitter. You may decide to be a teacher or to work in children's television. Your focus then will need to be on academic preparation for college. To be a teacher, you need to complete high school and four years of college. In some states, you also need to attend graduate school for at least a year. To prepare for work in children's television, you need to include drama, television acting, and production in your college coursework.

If you want to develop your aptitude for working with children by becoming a pediatrician, you will need to complete a pre-medical course in college, go on to medical school, and then specialize in pediatrics. All of these — working as a teacher's aid, as a teacher, in children's television or as a pediatrician — will allow you to develop an aptitude and interest in children. But each has its own special training requirements.

SECTION REVIEW
STOP AND REFLECT

1. What would you plan for your future if you had no limits on time, money, and other resources?

2. What makes planning for your future difficult?

3. List some of your aptitudes and interests that could lead to career opportunities.

4. Talk with two people who work in different jobs. Ask each one what skills they need to do their work.

5. How did each person you talked with decide on his or her career?

SECTION 2

PREPARING FOR THE FUTURE

If you are like most teenagers, you have some pretty high hopes for yourself. You dream of being independent and of having a good life. Most teens dream of having a family of their own. Some dream of being able to make the world a better place by helping others, perhaps by working in politics or government. These hopes and dreams are a result of all your values. If you want them to come true, you must act in ways that are consistent with your values, and you must set realistic goals.

Who Needs School?

The most important way to build for the future is by staying in school. Some teens never question school. Their families assume that they will complete high school and college. For other teens, whether or not to stay in school becomes a big question.

Why Some Teens Drop Out

There are some situations that seem to present strong reasons for dropping out of school:

- **Family needs.** Sometimes the family depends on income a teen can contribute. Sometimes a teen is needed to help care for young brothers or sisters so that the parents can work.
- **Opportunity to make money.** Sometimes teens can earn a lot of money by working instead of going to school. This money can come from a legal job. Or, it can come from illegal and unethical work, such as selling drugs.
- **Abuse or neglect at home.** Teens who are abused at home may be beaten by a parent. They be forced to have sex with a parent, relative, or neighbor. Or they may be abused emotionally and constantly told they are "no good." Teens who are neglected receive little or no attention or care from their parents. Abused or neglected teens often drop out of school and run away from home.

- **The desire to be out of school.** Schools are sometimes called "total institutions." Until age 16 students are not always there by choice. They cannot come and go whenever they feel like it. This causes some teens to feel as trapped as if they were in prison. They want to break away.
- **Failure or lack of success in school.** Teens who do not do well in school, particularly those who have trouble reading, often ask, "Why bother?" Sometimes very intelligent people have trouble in school because they have learning disabilities or other problems.
- **Pregnancy.** Sometimes teens who get pregnant are too embarrassed to stay in school. They may think there is not a place for them in school anymore. They may think that school won't matter since they are going to have a baby.

Dropping out of school is a ticket to disappointment and possible poverty in the future. Without the basic skills you learn as you complete school, you have little chance of getting a job that will pay more than the minimum wage. Most of the jobs that school dropouts can get offer very low pay. Also, these jobs often do not have benefits like health insurance, sick leave, or retirement plans. Illegal employment, such as dealing drugs or stealing, may bring the feeling of power that goes with instant money and all it will buy. But that kind of employment goes against the law and the basic values of society. Dealing and stealing are wrong!

Why Most Teens Stay

Some teenagers think there are strong reasons for dropping out of school. However, there are even stronger reasons for staying in school:
- **To provide options.** Many more jobs that offer above-average wages and benefits are available to the high school or college graduate. While the amount of money you can get for unskilled labor may sound like a lot now, you can't live on it with a family. A high school diploma is a basic requirement for most jobs, for college, and for many vocational training programs. If you drop out of school, you cut off your future opportunity to earn a decent living. You cut your chances of getting many kinds of jobs and of going to college, unless you are willing to go back and complete your basic education. You show prospective employers that you are not responsible. When you stay in school, it shows determination, a quality employers look for.
- **To break the cycle.** Even if you live in a home where you are abused or neglected, you can get out without running away and dropping out of school. If you have problems at home, talk with a social worker, a trusted teacher, a member of the clergy, or the parent of one of your friends. When you stay in school, you help to break the cycle of abuse or neglect that traps teens into being just like the people in the families they wanted to leave.

If you are pregnant, you can talk with your guidance counselor, a trusted teacher, or a member of the clergy about various programs for teen parents. You can learn how to have a healthy baby, how to care for your baby, and how to establish a strong, loving, respectful family. Community organizations are there to help you. And you can finish your schooling, so that your baby, too, can grow up caring and cared for.

- **To become knowledgeable and skilled.** The skills you get in school help you to understand and enjoy more of what you hear and see on television or in the movies, read in newspapers, and experience day to day. You will be a better citizen when you understand the issues in election campaigns and can decide for yourself on the best candidate. You can take a stand on social issues, such as providing care for the homeless or for the natural environment, when you understand the issues and know where to go for help.
- **For self-satisfaction.** You may be the only one in your family who cares whether or not you finish high school. If nobody cares, why bother? You are worth it! You may have to work harder. You may get less encouragement from home than anybody else in your class. But you can enjoy the satisfaction of knowing you did something very special and important for yourself and your future.

Dropping out of high school makes it very difficult to find a well paid job.

If your family needs you at home, it may not be easy, but you can finish school anyway. If you feel trapped in school and have a hard time dealing with the fact that you have to be there, talk with your guidance counselor, a teacher you can trust, or a friend who is staying in school. You don't have to carry these

Making school work means finding the right courses for you — ones that build skills in your area of interest. Of course, you have to find a way to make the required courses work, too.

feelings alone. You don't want to trade a feeling of freedom now for being trapped in low-skill, low-paying jobs the rest of your life. Many school dropouts, who said school was too much like prison, end up in prison.

Making School Work For You

Some people seem to have an easy time of it in school. They don't even study, and they make good grades. Others work and work, but still barely get by. People have different skills. Everybody can do some things well. If you find it hard to do school work, it doesn't mean that you are not smart. It may mean that the ways you are smart do not seem to be as useful in school. For example, you may be handy with car repairs, but that doesn't help you to learn history facts. Or, you may be a good athlete, but that doesn't help you learn math.

To get the most out of school, you have to

- **attend school regularly,**
- **make appropriate decisions about the courses you take,**
- **focus on learning,**
- **get tutoring** for a subject you have difficulty with,
- **take advantage of extracurricular**

activities,
- **use school resources.**

Attend School Regularly

Deciding to stay in school is only part of getting an education. You have to be there to benefit. Some teens are tempted to cut classes or skip school. When you miss school, you are missing out!

If you are caught in a situation where you can't go to school regularly, ask for help. Talk to someone—a social worker, teacher, the parent of a friend, or a member of the clergy—who can help you and your family find a solution to the problem.

Make Appropriate Decisions

In junior high or middle school, you probably have the chance to "elect" a course or two each year. In high school, you will have the opportunity to select several courses. Your selected courses can help you identify special interests and abilities. Sometimes they lead to career choices.

It is important for you to know what courses are available to you. Talk to your guidance counselor. Talk to your teachers. Tell them what you're interested in. Tell them what your goals are. Read the course selection booklet developed for your school. Then choose courses that will help you meet your

A high school co-op program allows students to work (for pay) half a day and attend school half a day.

goals. When there is time, select a course for fun!

If money is a problem for you and your family, you might want to consider the co-op or work-study program in high school. In this program, you go to school half a day and work the other half—for pay. While in school you complete your basic academic courses and learn how to get and keep a job. You also get high school credit and are supervised on the job by your co-op teacher. This program is particularly helpful to people who might otherwise drop out.

If you are fairly certain of going to college, you want to be sure you take all the required academic courses. You may not have time for a lot of electives in high school.

Whether you plan to go to work or go to college after high school, you must

Around the World

America: Your Right to Be Well Educated

As an American teenager, you have an opportunity to be well educated. Young people with different incomes and backgrounds can succeed in school by working hard and applying themselves. After completing high school, most students can attend a technical school, junior college, four-year college, or enter an apprenticeship training in a trade.

In many countries, however, the educational system doesn't offer such opportunity. Only the very brightest youth qualify for specialized training at a university. Others can receive only a basic education. They must learn the rest on the job.

Some education systems require students to make career decisions in their early teens. After testing, students attend a high school that prepares them for college or one that prepares them for a trade. It is nearly impossible for them to change direction later.

In many countries, one test can determine a student's future. A student's chances of attending a university may depend on one long, written exam covering everything ever learned in school. In the United States, you have many chances to prove your knowledge and abilities.

be aware of high school graduation requirements. Be sure you meet those requirements each year. If there is a problem, talk to a teacher or guidance counselor immediately.

Focus On Learning

Sometimes schoolwork seems to have little to do with the real world. How often are you going to need to know the names of the rivers in South America? Do you really need to know how to read a road map?

A lot of things you learn in school do become useful later. Others help you to develop skills and expand your thinking. It is true you may never be required to name the rivers in South America. On the other hand, when you read about the political situation in South American countries or see a movie set in South America, that knowledge will add to your understanding. If you take a job that involves considerable travel by car, you will be more than happy to discover you have map skills. You'll be able to find your way no matter where you are — as long as you have a map.

Many people find it difficult to learn to read. If school work is a struggle, get help. You might be surprised to know how many people there are who have had to overcome learning problems. Though most schools have programs for students who learn in different ways, teachers may not recognize that you need this special help. Sometimes people may think you aren't learning because you refuse to learn or because you have an attitude problem. Talk to your guidance counselor, a favorite teacher, or the reading specialist at your school. Don't wait around for help to find you. You have a right to a good education.

Take Advantage Of Activities

Most schools offer a variety of extracurricular activities. They might include sports, band, chess club, computer club, dance club, and many others. These groups share common interests and provide opportunities for expanding your skills or your knowledge as well as your circle of friends.

Use School Resources

Most school districts have a variety of resources available to you and your family. Some of these are recreational facilities such as a swimming pool, tennis courts, baseball fields, and so on. If your school has such facilities, find out when they're available to the community.

Professional resources available in many schools include a school psychologist, social worker, nurse, and dietitian, in addition to school administrators and teachers. Regardless of the problems you or your family may encounter, a teacher or guidance counselor can put you in touch with the proper professional.

Sports are just one type of extra-curricular activity. Find one that interests you. You'll have fun and build skills at the same time. What do you think these teens are learning?

Keeping Your Options Open

Dreams are only wishful thinking until you act. You can act in ways that make your dreams more likely, or you can act in ways that block your dreams.

Suppose you value responsibility and independence. You have a job after school working at a local supermarket. The manager tells you, "There are real opportunities here for young people with get-up and go!" You realize that if you worked full-time you would make more money and might advance within the store. This would give you the freedom to be on your own. You could assume more of the responsibility for your expenses and take some of the pressure off your family. You are tempted to drop out of school as soon as it is legal so you can work full-time.

If you take a full-time job before finishing high school, you risk getting stuck at one level in the store while those who have more education pass you by. But, you could get your high school diploma at night school. Or you could talk with your guidance counselor about a work-study program involving your school and the supermarket. Either of these would enable you to be responsible about your education and your future. You don't have to finish school the same way everybody else does it. You can make it work for you.

Many of the decisions you make represent trade-offs. You have to trade one satisfaction for another. Each decision you make reduces the possibility of certain other decisions, but it can open up new possibilities. If you keep the part-time job and stay in school, you give up the opportunity to make more money and gain financial independence now. But you can keep your future with the store open and explore other career options as well. If you take a full-time job with the store and go to night school, you give up freedom to be with your friends and the opportunity to explore many career options. But you begin to focus on a career and prepare for it.

Decisions can be changed, but changes become more difficult as time passes. You may decide to slide through school and study as little as possible, for example. Perhaps you even make good grades. Later, though, you may regret having missed out on so much education. You can still learn what you missed by studying on your own or by going through a special program, but you will have to spend extra time and make a special effort.

All of the decisions you make in your teen years should lead to more opportunities. Your decisions should not limit your opportunities. Planning for tomorrow is not easy when there is so much to think about today. But planning will help you achieve your goals and dreams about who you want to be, both now and in the future.

SECTION REVIEW
STOP AND REFLECT

1. If you could change something about school, what would it be?

2. What makes school work for you? Not work for you?

3. Which of your values can you strengthen by staying in school? Which of your values helps in your decision to stay in school? Is there any value that would lead you to leave school?

4. Describe what you would like your life to be like two months from now. In one year. Five years. Ten years.

5. What changes will you need to make to get where you want to be in two months? A year? Five years? Ten years?

CHAPTER 10 REVIEW

Putting Your Values To Work

STRENGTHENING YOUR VALUES

Reread *A Teen Speaks* **on page 201. Then answer the following questions.**

1. How do you think Twyla felt about herself before she left home? When Butch forced her to have sex with other men? Now that she's back in high school? In each instance, explain why you think Twyla feels this way.
2. What advice would you give Twyla if she were your friend and she told you that her mother drinks and hits her?
3. Is there any way the members of Twyla's family could have helped each other? Explain your answer.
4. If you were in Twyla's situation with Butch, how would you avoid being controlled by him? How would you get help for yourself?
5. Now that Twyla is off the streets and back in school, what problems do you think she faces as a single teen parent?
6. How do you think Twyla can help herself when she faces these new problems?
7. If Twyla were in your classes in school, now that she's a single parent, how would you react to her? How could you help her?

INTERPRETING KNOWLEDGE

1. Write a rap about your future and what you're going to do to get there.

SHARPENING YOUR THINKING SKILLS

1. Imagine that your class is having an important discussion. Everyone who speaks expresses a similar point of view, but you have an opposite opinion. Would you speak out? Why or why not?
2. Why is staying in school important for the future? How can students avoid dropping out?
3. What makes some dreams realistic and other dreams unrealistic?
4. How can you "make school work for you," even if you don't enjoy being in school?
5. How do your aptitudes and values help shape your dreams?
6. Explain how you can act in ways "to make your dreams more likely, or in ways that block your dreams."
7. Identify a dream you have for the future. Describe the steps you need to take to make your dream become a reality.

APPLYING KNOWLEDGE

1. Write a letter to a professional organization or trade union asking for information about a job or career you think you'd enjoy. Start a file in your classroom to share this information.
2. Get a part-time job or volunteer in your community in a field that interests you.
3. Join an afterschool club or start a new one.

CHAPTER 10 REVIEW

Putting Your Values To Work

PRACTICING DECISION-MAKING SKILLS

Read about each situation. Then answer the questions.

Situation A: Your best friend, Jessie, has been cutting classes in the afternoon to be with her boyfriend, Al, who is finishing up his senior year as a co-op student. He goes to school in the morning and works in the evening. Jessie loves Al, even though they often fight and sometimes hit each other. Jessie says Al is the only one who understands her and loves her.

Jessie lives with her grandmother, who often gets on her case about not going to school. Her grandmother has told you she's worried about Jessie.

Lately, you've noticed that Jessie's eyes are glazed, and she's getting very thin. You know Al deals drugs, and you think Jessie is on something. When you confront her, Jessie says she just does drugs once in awhile.

Now Jessie comes to you and tells you she and Al are going to get married when Al finishes high school this spring. Al wants Jessie to quit school. Jessie says that's fine. When you point out to her that she'll need a high school diploma to get a decent job, Jessie isn't concerned. She says if she doesn't get a job it won't matter because Al will take care of her. She says he can make enough money to support both of them. After telling you all of this Jessie asks you not to tell anyone.

1. What do you think Jessie's future holds for her? Explain your answer.
2. As Jessie's friend, how do you feel about her marrying Al?
3. If you were Jessie's grandmother, what would you say to Jessie? What actions would you take with her?
4. If you were in Jessie's place, what decisions might you have made differently?
5. In Jessie's current situation, list the choices she has already made and the options that are open to her now. State how each choice or option reflects her values. What do you think Jessie should do?
6. Where could Jessie turn for help?
7. Describe Al's position in this situation. Explain how the choices he has made in his life reflect his values.

Situation B: Jessie's guidance counselor stops you in the hall and tells you she's concerned about Jessie. The guidance counselor knows you and Jessie are good friends and asks you if Jessie is having a problem.

8. List several responses you might make to the guidance counselor. Beside each response, describe how it reflects your values.
9. What response would you make? Why?

Situation C: You notice that your friend, Carl, is getting into trouble because he doesn't do his homework. You have been with Carl when he has tried to do homework. He usually gets so frustrated that he just gives up. He tells you he doesn't even understand what he reads.

10. What would you advise Carl to do? Explain your answer.

UNIT 3

Caring For Others

CHAPTER 11	COMMUNICATING WITH EACH OTHER
CHAPTER 12	UNDERSTANDING CONFLICT
CHAPTER 13	BEING A FRIEND
CHAPTER 14	UNDERSTANDING YOUR SEXUAL IDENTITY
CHAPTER 15	MAKING SEXUAL DECISIONS

CHAPTER 11

Communicating With Others

A TEEN SPEAKS

Sometimes my dad is such a jerk! He makes me so mad. He'll be on my case about something and I try to explain what happened. But he won't listen. He just lectures at me. The more I try to tell him something, the more he lectures.

The thing that upsets me most is this friend of mine, Joyce. My dad just hates her. He thinks she's going to be a bad influence on me.

But that's not what it's like. I don't want to be like her—it's not me. We don't even like the same kind of music.

What I like about Joyce is that we can talk about things. She's got some real problems. Her grandma lives with Joyce and her mom—and she's getting senile. Joyce has a lot of pressures. She has a part-time job. Then she has to do a lot of the work at home. Her mom can be real mean. Joyce talks to me about her problems and I listen.

We don't just talk about problems. We talk about all kinds of stuff.

Everybody at school tells me I come out of the "Leave It to Beaver" family. My parents aren't divorced. My mom doesn't have a job—she's always home. My brothers are good kids. But it isn't always great. I love my family, but sometimes they really get on my nerves. Joyce understands.

My dad doesn't see any of that in Joyce. He's always on me to break off the friendship. I wish I could tell him how I feel about Joyce. But I feel like I'm talking to a brick wall.

Nancy

SECTION 1

WHAT IS COMMUNICATION?

Why is it that you can talk with some people and not with others? Why is it that, when you want so much to say something, you just can't seem to express it? And why is it that, when you do try to say something important, it seems to come out the opposite of what you meant?

Talking to others isn't always easy. Talk can help, and talk can hurt. There is a difference between talking with people and talking at them, too. And there is a difference between talking and communicating.

Communication: A Two-Way Process

Communication is more than talk. It is more than technology like radio satellites and telephone systems. It is a two-way process. Communication requires a sender, a receiver, and a message. **Communication** occurs when two or more people understand the message in about the same way. It occurs if the message has the impact on the receiver that was intended by the sender.

Suppose your mother tells you, "Be sure you're home by 3:30 this afternoon. I need you to run some errands for me." You hear her. You arrive at home by 3:20, ten minutes early. You're in time to run errands. The message meant the same thing to you and your mother.

Or, suppose your gym teacher sees someone throwing a baseball. It's headed right at the back of your neck. He yells, "Duck!" You duck and the ball whizzes past, missing you. Your teacher intended for you to miss being hit by the ball. He wanted to make you duck. Since you did, his message had the impact he intended it to have.

Your World Of Meaning— "What planet am I on, anyway?"

Good oral communication skills can be transferred to written forms of communication, such as letters, homework assignments — even business memos.

Have you ever noticed that sometimes people try to communicate with each other and completely miss the point? You hear one of them say, "That is beyond me! I don't understand what you're talking about!" It is as if they are both in their own little worlds. They can't seem to relate to each other.

In a way, you really are in your own little world. From the time you were born, you have been trying to make sense out of things. You've learned how to organize things in ways that seem reasonable to you. All experiences you have and your feelings about them get organized into your world of meaning. Your world of meaning includes

- **all the things that happen to you and the way you feel about them,**
- **what you believe,**
- **your wants and needs,**
- **your habits,**
- **the way you see life,**
- **your family,**
- **your friends,**
- **your values.**

In both cases — with your mother and with your gym teacher — communication was effective. Effective communication is essential to friendships, families, and society. Developing good communication skills will help you prevent misunderstanding, tension, and difficulties in all of your relationships. But it requires more than just talking. Being a good communicator requires that you learn how to say what you mean. It requires that you learn how to listen to others and see if you know what they mean. It is a two-way process!

No two people ever see things exactly the same way, not even people from the same family. Even identical twins have their own, separate worlds of meaning because they are not together every moment. Even people who are together when things happen often interpret those events differently.

Suppose Clarence, Kenny, and James are walking to school together when there is an automobile accident. They get to school and begin to tell the story:

Clarence says, "This car stopped in the middle of the street for no reason and this lady plowed right into it from behind."

Kenny says, "No, this dumb woman driver ran a stop sign and smashed right into that guy. She wasn't watching."

James says, "A gray dog ran out into the street and nearly got killed."

So who is telling the truth? All three, but each from his own world of meaning. Clarence didn't see the dog. Kenny, whose dad is always putting down women drivers, focused on the woman. James had been watching the dog and saw it run into the street. He was afraid it would be killed. His dog was run over several weeks ago.

Their World Of Meaning— "Do I understand you?"

When you try to communicate with someone, you construct your message from your own experience—your world of meaning. When someone tries to communicate with you, you figure out what they mean from your own experience—your world of meaning.

Communication is more than simply sending a message. It is more than receiving a message. The person who receives a message has to figure it out, based on how he or she understands things. The only way to know if your message has been received the way you meant it is to get some feedback. You have to ask the receiver to let you know what he or she thinks you meant. The same is true if you are the receiver. You need to know if what you think somebody means is really what that person means.

Suppose you said, "I'm so hungry I could eat a horse." If the person you are talking to is from a culture in which horse meat is eaten, he or she might

Asking questions will enable you to get feedback and to determine if you have understood correctly.

A sense of humor goes a long way in getting your point across.

think that you are craving a horse steak! The way that person can find out is to ask. "Do you eat horses in this country?" your friend might ask. "No. That is an expression that means I'm really, really hungry!" you could reply.

Let's Talk

Communication is more than talk. But it does include a lot of talk. How good a talker are you? Do you talk too much? Or do you wish you could talk a little more? Sending messages through words is an important part of communication.

Talk That Helps — "Tell me more!"

Talk that helps contributes to understanding. It also enables people to respond to one another with empathy.

Encouraging Messages

"You made a good point." "Interesting idea!" Comments like these build people up. They open communication lines and build self-confidence, and good relationships. Don't be afraid to ask for more information. People who feel encouraged and supported open up.

In almost every situation, you can find something positive and constructive

> **Talk Starters**
> - Tell me more.
> - Are you saying that. . . ?
> - Do you mean that. . . ?
> - Are you feeling. . . ?
> - Help me understand.
> - Can we talk?
> - Tell me if I'm wrong.

> **Talk Stoppers**
> - That's really dumb!
> - How would you know?
> - You're wrong.
> - You don't know how I feel!
> - Shut up.
> - You don't know what you're talking about!
> - I don't believe it!
> - Nobody your age would understand.
> - I don't want to talk about it.
> - I don't want to listen.
> - There's no problem—everything's okay.

to say. Take time to offer words of encouragement, support, and empathy. A sense of humor, also, helps when it is used to laugh with, and not at, someone.

Clear, Specific Messages

What is the point of your message? Being clear and specific in what you say helps you get your points across. When you can't say what you mean, you can't communicate well. In fact, if you can't put a message into words, you don't understand the message yourself.

Think before you talk. Fuzzy thinking produces fuzzy messages. You can't expect others to understand what you want to say if the message isn't clear in your own mind.

Consistency is also part of being clear. If you say one thing one time and something different the next time, your message won't be understood. People may not believe or trust you, either.

Open, Honest Communication

If he can't figure out that I'm mad at him, it's too bad!

Have you ever felt that people should know that you feel hurt, angry, or sad? You might have decided not to say anything about your feelings. Or perhaps you shared only part of a problem with a friend. You couldn't bear to tell the whole story! Or maybe you locked everything up inside, even though you wished you could talk. Not talking or holding your feelings inside stops communication and often creates misunderstanding.

Your listeners aren't mind readers. You can't expect them to know what you don't say. And don't expect them to guess feelings that you're hiding. Not even best friends can always do that. Openness builds understanding.

Say what you mean. Honest communication is the foundation of good personal relationships. But honesty without respect and sensitivity can be cruel.

Speaking with Respect and Sensitivity

Comments that show respect and sensitivity are caring and keep communication lines open.

> Please, could we talk? I'm really feeling angry. I really think we can work this out.

Being tactful is another sign of respect and sensitivity. **Tact** is communicating something difficult without offending or hurting. To be tactful, you need to be aware of how others might respond to your words or actions.

> How about if you call me sometimes? When I'm always the one to call, I start to think you don't care.

Keeping Confidences, Building Trust

As people develop trust, they begin to communicate in a deeper, more personal way. They trust each other with their feelings, and ideas.

To build trust, people communicate in a supportive way. For example, they keep promises to each other, avoid lying, and avoid gossiping about one another.

"I Messages"

Sometimes "I messages" are more effective communicators than "you messages." "I messages" focus on you and how you feel instead of sounding like a verbal attack on someone else. For example, "Why do you always have to be late?" sounds like a verbal attack. An "I message" sounds different: "When you're late, I get upset. It makes me feel like you don't care that much."

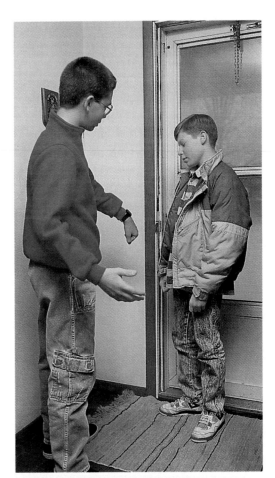

"You're always late and it really ticks me off. Why can't you ever be on time?"
Tom's lack of tact does not promote good communications.

Learn the three parts of "I messages."

"I messages" focus on the consequence. An "I message" places the problem where it belongs. When you send an "I message," you are saying "This is a problem for me," instead of "You've got a problem."

Mel did this when he tactfully told his friend about calling, "When I'm always the one to call, I start to think you don't care." For this reason, these statements are less likely to create bad feelings and rebellion. Others feel less threatened when you explain the effect of an action on yourself. They know what you think, and why.

"I messages" aren't always easy to state. They must be said in an open and friendly manner. "I messages" that are said in a hateful, demanding, or accusing way close communication. "I messages" require the sender to be open and willing to share feelings. But when you are open, you develop closer relationships with others.

To give "I messages," remember these three parts:

1. **When you . . .** (state the action),
2. **I feel . . .** (state the feeling),
3. **because . . .** (state the consequence).

Talk That Hurts — "This is your fault!"

Some messages discourage, rather than encourage, communication. Talk that hurts may even destroy relationships.

Gossip, Lies, Insults

Gossip, lies, and insults are signs of insecurity. Some people use this talk if they want power over others. Maybe they have low self-worth, so they build themselves up by tearing others down. Perhaps they can't accept responsibility for their own mistakes. Gossip, lies, and insults show neither respect nor caring; instead, they destroy relationships.

Blaming and Accusing

Some people try to give the impression that they're always right. They blame others for mistakes instead of accepting responsibility for their own words and actions. By blaming or accusing others, insecure people try to feel in control.

Some of the most destructive communication is full of "you messages" that blame, accuse, and criticize. For example, "Can't you ever do it right?" or "You're never home on time." These "you messages" shut off communication. They belittle the other person, making him or her feel unworthy. People often respond defensively with silence, anger, or equally blaming statements.

Rather than using "you messages" to blame, give "I messages" to say how the other person's words or actions seem to you or make you feel. "I messages" keep communication open and don't assign blame.

Threatening

> I'll never speak to you again if you go out with him.

> If you don't come with us, you won't have another chance.

> I bet I can't trust you with anything.

This kind of threatening language creates barriers to relationships. Like other negative communication, it is controlling language that shows no respect or caring for others.

How do you handle gossip?

Nagging, Preaching, and Complaining

People usually don't respond positively when you bombard them with "shoulds." For example, "You should get a job after school." This language sounds controlling and doesn't show respect for another.

Sometimes you have to complain to change something. An "I message" rather than a "you message" is usually the best way to get action. Complainers can be tiring after a while. In fact, people who complain a lot often find that others ignore them.

Interruption

Interruption short-circuits good communication because the complete message isn't sent. The receiver responds to only part of the message, so the response may be inappropriate. Listen to the whole message before you respond.

On the other hand, people may interrupt because someone dominates the conversation. Communication isn't a two-way process when one person does all the talking.

If someone interrupts you, repeat your message in a patient, yet assertive way. Look at the listener to engage him or her in your conversation. But also be sure you are giving others an opportunity to get in.

Being Opinionated

People are entitled to their opinions. Usually, however, there are many ways to look at a situation. Each way might be valid. A "know-it-all" stops communication because he or she isn't open to other people's thoughts or ideas. Speaking without always judging keeps communication lines open.

Sarcasm

You've heard people say one thing when their nonverbal expressions send a different message. For example, "Those singers are really talented," but the tone of voice says just the opposite. Sarcasm stops communication and destroys relationships. Instead, good communication occurs when people express themselves without fear of being put down.

Communication may be more than talk. But by learning to use talk that helps rather than talk that hurts, you will become better at communicating with others.

SECTION REVIEW
STOP AND REFLECT

1. Describe a time when you tried to tell someone something important and couldn't get your message across. Tell how you felt.

2. What was the message? Tell what impact you wanted it to have on the receiver.

3. Why do you think that the message didn't get across?

4. Tell about a time when you and someone else (family or friends) came up with different versions of the same experience.

5. Make a list of things about other people that really bug you. (Use fictional names instead of each person's real name so you won't risk having your list create communication problems!)

6. Next to each item on your list, write an "I message" that you could use in communicating with the other person.

SECTION 2

COMMUNICATING IS MORE THAN WHAT YOU SAY

When you communicate with another person, you put your whole self into communicating. You talk in ways that are helpful or hurtful. And you send messages with your body language and tone of voice.

You also listen in order to receive messages. And you listen for more than words. You try to pick up on exactly what the sender means.

The way you communicate with others says as much about your message as the words you use. It also says a lot about your values.

Communicating Your Values And Beliefs

How do you say *NO* if someone offers you a beer or a joint? How do you speak out when others toss cans on the highway? How do you avoid sex on a date? Standing up for what you believe and value isn't easy. It takes conviction, strong values, self-confidence, self-control, and a clear understanding of why you think the way you do.

Talking About Values — "What I say matters!"

"I messages" are often the best way to express your beliefs and values. For example, "When you push me to drink, I get mad because I feel you don't respect my values; I like being in control of myself." Or, "When you toss cans on the highway, I get upset because I feel responsible for keeping the community clean for everyone." Or, "When you try to push me to have sex, I resent it because I don't want to lose my respect for you or myself."

"I messages" help you explain why you think or feel the way you do; others can respond to your reasons instead of your verbal attack on them.

Showing My Values — "It's what you don't say!"

Words are just one way to send and receive messages. People also communicate without words, using nonverbal behavior. Some nonverbal expressions are word substitutes. For example, a nod or a shake of your head usually means *YES* or *NO*.

Body movements, silence, your tone of voice, and your appearance also indicate how you feel about what you're saying. You may be unaware that you send these messages, but an effective listener looks for meaning beyond words.

People can express many feelings without saying a word. What message is this teen communicating?

Body Language

You've probably heard the phrase, "Actions speak louder than words." In fact, your actions may speak so loudly that what you say about your values isn't heard!"

Body language is the way you express yourself through movement, posture, and facial expressions. You can say *NO* when somebody offers you a cigarette, but look like you aren't sure you mean it. Or you can say you have a problem with littering, but act like you don't care that much. You can say you don't want sex, but act in a way that says you do.

When you smile, hug, kiss, cry, or clench your teeth, you "talk" with body language. A smile, for example, may express happiness, warmth, friendliness, or self-control. A clenched fist might express anger or stress. What messages might these motions send—slamming a door; slouching in a chair, putting your arm around someone?

People send mixed messages when their words don't match their body language. What they say is different from what they feel or think. The difference shows in their actions. For example, you can see tension when a friend nervously bites her nails, even if she says everything is fine. What message would you get if someone said "It's great to see you," but at the same time he or she was looking at someone else?

Body language can be misinterpreted among people of different cultural backgrounds. For example, a pat on the head to an American child may show affection. In Japan it shows disrespect.

What is the use of silence doing to this communication process. Are there times when silence can help the process?

Silence

Silence communicates even though no words are spoken. Silence is neither good nor bad. The way silence is used, however, is good or bad.

If you disagree with someone, or if a value is being challenged, moments of silence can be helpful. A little time lets you think so that you don't say something you will regret later.

Not talking can be destructive when you keep your problems—and your joys—to yourself. It can also be destructive when you don't stand up for what you value.

Sometimes you may feel afraid to talk because you think you will be ridiculed. Maybe you worry about saying the wrong thing or being misunderstood. Or maybe you don't want to talk about or admit a problem. This kind of silence creates stress, keeps you from working through your problems, and becomes a barrier to good relationships.

The Tone of Your Voice

Say this sentence aloud four times: "I would be glad to do that for you." Each time, emphasize a different word—*I, glad, that, you*—in the sentence. How does the meaning of the sentence change?

Your tone of voice can communicate as much as your choice of words. For example, you can hear anger in your parents' voices even when they are trying to control their words. You can detect insincerity even when the "right" words are spoken. You hear the soothing sound of caring words and the bouncy sound of humor. When you are taking a stand for your values, the tone of your voice can show how strongly you feel.

Your appearance tells others a lot about what you value.

Your Appearance

As a teenager, you're probably well aware of personal appearance. You might follow a fashion style to be accepted by others or purposely dress differently to show your independence. You might be attracted to certain other people because you like their looks. And you might groom yourself and dress so you attract others to you.

Your grooming, your clothing, and your posture tell others a lot about what you value. The way you are groomed and the way you hold yourself can show confidence and personal pride.

Confronting Problems — "Let's face it."

When you communicate your values to others, you may run into conflict. **Conflict** means a struggle or fight. You may experience conflict because you admit problems to your family or friends — "I got a D in math," "I just got into a fight," or "I'm pregnant."

You can have conflict, too, when you try to act in a caring, responsible way toward others — "I notice you've started drinking a lot," or "I'm not ready

Sharing your problems with someone you trust can be helpful — particularly if you are willing to accept their guidance.

for a heavy relationship." It is hard to tell people things they don't want to hear. You may be afraid of what will happen. Or you may not want to admit you have problems that you need to confront.

Most problems don't just go away. You need to address them. If you can, share your problems with someone you trust. Choose a quiet time and place to talk so you won't be disturbed. Accept responsibility for your problem. Don't lay blame. Openly seek guidance by signaling that you will accept help.

Confronting others with their problems — perhaps their selfish or abusive behavior — can be even harder than revealing your own problems. Show you care by being open, accepting, and understanding. Encourage others to share by saying, "Tell me what's wrong," or "You're my friend. I'm concerned about you," or "It hurts me when you say . . . Can we talk?" Then be a good listener.

Constructive Criticism

No one likes criticism. But everyone needs to give or receive it on occasion. Negative criticism can decrease self-worth and result in conflict. Given in a constructive way, criticism can be caring, and supportive. When received openly, constructive criticism is a good way to learn from one another.

Learn to give constructive criticism and to receive criticism without becoming defensive.

- Focus on the action, not the person. For example, instead of "You'll never learn," say "That approach didn't work; you might try it this way instead." Use "I messages."
- Offer constructive criticism when the person isn't angry, visibly upset, or busy with something else.
- Offer suggestions; don't belittle or ridicule. Indicate your willingness to help.

Communication—Cultural Differences

Communication may have very different meanings in different cultures. This often leads to serious misunderstanding between people and among nations.

- In China, a letter written in red ink means the relationship between writer and reader is being broken.
- In Thailand, patting a child on the head is offensive because the head is sacred.
- In Israel, Egypt, and Singapore, showing the sole of your shoe to someone, perhaps when crossing your legs, is an insult.
- In Bulgaria, head motions for "yes" and "no" are the opposite of those used in the United States.
- In Germany and Austria, a person's title is very important, even to close friends. For example, you might call a man "Herr Professor," which means Mr. Professor.
- In Saudi Arabia, admiration of someone's possessions makes the person honor-bound to give them as a gift. Refusal of such a gift is insulting.
- In some Asian countries, the American hand gesture for calling someone is used only for dogs.
- In Greece, people stand very close when they talk face to face. Americans prefer to be further away, about 30 inches or an arm's length.
- Among the Maori tribe in New Zealand, people touch foreheads as a greeting instead of shaking hands or hugging.

Using Feedback — "What I hear you saying..."

At 4:00, Christina phoned her brother Jason and asked, "Could you pick me up soon in front of the school?" He responded specifically, "I'll be in front of the school in 15 minutes." Then she said, "Thanks—in front of school in 15 minutes." Jason was there with the car at 4:15 sharp.

Christina sent a clear message. Jason received the message, interpreted it correctly, and sent a message back. Christina gave him feedback so they knew they meant the same thing. He picked her up as promised.

> Tad asked Trisha, "Could you please buy me a blank videotape while you're out shopping?" When he asked, Trisha wasn't paying much attention and absently said, "Okay." She returned home later—with a cassette tape. Tad was angry with her. They argued.

Tad sent a message, but Trisha received the message incorrectly. She bought the wrong kind of tape. They did not communicate well, so the end result was conflict.

Simply sending a message is not communication. The message must be received and accurately interpreted. In fact, communication is a cooperative effort. Both sender and receiver share equal responsibility for the message.

Responding with a question can help you interpret what you hear.

In the first example, Jason responded with a clear message. Then Christina let Jason know she understood his response. Both accepted responsibility. In the second example, neither Tad nor Trisha checked to be sure the message sent was the same message received. Neither accepted responsibility for the message.

Let's Listen

Did you know that 60 percent of communication time is spent listening, not talking? In fact, your role as a listener is as important to communication as your role as a talker. However, you've probably observed conversations in which people were talking but not one seemed to be really listening.

Active Listening — "I hear you."

Listening is more than not talking, and it's more than hearing words. Active listening is focusing on the message and checking with the sender to be sure the message sent is clearly understood. Active listeners give both verbal and nonverbal feedback to the sender.

Being an active, fair-minded, and open listener encourages communication. In fact, people like others who listen well. They enjoy sharing their ideas and feelings when they don't feel the listener will judge, interrupt, or correct them. Good listeners are often well liked.

A good listener doesn't have to agree with everything the other person says. He or she will also be fair, taking personal biases into account. Acceptance—through tone of voice and words—shows respect and caring.

Good listening skills have other benefits. You learn by listening and observing. Success often depends on hearing directions. Self-worth develops

You can't be an effective listener if you try to carry on two conversations at the same time.

when you hear how others respond to you. Misunderstandings are more likely avoided when you take time to listen well.

Listening Is a Skill

Listening is hard work. It's a skill that takes practice and can be developed. However, many people don't know how to be good listeners. You can become a better listener if you

- **keep eye contact as you listen.**
- **concentrate on what people are saying,** not on how you'll respond. Don't change the subject or let your mind wander.

- **eliminate distractions,** such as television or a loud radio, as you talk. Excuse yourself if you must be distracted for a moment.
- **carry on just one conversation at a time.**
- **show your interest.** Use gestures, such as nodding your head, to encourage communication.
- **listen for the main point** of the message.
- **listen for feelings** as well as words.
- **be sensitive.** Empathy, or imagining yourself in the other's place, can help you understand the message.
- **"hear" nonverbal expressions,** too. Do gestures and personal appearance match the person's words?
- **listen for silence and tone of voice.** What might someone be saying without words? Can you detect meaning from the tone of voice?
- **wait until the person has finished talking before you respond.** Don't interrupt.
- **ask questions,** or rephrase what you heard, to be sure you understand. For example, "I understand you are saying...," or "Do you mean...?"
- **avoid judgmental or closed responses.** Instead, give open responses. A closed response might be, "You can't be unhappy. You have a lot of friends, good grades...." An open response might be, "You seem unhappy. Let's talk about it." Another closed response might be, "things don't always go as you'd like them." A more open alternative might be, "How might you change that next time?"

Learning how to be an effective talker and listener requires work. And matching your body language and actions to what you say can be just as much of a challenge. But it is through communication that we express our values and connect to the world and others in it. Even people who are unable to hear or speak learn to communicate. Without the ability to connect in one way or another, we would know very little about ourselves or the world around us.

SECTION REVIEW
STOP AND REFLECT

1. Think of something you need to say to others about a value. Say it as an "I message."

2. Describe someone else who has a habit that really annoys you. How can you tell them with an "I message?"

3. Describe someone who is a good listener. What makes you think so?

4. Think of a person you know who needs someone to listen? What makes you think so? How could you show care by listening?

5. Tell about a time you didn't give or receive a clear message. What happened? How could you have made the message clear?

CHAPTER 11 REVIEW

Putting Your Values To Work

STRENGTHENING YOUR VALUES

Reread *A Teen Speaks* **on page 227. Then answer the following questions.**

1. What kind of image is Joyce communicating to Nancy's dad? What makes you say this?
2. Should Joyce try to communicate a different image to others? Why or why not?
3. How might Nancy help her father get to know the "real" Joyce?
4. What kind of communication problem do Nancy and her father seem to have?
5. What would help Nancy and her father begin to communicate better?
6. What would you do if you were in Nancy's situation?
7. What would you do if you were Joyce?
8. When a person is misjudged by others, whose fault is it?

INTERPRETING KNOWLEDGE

1. Make a poster about good communication. Share it with your classmates.
2. Look through some magazines for pictures of people. Collect pictures that illustrate the different ways people communicate without using words. Label each picture you choose with a caption that describes what is being communicated. Create a display of your labeled pictures.

SHARPENING YOUR THINKING SKILLS

1. How would you define **good communication**?
2. Give as many examples as you can of times when you used good communication strategies.
3. What do you think your words and actions communicate about you? Do they present an accurate picture?
4. Think about a time you hurt someone's feelings unintentionally. Identify three ways you could have avoided the hurt feelings.
5. How do "talk stoppers" hurt communication?
6. Describe a time you used silence to help communication. What was the result?
7. Now describe a time you used silence in a way that blocked communication. What was the result?
8. Name three ways you can become a better listener.

APPLYING KNOWLEDGE

1. For a week, keep a list in your journal of all the times you use either "talk that helps" or "talk that hurts." During the following week, try to increase the number of times you use helpful talk and decrease the number of times you use hurtful talk.
2. Think of an issue, a project, or a person you care a lot about. How might you communicate your caring without using words?

CHAPTER 11 REVIEW

Putting Your Values To Work

PRACTICING DECISION-MAKING SKILLS

Read about each situation. Then answer the questions.

Situation A: One Monday, Anne arrives at school and quickly goes to her locker, where she usually meets Lucy. When she gets there, Lucy is nowhere to be seen. Anne waits until the first bell rings and still does not see Lucy. "That's odd," she thinks as she rushes to class. She gets there late and sees that Lucy is already sitting at her desk. Lucy doesn't even look her way. Anne wonders what is going on. When the class is finally over, Anne waits for Lucy outside the classroom. But Lucy dashes right by her without so much as a look.

At lunch, Anne finally corners Lucy, who has continued to avoid her. "Lucy, what is the matter? Are you mad at me or something?" Lucy just looks at her and says, "You know what's wrong. I thought you were my friend, but I've found out that you can't be trusted." Now Anne is not only hurt, she's angry. She starts yelling at Lucy, "Who do you think you are? I don't know what you're talking about!" This makes Lucy turn around and sneer, "Oh, you don't know about telling John that I like him, even after you swore you wouldn't? I could have died when he and his friends teased me." Anne shouts back, "I never said anything!"

1. Describe this situation. Does Anne need to make a decision? Explain your answer.
2. What are some actions Anne might make?
3. What might the consequences of these actions be?
4. How do you think Anne feels? How do you think Lucy feels?
5. Do you think there is ever a good reason for not keeping a friend's secret? Explain your answer.

Situation B: Same as Situation A, except that the secret Lucy told Anne was that she has discovered she is pregnant. Anne tries to persuade Lucy to tell her parents or a teacher, but Lucy refuses. Anne is really worried about her friend and doesn't know what to do. She goes to a favorite teacher to get some advice. The teacher decides to talk to Lucy about what she ought to do. Lucy is very angry that Anne has betrayed her trust.

6. How would you describe the situation now? What decision does Anne need to make?
7. What are the possible actions Anne could take?
8. What are the consequences of each action?
9. If you were Anne, what decision would you make?
10. Imagine that you take the action you have decided is best. What might the results be?

Situation C: Same as Situation A. One week later, Lucy finds out that another person had overheard her telling Anne about John. Anne was telling the truth—she hadn't told anyone what Lucy said. By now, though, the girls are not speaking.

11. What challenge faces both girls?
12. What actions might Lucy take to repair the situation?
13. How can this kind of miscommunication be avoided?

CHAPTER 12

Understanding Conflict

A TEEN SPEAKS

I've been taking care of myself and my little brother and sister after school since I was in fourth grade. They're usually pretty good—but sometimes they fight. Mostly they fight just when Mom comes in the door. They're hitting and pounding on each other and calling each other names. Then Mom is all over me about letting them get started.

Sometimes I watch the soaps for a while when I get home, just to unwind a little bit. But I'm really good about starting my homework and seeing that my brother and sister get going on theirs. Once in a while I talk with my friend on the phone or maybe watch a little extra TV. It seems like every time I do, Mom comes in and finds me not doing my homework, and she's at me again.

It's tough to be responsible all the time. I know I let her down sometimes. But I really wish that once in a while she'd just walk in and give me a hug and say she's glad to see me and she's proud of me for doing such a good job. It seems like all she ever does is nag me or start giving orders about what I need to do that I haven't done.

Willie

SECTION 1

RECOGNIZING CONFLICT

When was the last time you got into a fight at home? Who started it? How did it end? Fighting at home is one of many kinds of conflict. No matter what kind of family you live in, chances are you experience some conflict at home.

What Is Conflict?

Conflict means a struggle or fight. It happens when two people disagree strongly about what to do or what to believe. Conflict often involves shouting and expressing strong feelings. It can lead to physical violence, as when two people get into a fist fight.

Struggles or battles do not always show, however. You can have an inner battle with yourself.

> Suppose your family expects you to attend your grandmother's birthday party. But the party is on the same night as an important school dance. While your parents would like you to go to the birthday party, they tell you to make the decision. You experience an inner struggle. You want to please your family. At the same time, you want to go to the dance.

Kinds Of Conflict — "Who is the enemy?"

We usually think of a struggle or fight as a something between two or more people.

> One group at school claims part of the school cafeteria as its territory. The word is out that anybody else who sits

there will really pay for it! Your group decides to move into their territory. Sure enough, fights break out.

Conflict can also happen between ideas or values. For instance, telling the truth and loyalty are both important values.

> What if your friend's mother calls you to find out what happened at the party last Saturday night? Your friend has confided in you that she had been drinking.

What do you do? Do you tell the truth, or do you refuse to betray your friend? In this case, truth and loyalty are conflicting forces in your inner struggle.

> Suppose your grandmother believes that children ought to be spanked when they misbehave. Your dad thinks it is better to take away privileges than to spank you.

Your grandmother and dad have conflicting ideas about discipline which could create problems.

Conflict can lead to a lively discussion. When it does, people who disagree learn to respect their differing viewpoints. Conflict can become long, drawn-out bickering or quarreling—and even lead to actual blows or fighting.

Conflict may be short and personal, as in a brief afterschool argument with a friend. Or, conflict may be a long, continuous struggle, as within a family or between nations.

When you can't resolve an inner conflict alone, seek advice.

Positive And Negative Conflict

You probably know that there is long-standing conflict between the two major political parties in the United States. Each party has its own ideas about how the country ought to be run. Each tries to convince the voters that it is right and the other party is wrong. This kind of conflict is healthy. It seems to result in a better country for all of us.

Two countries may get into a conflict over what is best for the rest of the

If you don't talk to each other, there is no opportunity to solve your problems. Unsolved problems tend to decrease your feelings of self-worth.

world. The clash of ideas can result in arguments and debates in the United Nations or in letters and meetings between the two countries. In that case, the conflict can be healthy.

Improved relationships between the countries might lead to cooperation that is good for the rest of the world. But in other cases, the conflict can become so intense that neither side is willing to change or modify its position. War is sometimes the result. This is negative conflict.

Negative conflict is just the opposite of positive conflict. It leads to more conflict and bad feelings. Instead of helping to solve a problem by bringing out differences of opinion, negative conflict makes the problem worse.

Conflict becomes negative when people begin to use destructive tactics in dealing with normal, healthy conflict. These tactics "add fuel to the fire." They increase the hostility people feel toward the problem and toward those who are involved in it.

Overinvolvement in Conflict

People can become so anxious about conflict that they actually become conflict-prone. That is, they keep getting involved in conflict in order to tell themselves that they aren't afraid. The boy who goes around with a chip on his shoulder acting macho may be trying to appear tough because he is really afraid of conflict. The person out looking for fights isn't learning how to deal with conflict in a positive way.

Clamming Up

Conflict is destructive when people refuse to talk to one another.

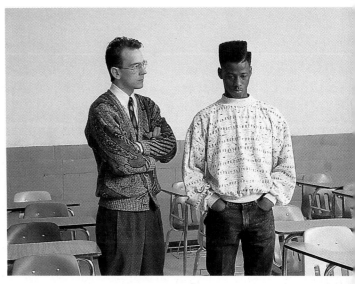

Don't let things get out of hand. Talk!

> Suppose you find out your best friend is gossiping about you. You get so upset that you refuse to talk to her. Since you won't speak to her, you have no way of knowing her side of the story. You begin thinking about how she betrayed you. The more you think about it, the angrier you feel. You become so angry that you turn around and go the opposite direction in the hallway when you see her coming.

If you act this way, you prevent yourself from having any opportunity to find out the truth.

Expecting the Worst

You can bring people into a conflict by the way you treat them. When you expect the worst, you often get it.

> Suppose one of your teachers, Ms. Bloom, snaps at you when you ask her a question on the first day of class. The truth is, she's nervous about meeting new students. She doesn't realize she sounds cross. But you think Ms. Bloom doesn't like you. "Well, I don't like you either!" you say to yourself. You begin treating her as someone you don't like. When you see her outside of class, you ignore her and look away. You don't speak up in class unless she asks you a direct question. Then you answer in a sullen way. Ms. Bloom begins to wonder about the way you are acting and decides that you must not like her. She starts treating you like a "difficult student."

How might this be negative conflict?

Your reaction to Ms. Bloom's behavior has caused her to become the way you said she would be. She has become someone who doesn't like you very much.

Hanging on to Conflict

What if one of the boys in your class thinks girls aren't as good at basketball as boys? He insists that girls are not as strong and that boys are faster. "Some girls are good," he says, "but when the press is on under the basket, they're afraid they'll break their fingernails!" The girls challenge the boys to a game, and the girls win by 9 points. Your friend still refuses to admit that girls are as good at basketball as boys. He says it was dumb luck and makes all kinds of excuses.

Conflict can be destructive when people are unable to admit they are wrong. This boy is afraid he will look silly if he admits he was wrong. He is hanging on to the conflict.

You can't be rid of all struggles and probably shouldn't try. But you can learn how to keep conflict from becoming destructive. Recognizing and avoiding tactics that "add fuel to the fire" help keep conflict positive.

Whether conflict ends up as a positive or negative experience depends on
- how serious people believe the conflicting issues are,
- the way they choose to handle these issues.

SECTION REVIEW
STOP AND REFLECT

1. Describe a positive conflict that has helped you to grow in some way.

2. Why do you think the conflict was positive instead of negative?

3. Describe a negative conflict that you have been involved in.

4. What destructive tactics were used by people involved in the conflict?

5. If you were involved in the same conflict again, how could you help to make it positive?

SECTION 2

WHERE DOES CONFLICT COME FROM?

Sometimes it seems as if life is just one struggle after another. As soon as you get one problem solved, there is another to take its place!

Believe it or not, most people seek out conflict in everyday life. Sports, competitive games, and teasing friends are all ways of getting involved in conflict. So are family feuds like arguing with your dad, just to get his attention, or picking a fight with your brother just because you're bored!

Sources Of Conflict

Conflict is a normal and natural part of our lives. It is part of growing and learning to live with yourself and others. Conflict reflects the fact that you have differences in values—or believe you have differences in values. Those differences matter to you. Conflict can and should lead to healthy growth and greater maturity.

In fact, a certain amount of conflict may be necessary for your healthy intellectual, emotional, and spiritual growth. Conflict teaches you how to deal with people and ideas you don't agree with. For example, you and a friend may strongly disagree about which music group is the best, but you can still be good friends.

Conflict also helps you to establish your personal identity and values. By expressing your own ideas and beliefs, you show your friends just who you are and what you stand for. Your friends may disagree and challenge your values. This gives you the opportunity to test and defend your values.

Conflict can lead you to ask, "How come she thinks that way and I think this way?" This kind of conflict often reveals just how much information you have—or don't have—to back up what you believe. It can lead you to revise or even change your ideas about things. When you're open to new information you may think, "I didn't know that! I see it differently now."

When conflict brings a problem or difference out into the open the problem can be discussed and solved. And growth

Sometimes you have to make hard decisions about what is important. Often, you have to look at long-term goals to remind yourself that sometimes it's okay not to go along with the crowd.

takes place. When conflict is left hidden and unexpressed inside us tension builds. Tension becomes a barrier to our relationship with ourselves and others.

We can't avoid conflict. But we can learn how to deal with it. The key is learning to keep conflict positive, or lively and productive.

Most conflicts grow out of differences that are related to territory or space, property, values and beliefs, power and authority, and who gets rewards or privileges.

Conflict For Teens

Teens most often experience conflict as they try to relate to themselves, their family, their friends, and their school. Some teens also have conflict with their community.

Your greatest conflicts may be with yourself. Even when you get into a conflict with someone else, you have to figure out how it fits with the kind of person you choose to be. You may not be aware of it, but you are always struggling to decide whether the way you deal with conflict fits with your values. When you resolve a conflict in a way that fits with what you value, you end up being most satisfied.

Conflicts At Home

Most teenagers say that their biggest conflicts occur at home. Since family is one of our most basic values, this shouldn't be surprising. You want to feel that you belong in your family. You want family members to like, respect, and love you. But living in a family isn't easy. Teenagers have many complaints about the way their family treats them.

> My parents are always
> nagging me.

Around the World

Saving Face

How do you handle conflict? Do you talk things through in a positive way? Do you talk back, yell, or refuse to talk? Or do you avoid conflict at all costs?

In the Japanese way of life, people are taught to "save face"—for themselves and for others. There is a strict code of manners and conformity that helps people avoid conflict and misunderstanding. For example, children are likely to follow the expectations of their elders, so there may be less parent-child conflict.

Japanese people often avoid open conflict even when you may think it's justified. If someone used crude language, a Japanese teen might remain expressionless, pretending not to hear. An American teen might show anger. Both may feel angry or insulted, however.

At school or work, Japanese people don't talk back or challenge teachers or bosses who correct them. Talking back shows lack of respect. Instead, the person may smile as he or she listens. You might interpret the smile as insolence. But to the Japanese, that smile helps take away the pain.

When I'm at school my mom goes into my bedroom and snoops around. She says she's just cleaning, but my stuff is never where I left it.

My dad can't stand my friends. He's always finding reasons why I shouldn't hang out with them.

I feel like I'm in the army. My folks order me around all the time: "Do this." "Don't do that." Why can't they just leave me alone?

Conflicts With Friends

Teenagers also experience conflict with their friends. Conflict with friends reflects the fact that friendship is an important value. It's not unusual to hear teens complain:
- "My friends take advantage of me."
- "My friends don't respect my opinion. When I make a suggestion, they put it down."
- "My friends aren't loyal—I'm always getting dumped and then they expect me to make up like nothing ever happened."
- "My friends take me for granted. They think I'll always be here for them, no matter what."

Conflicts occur at school, too. At school you relate to many groups of people in one place. You're part of a crowd. And, you are always being evaluated by teachers and other students. You want to get along with your friends and with your teachers at the same time. You may want to succeed at school, but you may be afraid of being too successful. You know your parents want and expect you to get good grades. But some of your friends may not value good grades at all.

Conflicts In The Community

Conflicts with your community or society can occur when you break the law or go against **social norms,** what your community expects of you. Social norms are not laws, but they do tend to operate a lot like laws. When you go against them, you are made to feel as if you had committed a crime.

Social norms are not the same in every community. One social norm in our country is that you graduate from high school. If you drop out of school, society makes it difficult for you. Jobs are harder to get. The jobs that are available usually pay only the minimum wage. Like laws, social norms are based on values.

How intense your conflicts get will depend upon
- how strongly you feel,
- what you understand the consequences of the conflict to be,
- your previous experience with conflict,
- the ways you deal with conflict.

SECTION REVIEW
STOP AND REFLECT

1. Describe a conflict you've been in.
2. Who was involved?
3. What was the source of the conflict?
4. How did you feel when it was over?
5. What are some values that have caused you to have conflict with yourself? With others?

SECTION 3

DEALING WITH CONFLICT

It is important that you know how to deal with conflict. Even if you do, you will not always be able to resolve a conflict. You cannot control how other people will act. They also have to want to resolve the conflict and help you resolve it.

Handling Conflict

It's not always easy to resolve conflict. In fact, for some people it is so difficult or frightening that they simply avoid conflict altogether. They pretend it doesn't exist, run away from it, or make excuses for not dealing with it. Other people meet conflict head-on by confronting it and making use of some very simple steps for solving problems.

Confronting Conflict — "Let's face up to it!"

It is always best to admit there is a conflict, even if you choose not to deal with it right away. Sometimes it's a good idea to avoid conflict. Some problems may even work themselves out if you leave them alone. For example, a friend has accused you of being dishonest. Instead of trying to convince him that he is wrong, you could wait until he cools off. He may realize that he is wrong when he has a chance to think about it.

Confrontation is usually the best way to deal with conflict or with problems that will lead to conflict. However, most problems—or puzzling, frustrating situations we face—will simply not solve themselves. They are often the source of conflict. It helps to know when and how to confront problems so that the conflict they bring can be productive and healthy.

For example, you may think that your curfew is too early. So you decide

Timing can influence the result of a serious discussion. It is better to approach people when they are relaxed and happy.

to talk with your dad about it. What if he comes home from work and announces that he's completely worn out because he's having some big problems on his job? It is probably better to wait to confront him with your curfew.

By confronting conflict, you can usually avoid harmful consequences. You can prevent a problem from getting worse and leading to greater conflict.

Imagine you have failed a math test. You decide not to tell your family. "It's really only my business, anyway. There's no use upsetting them." You feel guilty and are afraid that your family will find out. You're storing up fear and defending yourself so you can keep your secret.

When your dad asks if you need any help with your homework, you accuse him of jumping on you all the time. You dump some of your feelings on him. Later, your parents get a notice about your math grade, and they are really upset.

Most parents like to help their children be successful, but they need to be told when help or advice is needed.

You could have dealt with the conflict head-on and just told your parents about the test. Then you would have released mental and emotional energy. You might have gotten some good help from your dad and improved your relationship with your family. You might have won respect for being honest, too. When you try to avoid conflict, you usually increase conflict.

Sometimes conflict creates so much tension that you feel you must solve the problem immediately. You become willing to accept almost any solution just to ease that tension. But new problems may arise that create even more conflict when you don't look carefully at all the issues.

Two sisters, Darla and Kate, have been having terrible fights over clothes. Darla often wears Kate's clothes without asking. The last fight began when Darla was caught wearing Kate's favorite blouse. Before long the sisters were screaming and slapping each other. Darla ended up with a bruise on the side of her face, and Kate had a deep scratch on her arm. Both girls were very emotionally upset at what they had done to each other. They tearfully agreed never to argue over clothes again.

Deal with the causes of a problem, not just the symptoms.

Kate and Darla's agreement is doomed to fail because they have not dealt with the basic issues: property rights, caring, and respect. Any solution to a problem must deal with the causes of the problem, not simply the symptoms. Kate has strong feelings about what belongs to her. She feels she has the right to decide who can borrow her clothes and who cannot. She also feels that no one has the right to take what belongs to her—even her own sister.

Darla does not seem to respect Kate's property rights. She acts as if anything that belongs to her sister is hers without asking. In reality, she admires Kate and wants to be just like her. But their fighting reveals little caring. Only after the damage has been done did they agree to stop slapping and hitting. Even that may have been more out of self-interest ("I don't want to get hit anymore.") than out of caring for the other.

What might have happened if Kate and Darla had tried to resolve their conflict over clothes? Instead of fighting and then jumping to a solution, they might have developed more respect for each other's feelings. They could also have developed a workable plan to solve the problem.

SECTION REVIEW
STOP AND REFLECT

1. What tactics do you most often use to deal with conflict? Which are positive tactics? Why?

2. Describe a time you tried to avoid a conflict. What tactic or tactics did you use? Tell what happened.

3. Why did you try to avoid conflict? Do you think this was a positive or negative decision?

4. What conflict do you know about that you could have a part in solving?

5. How might you go about trying to solve it?

SECTION 4

USING CONFLICT FOR PEACE

a *peacemaker* is a person who works to bring about peace through positive conflict and conflict resolution. *Conflict resolution* is the process of ending a conflict by problem solving and cooperation. Learning how to live as a peacemaker is an important contribution you can make toward world peace.

How can one person do anything about world peace? When you master the skills of conflict resolution, you are mastering the skills that can help establish and maintain peace throughout the world. Although you may feel powerless to influence world conditions, you can make valuable contributions to peace in your own part of the world.

Being A Peacemaker

When you value caring, you are already committed to being a peacemaker. Peacemakers show their care for other people. They work to help others show respect and caring.

If you're going to be a peacemaker, you have to learn how to deal with attitudes and behaviors that block conflict resolution through problem solving and positive conflict. You have to learn to see the options and to determine the facts as clearly as you can without jumping to conclusions. You have to remain calm and objective. You need to see both sides of an issue.

To be a peacemaker, you need to understand the difference between competition and cooperation. When you compete you are measuring yourself against some standard or another person. When you cooperate you are working with others toward a goal that is good for everyone involved. When faced with a conflict, most people react as if they were in a competition. It never works when they do. In competition there is almost always a winner and a loser. The object in competition is to win. The object in resolving a conflict is to solve a problem in a way that satisfies everyone involved.

Competition can be fun. Learning how to win or to lose graciously can build character. One of the ways you grow is by comparing yourself to others. Then you test your growth through competition. Like conflict, competition is a part of life. And, like conflict, competition can be healthy or destructive.

Competition is one of the ways you test your growth, but it should be approached in a spirit of fun.

Healthy competition
- helps you discover more about your own strengths and limits,
- is done in the spirit of fun,
- helps you to look out for your own interests and see that you get your fair share of things,
- helps you achieve goals.

Competition isn't healthy when it stops being a route to self-discovery and becomes a means of proving your self-worth. When you need to win over others to feel better about yourself, competition is no longer healthy. No matter how many prizes or congratulations you win, you can't win good feelings about yourself.

Competition isn't healthy when it is hostile, either. A hostile competitor gloats. Hostile competitors are glad when someone on another team gets hurt or when the people they are competing with have problems. Hostile competitors seem to take over every activity—even conversation! They have to be the best (or the worst, or the most dramatic) at everything.

Competition is no way to resolve a conflict. You can't have healthy competition in a conflict.

Competition — "May the best team win!"

Think about a big football game between two rival schools. The two teams practice so they can win. They don't practice to lose! Competition between schools can go on for years. Every year people from each school root for their team and hope it will win. This can be fun for everyone involved.

But when you treat a conflict like a competition and set out to win, you make the conflict even worse and keep it going. It isn't fun for anyone involved. By looking at the characteristics of competition, you can see why it doesn't work in conflict resolution.

Once you experience the thrill of winning, you want to win again. Why not? If you are competing with yourself, you want to set new goals and do even

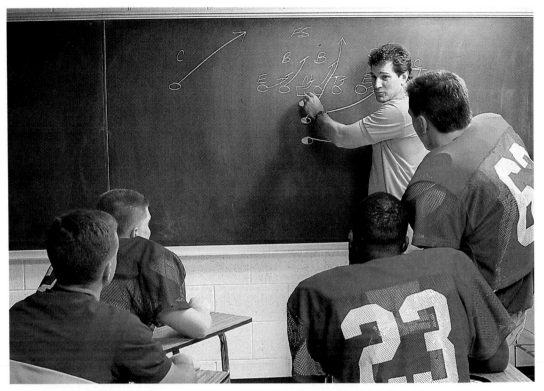

Secrecy adds excitement to games, and it pulls team members together. In other relationships secrecy causes serious problems. Can you give an example of this?

better. If you are part of a losing team, you want to lose by fewer points and keep striving to win. It is natural for competition to make you want to compete more. It is part of wanting to do better.

Competition can bring people together. When they play together they agree to follow the same rules. They often find out how much they have in common and gain new respect for each other. This opens communication.

Competition can close communication, too. In a competition between two groups, it is often important to keep plans secret. For example, the football players preparing for the game are sworn to secrecy. The coaches don't want any part of their game plan to get out. Playbooks are kept under lock and key. Secrecy is not a bad thing in a competition. It adds to the excitement and surprise in a game. It tests responsibility and trust.

When you are trying to settle a conflict, keeping secrets is destructive. This is one reason you can't settle a conflict by reacting in a competitive way. Unless ideas, goals, plans, and motives are brought out into the open, you can't get at the problem or issue. When you can't get at the problem or issue, you can't settle the conflict. Open communication is very important.

Rumors are not reliable sources of information. Can you give examples of this? How can rumors be damaging? How can they get you in trouble? How do you think rumors get started?

Rumors are a part of many competitions. People want to know what is going on. So people talk and rumors often get started. Usually rumors are annoying. But they are destructive when they interfere with effective communication. They are always destructive in conflict. Secrecy and rumor in a conflict make it harder to find out what needs to be done to resolve the conflict.

Lack of communication and rumors make it hard to know what is actually going on in a competition. Information can be critical to making plans for a game. Coaches usually send out scouts to watch other teams. They want to find out all they can before they compete. They expect that the other team will send out scouts, too.

An open exchange of information is essential to resolving a conflict. When you send out a scout to find out what somebody is thinking or doing, instead of asking him or her yourself, you risk getting the wrong information. In a competition that is part of the fun. But it isn't fun to try to outguess someone when you're trying to settle a conflict.

Destructive Competition

Unhealthy, destructive competition can get violent. It can become an excuse to act out bad feelings. Healthy competition never includes violent tactics. When people take a competition too seriously, they can get so emotionally involved they use poor judgment.

For example, some teens might announce that, if your team loses the big game, they will be waiting for the fans from the other school after the game. They may also threaten to beat up everybody from that school who happens to come into your neighborhood. If teens from the other school act in the same way, you can expect violence.

When conflict is treated like a competition, it automatically becomes destructive. Things get worse. People get frustrated, and violence is almost certain to happen. Each time a competitor does something to the other side emotions run higher and a negative situation escalates. On the other hand, a peacemaker or negotiator may turn a situation around and bring about a positive outcome.

In destructive competition people try to weaken their competitors. They want to make themselves seem more powerful. Some students might wait outside the practice field of your rival school. They pick a fight there with one of the star players, hoping he will be too bruised and hurt to play or that he will be suspended for fighting.

When conflict is treated like a competition, this is what happens. If you think you have to be the winner in order to get rid of a conflict, you'll find yourself doing what you can to weaken the person you see as an opponent. This will create more bad feelings and greater conflict. You may be tough enough to win, but you don't solve the problem. The issues will come up again, later.

People who take competition to the extreme tend to view their opponents as holding opposite values or at least different values. This means that they see themselves and their opponents as more different than alike. They assume that their opponents will disagree with them on most issues. They see their opponents' values as a threat to their own values. The word is out at your school, for example, that the students who attend the rival school are weird. None of them could possibly fit in at your school!

Conflict, treated like a competition, makes people's differences more important than what they have in common. This is because you have to think about people in ways that will make you want to fight with them. If you start thinking about how much you're alike, you don't want to fight so much.

When people treat conflict like a serious competition, they end up feeling hostility and suspicion. They assume that their opponents will not be friendly and that they cannot be trusted. This happens in competition when winning becomes the only thing that really matters to a team. For example, when some of the students from the rival school show up at the skating rink in your neighborhood, you and your friends leave early because you're sure they came just to start a fight. Because you feel hostility and suspicion, you automatically imagine they do, too.

Cooperation — "I win, we win!"

Conflict is not a competitive sport. To resolve a conflict you have to be able to cooperate instead of competing. Cooperation can make the difference between a conflict that is destructive and a conflict that is constructive.

Constructive conflict resolution has the same characteristics as cooperation. These are almost exactly the opposite of the characteristics of competition.

Cooperation Leads to More Cooperation

People who treat conflict like a competition find that there is more to compete over. When you approach a conflict in a cooperative way, you set the stage for more cooperation to take place. The more you cooperate, the more likely you are to continue to cooperate.

Suppose the conflict between your school and its rival has really gotten out of hand. Not only is there fierce, destructive competition in sports, but there have been outbreaks of violence between rival gangs from each school. Students and faculty become so concerned that they decide to do something.

They send a group of students and faculty to a meeting with students and faculty of the rival school. The groups discuss how they can deal with the growing conflict. Everyone agrees that both schools need to cooperate. The problem is, they don't want to cooperate.

The two groups have to search for something that makes cooperation important for both. People in the two communities are concerned about the conflict and violence, too. A group of businessmen and women announce that if the two schools will work together to clean up the park between the two communities, each school will be given new athletic equipment for its physical education program.

They both see the value of getting more athletic equipment. So they agree to take on the project. Their conflict has not been dealt with directly. But as the two schools begin to cooperate, they will find that cooperation on this project will lead to conflict resolution in other areas.

Cooperation Encourages Open Communication

When you begin to cooperate, communication opens up. It has to be open or you can't work together. At first, efforts to open communication may be cautious. But the more you cooperate, the more you can cooperate.

Representatives from each school meet to make plans for the project. Ideas are shared at both schools. Enthusiasm runs high. When you see some kids from the other school, you ask them about the project. You see each other as persons working toward the same objectives. You feel free to talk to each other without trying to mislead the other.

You discover that you're both talking honestly and that you want to share information with each other. All the stu-

Healthy competition never includes violent tactics. Competition is not an excuse to act out bad feelings.

dents want their school to benefit from the cooperative project.

When people cooperate, they may still try to persuade other persons to see their viewpoint. But they do so in ways that are acceptable to both. For example, when the two student groups meet to plan a joint project, disagreements will most likely arise. Rather than quitting, they continue to work on the project, trying to convince each other without upsetting anyone. They know if they get hostile the project will break down.

Cooperation Strengthens Values

When you are cooperating with others, you benefit when they are strong. The more power one group has, the more that group can contribute to the cooperative project. Your school, for example, discovers that the project at the other school isn't getting enough support due to the lack of publicity. You need the project to be strong at the other school, too. So your school agrees to send one of your best students to help plan ways to publicize the project in the

It's okay to disagree. The important thing is to keep talking.

other school. It has become important that both schools succeed! In a competition, you would never send your best players to help the other team win the game!

Persons or groups that cooperate tend to see each other as sharing the same basic values. They think of others as being more like them than different, even though they may disagree about some things. The other school may be your rival in football, but the students there are okay people!

Persons and groups that cooperate assume that others are trustworthy. And even if the individuals or groups do not know each other, they act as if they can count on each other to be friendly and trustworthy. For example, an argument breaks out at the skating rink. It involves students from your school and its rival. Everyone listens to both sides of the story. You all agree to leave each other alone for a while and to cool off. You trust the kids from the other school to keep their word. They trust you and your friends.

Conflict Resolution Works

Using cooperation to resolve conflict is not a series of steps. But it does involve attitudes and behaviors that open communication and help to strengthen everyone involved. Strangely enough, the way you learn to do this is by jumping in and doing it.

The key to conflict resolution is getting the idea of winning and losing out of your head. It just doesn't work. Healthy competition is great, but there is no such thing as healthy competition in conflict resolution.

The concept of conflict resolution through cooperation must be more than merely a topic for discussion in school. It is a process that really works. You can use it at home, in relation to your friends, and in school. You can help to produce constructive conflict resolution, helping people learn to cooperate and to use the problem-solving process. By acting towards others at home and at school in the way you would act as a result of cooperation (trustful, friendly, open), you can actually create cooperation.

Sometimes, however, it is impossible to work through a problem, even when you cooperate and use the problem-solving process. Deadlocks, in which nobody will budge, occur when people involved in a conflict can't find a solution that both can agree on.

To solve a deadlock, they need a mediator, or somebody who is not involved, to help them look at the problem. Usually, with a mediator, people can agree on a solution. Teens can seek mediation to help undo destructive conflict-resolution tactics and build cooperation.

In fact, there are some schools where students are involved in special programs of peer mediation. The mediator becomes the person in the middle between two students who are experiencing conflict. The mediator has been trained to suspend judgment, to be a careful observer and listener, to ask questions, and to respond to both parties involved in the conflict. The mediator tries to resolve the conflict by making use of the problem-solving process.

Schools where peer mediation is practiced report a good level of cooperation and harmony in their schools because peer mediators have been peacemakers.

Many schools have problems with student violence. Where students begin working together to create cooperation, student violence decreases.

You can't expect world peace if you aren't able to live in peace with your family and friends. Being a peacemaker shows how much you value caring, family, trust, respect, and responsibility. What you do today can greatly influence the world you inherit tomorrow!

SECTION REVIEW
STOP AND REFLECT

1. Which destructive tactics have you used in dealing with conflict? What was the result of each?

2. What cooperative tactics might have made the conflict a healthy one?

3. Tell about a time when you or someone you know treated a conflict as if it were a competition. What happened as a result?

4. Describe a conflict you are having with someone (or a recent conflict). How might cooperation help you to resolve the conflict?

5. What could you do to create cooperation at home? With friends? At school?

6. What types of conflict do you experience? Can you think of ways to resolve these conflicts?

CHAPTER 12 REVIEW

Putting Your Values To Work

STRENGTHENING YOUR VALUES

Reread *A Teen Speaks* **on page 249. Then answer the following questions.**

1. What are the conflicts in this situation?
2. What values are present in this situation? What values are lacking? Explain your answers.
3. Choose the conflict that you think is most important. Describe a negative way to resolve it. Describe a positive way to resolve it.
4. Explain why the negative approach will probably lead to more negative conflicts in the future.
5. Explain why the positive approach will probably lead to more positive conflicts in the future.
6. What do you think will happen if Willie and his mother never confront their conflict? Do you think this is good or bad? Explain your answers.

INTERPRETING KNOWLEDGE

1. In your journal, describe how you would feel if you always won every competition. Then describe how you would feel if you always lost every competition.
2. Choreograph two dances with two dancers. The first dance should symbolize negative conflict. The second dance should symbolize positive conflict. If possible, arrange for the dancers to perform the two dances for the class.

SHARPENING YOUR THINKING SKILLS

1. In your own words, tell what conflict is and give an example.
2. Conflict can occur between people and between ideas. Give an example of each.
3. Some conflict is good and some is bad. Explain the difference.
4. Can you deliberately create conflict? Explain your answer.
5. Why is it important for people to learn to deal with conflict?
6. Is it possible to avoid conflict completely? Explain your answer.
7. Why do some conflicts seem much stronger than others?
8. Describe an inner conflict you might have.
9. What is a peacemaker? Give three examples.
10. Describe ways in which competition can be healthy.
11. Describe ways in which competition can be destructive.
12. How does cooperation differ from competition?
13. Why is cooperation important in managing conflict?

APPLYING KNOWLEDGE

1. The next time you watch a sporting event, compare the actions of the coaches, the players, and the fans. Think about how the actions of all three groups affect whether the competition is healthy or unhealthy.

CHAPTER 12 REVIEW

Putting Your Values To Work

PRACTICING DECISION-MAKING SKILLS

Read about each situation. Then answer the questions.

Situation A: You have science during the last period of the day. Mr. Bridges, the teacher, seems to be picking on you. Whenever several students are talking, it is you he asks to quiet down.

One day you have to go home right after school to watch your little brother because the babysitter is sick. In science class that day, the whole class is goofing off, but he singles you out. You stop talking, but everyone else keeps on, so you touch the student in front of you and ask her to quiet down. She turns around and accidently knocks everything on your desk to the floor. Mr. Bridges says, "That's it! You all have detention after school."

When school ends, most of the students stay in their seats. You start to leave. Mr. Bridges says, "Where do you think you're going?"

1. What decision do you need to make?
2. What is the challenge?
3. List several actions you might take.
4. For three of the actions you listed, list the possible positive and negative consequences.
5. What decision would you make? Why?

Situation B: Same as Situation A, only this time as you head out the door Mr. Bridges quietly asks to speak with you. Out of earshot of the other students, he tells you that the other students follow your lead; that is why it might seem he is picking on you. He thinks that if you behave properly, the others will too. He then asks you what you think.

6. What decision do you need to make before you answer his question?
7. What is the challenge?
8. Do you think your opinion of Mr. Bridges will change? Explain your answer.

Situation C: Same as Situation A, except that on your way out you tell Mr. Bridges that you have to take care of your little brother today. You explain that you will stay after school tomorrow instead. Mr. Bridges says, "You should have thought of that earlier. A rule is a rule."

9. What is your opinion of Mr. Bridges now?
10. What will you do? Show the steps of the decision-making process and how you arrived at your decision. What do you expect the consequences will be?

CHAPTER 13

Being A Friend

A TEEN SPEAKS

I never really had a friend all the way through grade school. I guess you could say I was a loner. I didn't really want to be. It just seemed like I got left out a lot. You know how kids are—they form little cliques. If you don't happen to fit, it's just too bad.

Anyway, when I got to junior high I met Diedra. We hit it off instantly. She had a lot of friends, but not a best friend. I guess she hadn't ever let herself get really close to anybody. We sat together in homeroom. Then when I found out we had a couple of classes together, I got a seat next to her. And we started getting together for lunch. She's so much fun to be with.

Then I asked her over. That's when we really got close. I needed a best friend, and I guess she did, too! Diedra says most of her friends are kind of superficial. They're nice and everything at school, but they never seem to want to do anything outside of school or be that close.

That's hard for me to imagine, because Diedra is the greatest. My mother says maybe it's because Diedra is physically handicapped. She was born with a birth defect. She has to wear these big braces on her legs and walk with metal crutches. But that doesn't keep us from being friends. We do lots of stuff together. We tell each other everything—and I mean everything. I'm really lucky to have a best friend like Diedra.

Mary

SECTION 1

WHAT IS FRIENDSHIP?

have you ever thought about what life would be like without friends? It would be empty. Friends are fun to be with. They give you an opportunity to test out your ideas and theories about life. They help you sharpen your thinking. You have to figure out how to express yourself so your friends can understand you. They make you feel better about yourself. Friends help you clarify your own values and stand up for them.

The Value Of A Good Friend

Life can be lonely, even in a crowd. You can be surrounded by people, even at a party, without feeling that anybody cares about you. You can dance with somebody all evening and still feel as if you are entirely alone. What passes for friendliness—smiles, saying "Hey!" and pleasant conversation—can be very **superficial,** or on the surface. Life in a crowd often reminds you of how much you really need a close friend.

Arno talks about being a writer. He wants to write for television or maybe the movies. He's good at it. He writes stories and keeps them in a notebook that nobody but Cliff has ever been allowed to read. He has always done well in school and gets A's on writing assignments.

Arno and Cliff have been friends for a long time. One of the reasons Arno has stuck with his writing is that Cliff has always supported him. And sometimes when Cliff hasn't gotten along so well with his family, Arno has been there for him. They've never made a big deal about their friendship. They just like being together.

One day Penny, one of the girls in their class, asked Arno if he would write her term paper for English. Penny's grades are low and her parents have threatened to ground her unless her grades improve. Penny's the kind of girl Arno has always dreamed about. She's beautiful and very popular.

Cliff can't believe Arno would even think of writing someone else's term paper. "But Penny is everything I could want in a girl," Arno explains. "If I tell her I'll write the paper, she'll have to spend some time with me and maybe she'll get to like me. Besides, who's going to know?"

"She just wants to use you," answered Cliff. He was really upset. "As soon as she gets what she wants, you'll be history! You don't believe in cheating. Promise you won't write her paper."

Arno wouldn't promise. He started avoiding Cliff and spending a lot of time with Penny. Cliff figured he'd lost his best friend. Then he got a call. "You were right," Arno told Cliff. "Penny dumped me. At first I tried helping her with the paper. But she kept pushing me. As soon as she found out I wasn't going to actually write her paper, she split. I know I acted like a jerk. I should have listened to you."

"Hey, are you going to be home for a while? Maybe I'll drop by."

Good friends provide advice, understanding, affection, and approval.

Can you identify the strengths of this friendship? To have a true friend, like Cliff, is to have someone who cares about you even when you make mistakes. A true friend helps you to learn more about yourself. Friends tell you when you are making a big mistake and hold you accountable for living up to your values. True friends, like Arno, will apologize when they've made a mistake. They'll appreciate the fact that you held them accountable for living by their values.

The best friendships are among equals—that is, where both friends give and take. Sometimes you see a friendship where one friend leads and the other follows. Or, you may know of friendships where one, like Penny, is a user and the other gets used. Or, you may see one friend who does all the giving and the other who does all the taking.

All of your friends don't have to be just like you. Older people, children, and people of different cultures make good friends, too.

Patterns Of Friendship — "Who are my friends?"

Most of your friends are probably other teenagers. You have similar interests and backgrounds. You may have the same classes, like the same music, enjoy the same activities, and share some of the same family problems. You are more likely to see things the same way as people your own age.

Other people who are both younger and older make good friends, too. Teachers, people you meet at work, a child next door, a member of the clergy, an elderly neighbor—many others in your community could be your friends. Each of these friends would have different interests and offer a special way of thinking about things.

Family members can be friends, too. Brothers and sisters, especially twins, are often very close. Husbands and wives often describe each other as "my best friend."

Boy-Girl Friendships

Mark and Sandy have known each other for years. As children, they were best friends. By fourth grade, each had found friends of the same sex. But their families did a lot together. So, as teens, Mark and Sandy are still close. But their relationship is **platonic**, without romantic or sexual involvement. They'd never think of dating each other.

There are advantages in having close friends of the opposite sex. They can help you understand the feelings, actions, and thoughts of the other sex. But you don't have the pressure of dating. For girls who don't have brothers and boys who don't have sisters, boy-girl friendships are often very important.

Friendship Groups

If you stop and think about it, you'll probably realize you are part of many different groups. Some are groups of people your age, like clubs at school or teams. Others may include children, teens, and adults, like community or religious groups. In some groups, people have a lot in common. Other groups have people with all kinds of interests. People in some groups are close friends. In other groups, people barely know each other.

You can get to know a lot of different people in friendship groups. Sometimes close friendships develop with people you get to know over time in these groups.

Cliques

Cliques are exclusive friendship groups. They are often organized along social class lines—like rich kids, kids from a certain neighborhood, kids in a particular ethnic group. Sometimes these are groups like the jocks and the kids who get good grades. It is a rare clique that includes people of all different kinds and interests.

Being in a clique helps you feel like you belong. On the other hand, being rejected by a clique you want to be in can be hard to take.

Usually cliques have certain unwritten rules. People who are "in" dress and act a certain way. When you are in a clique, you know that others will see you as they see the clique. That can make you feel safe, but it can be tough if you go against the clique's unwritten rules. You could get dropped.

The clique can come down hard on members who try to be different. They seem sure of themselves and in control. In reality, everyone sometimes feels awkward and afraid of being different. All people wonder if they look and act like they have everything under control—even the people who look as if they have it all together.

If the clique you are in doesn't share your values, you can be under a lot of pressure. Then you have to decide what you will value most, your own identity and freedom or the crowd.

What Makes A Good Friend?

When a group of teens were asked to list the characteristics of good friends, they reported that the people they like are usually
- friendly.
- trustworthy. They won't tell secrets.
- accepting of others.
- willing to say what they think without being unkind.
- willing to have fun and join in what is going on.
- full of good ideas about what to do.
- cheerful, enjoy a joke, and have a sense of humor.
- a good sport.
- fair
- willing to be themselves. They aren't phony.

Characteristics of Real Friends

Friends are
- caring and helpful. Friends go "the extra mile" for one another.
- cooperative. Friends cooperate with each other. Friends feel equal.
- willing to share. Friendship requires give and take on both sides from time to time.
- accepting and respectful. They don't need to pretend.
- trusting and trustworthy. Friends trust each other. They don't have to be on guard.
- loyal and committed. Friends stick up for each other.
- open. Friends are open with each other. They clear up misunderstandings instead of holding grudges or refusing to talk.
- honest. Friends don't lie to or cheat each other. And they don't spread rumors about each other. They talk about each other's faults without making judgments.
- unselfish. Friends take time to think about and to do things for one another. They aren't wrapped up in their own problems or selfish needs.
- kind. Friends don't talk about each other "behind their backs." They lend a helping hand when they can.
- understanding. Friends try to understand why the other thinks or acts a particular way.
- forgiving. Friends are willing to forgive mistakes. They are also willing to ask forgiveness and say "I'm sorry," even when they need to swallow their pride.
- considerate. Friends often spend time together without talking. They aren't afraid to be silent. They respect each other's need to talk or not talk. They don't pry.

Making Friends — "Will you be my friend?"

The people you know best are usually those you come in contact with frequently. You can find friends who share your values by spending time in the same places they do. Likewise, you can avoid dangerous or illegal activities, such as drugs or alcohol, by not making friends with people who choose this behavior and by staying away from places where they hang out.

Most friendships develop over time. Some relationships never grow beyond a passing "Hi! How are you?" Others become deep and lasting.

One way to avoid drugs is to make friends with others who are not into drugs.

Stages of Friendship

Do you know the difference between an acquaintance and a friend? An **acquaintance** is someone you see and talk with but do not know very well. An acquaintance may also be someone you know well but don't like or don't want to spend time with. Someone you know well and like to spend time with is a **friend.**

Sometimes you meet someone you instantly like. In that case, acquaintance turns to friendship almost immediately. For example, you might meet someone the first day of school and be friends from that moment on. Can you think of a friendship that developed very quickly?

You can have relationships with a lot of people. A **relationship** is a bond between people who share some of the same interests, who exchange information, who discuss their feelings. Acquaintances, friendships, and family members are examples of relationships.

Relationships may become close for a while, and then casual again. Close friendships develop in the following stages:
- getting acquainted,
- feeling comfortable together,
- sharing yourself,
- depending on each other,
- filling each other's needs,
- building intimacy.

Good friends can count on each other for support — in bad times and good times.

Getting Acquainted

Acquaintances are as important as friends. They help you to form networks to meet your personal goals. A **network** is all of the people you know and all of the people they know.

This is how a network works. Imagine that you'd like to write a sports column for a large city newspaper someday. Your science lab partner edits the school newspaper and asks you to cover the cross-country meet as a guest journalist. At the meet, one of your friends introduces you to his dad, who works on the city paper. He invites you to spend a day watching how a newspaper gets produced. Your network is moving you toward your goal of writing a sports column for a large city paper.

Feeling Comfortable Together

> I've had my locker next to Margaret's all semester. We've never been friends. But today we ended up as partners in PE. I couldn't believe it — she's really fun!

Friendships grow as people are in situations where they feel comfortable with each other. They begin to feel at ease. Some people say this is "good chemistry."

You can develop good relationships with others by being caring, open, and accepting. If you have feelings of self-worth, you are probably interested in others. You aren't likely to be wrapped up in yourself.

Sharing Yourself

As a friendship grows, people begin to talk about their feelings. They discuss their dreams — what they want, whom they'd like to date. Or they may talk about how they get along with their parents. This stage in friendship is self-revealing.

Revealing yourself to another person can be risky. Some things people reveal may cause bad feelings or conflict. True friends are able to work through those feelings, and the friendship grows even stronger. If friends don't share their feelings with each other, the friendship won't grow.

Foreign Students—A Chance for Friendship

Do you have foreign exchange students in your school? If so, you have a special opportunity for friendship. These students have left family and friends thousands of miles away and have come to experience family life and school in America. Often they have come alone, anxious to make new friends. They have come to share their unique culture and to learn about yours. How can you make them feel welcome?

- Go out of your way to be friendly. Don't assume they will introduce themselves.
- Help foreign students feel at home. You might arrange a school tour when students first arrive.
- Ask them about themselves and about life in their countries: What is school like? How do teenagers use their free time? Do they date, drive, enjoy sports, eat pizza? How do families spend time together? Listen to what they tell you.
- Introduce them to people you know. By sharing your circle of acquaintances and friends, you'll help them find friends of their own.
- Be sensitive. Put yourself in the foreign student's shoes. Life in a different culture and in a new family is exciting, yet overwhelming at times. Often the foreign student will feel homesick.
- Be understanding as you help foreign students fit into your community. Correct them in a positive, sincere, and non-judgmental way. For example, you might say, "I think this is what you mean, but here's how we say it. . . ."
- Be open-minded. Be careful not to stereotype. People from different places are like you in many ways.

Depending on Each Other

I knew you'd be here when I needed you!

I knew you'd love it. That's why I bought it for you!

As friendships grow, people learn to depend on each other. They know each other well enough to anticipate a thought or action. They can count on one another.

Filling Each Other's Needs

Close friends help fulfill each other's needs. They respect each other's need for independence or space. They don't depend on their friends to tell them what to do. They don't cling to friends in order to control them. True friendship is freely given and freely received.

Each person has different needs that a friendship can fulfill. Affection, self-confidence, and recognition are some of the needs that can be met.

Building Intimacy

Have you noticed how some people keep their distance from others? There are cultural differences in how much space you should leave between yourself and another person. Usually, though, people stay about an arm's length from others unless the area is crowded.

As a relationship develops, you feel more closely involved with each other. Friends aren't as sensitive about keeping their distance. Friends often sit close together when working on a project. They may pat each other on the back, exchange friendly punches, or hug.

People who feel close to each other share their most private feelings and ideas. For example, they might share an embarrassing experience or talk about something that makes them feel ashamed.

Becoming A Friend — "May I be your friend?"

You've probably heard this saying: "In order to have friends, you must show yourself to be friendly." In order to be a friend, you have to develop the qualities that build friendship. They are the qualities that show the value you place on yourself and others.

SECTION REVIEW
STOP AND REFLECT

1. Make a list of your friends. Next to each name, write the stage at which you think your friendship is.

2. List the names of friends you have or friendships you might develop among people who are younger or older than you are.

3. Describe the friendship groups you belong to.

4. Interview an elderly person and a child. Ask what they look for in a friend. Compare what they think.

SECTION 2

BUILDING FRIENDSHIPS

good friends often say that their friendship seems almost effortless. But friendship isn't something to take for granted. Sometimes a friendship never has a chance because there are too many barriers. Sometimes a friendship is doomed because the people involved do not treat each other in a caring way. But the friendships that work are worth the effort. A true friend is somebody special.

Barriers To Friendships

Some friendships don't have a chance because people create barriers that get in the way. And some people never get past the barriers to know one another.

Prejudice

I already know I won't like her. You can't trust people from that country.

You'll like her. People from that school are the greatest.

Prejudice is a feeling toward someone or something that is not based on experience. It means to judge before knowing the facts. It can be favorable or unfavorable. Favorable prejudice causes you to accept people or ideas without examining them. Unfavorable prejudice causes you to think badly of people or ideas without giving them a chance. Can you think of an example of prejudice?

Sometimes people make a mistake in prejudging somebody or something. When you make a prejudgment, you often change your mind once you learn the facts. Prejudiced people ignore the facts. When you try to help them understand the facts, they get emotional. They become upset or angry when their prejudice is challenged.

Describe these teens. Would your description change if you knew one was a champion athlete?

Prejudice is a barrier to friendship. When you are a victim of prejudice, you may withdraw in order to avoid people. You don't want to be rejected. Sometimes you miss out on sincere offers of friendships.

Prejudice is a learned behavior. We pick it up from our families and communities. It is difficult to overcome. Learning the facts helps. Then you can begin to identify your own prejudices. When you have done this, it helps to look for positive ways to get to know those people you are prejudiced against.

For example, if you are prejudiced against students from another part of town, work together on a committee with some of them. When you have positive experiences with people, you find that you have more positive feelings about them. Prejudice begins to fade away.

Stereotyping

> He's a jock. Don't expect him to volunteer to work at the nursing home.

> Look at those glasses. You know she's got to be a real nerd.

A **stereotype** is an exaggerated belief about an entire group, such as race, religion, jocks, or intellectuals. We use it to justify the way we treat all people in the group. When we stereotype people, we don't see them as individuals. We see them as their category.

When you accept each person as the unique individual that she or he is, you avoid being controlled by stereotypes.

Competition

> I'm not going to be friends with Elaine! She's too pretty.

Nobody wants to be second best. You may be tempted to look for friends who seem like you as much as possible.

But if you always hang out with people who are just like you, you miss out on learning new things. And you might be cheating yourself of a true friend.

Self-Centeredness

It doesn't matter whose idea it was, he always takes the credit. And it makes me so angry!

Most people don't like to be around others who are stuck up, or think they're always right, or act self-centered. If you always have to have things your way or hold yourself above other people, you aren't likely to have or keep friends.

Gossip

"Jane was using a needle in the bathroom." One bit of gossip like this can end up an hour later as "Jane got caught in the bathroom shooting up." By the end of the day, people could be avoiding her. In reality, Jane was giving herself an injection of insulin for diabetes.

Gossip can hurt and destroy. It is a barrier to new friendships, and it can destroy a good friendship. For example, people who don't know Jane may avoid her because they don't want to be involved with someone on drugs. Jane might avoid her friends because she feels so hurt that they believed the gossip.

Responsible caring people don't gossip.

How can conformity cause stereotyping? How might this type of stereotyping cause problems?

Conformity

It is easy to care for people who are like you. If you are like most teens, you are probably part of a clique or group, all of whom think, feel, and act a lot like each other.

But **conformity,** acting exactly the way others in a group act, can be a barrier to friendship. It can cut you off from people outside your group. And it can cause people to stereotype you.

A good friend won't push you to do something you don't want to do.

Popularity

Some people dream of being popular. Others want to be well liked, but they don't need a lot of attention. You don't have to be popular to have people like, accept, and respect you. Popularity isn't a guarantee of success in school or later in life.

Popularity can disappear as fast as it appears. For example, you may be instantly popular if you score the winning touchdown at a football game. But what happens when you haven't played so well? To stay popular, people often have to spend a lot of energy to keep up an image.

If you focus on being popular, you may not have time to give to deep friendships. At some point, you will probably realize that the quality of your friendships is more important than the quantity.

Besides, the things that make you popular in school aren't especially important later on. As you grow older, you find that people seek out those who can be true friends.

When Friendship Doesn't Work

Sometimes friendships don't work. This can happen when people abuse the trust, respect, and caring a friend gives. When this happens, friendships can become destructive.

There are many negative influences on friendships. Some of them are
- peer pressure,
- feelings of jealousy and possessiveness,
- using others,
- power plays.

Peer Pressure

> Come on. One cigarette won't hurt.

> Your folks won't be home for a week. They'll never know we had a party at your place.

> You're the only person I know who isn't having sex. If you really love me, you'll have sex with me.

Sometimes people abuse friendships by pressuring others into something that goes against their values and beliefs. A true friend won't push you to do something you don't want to do. In the same way, your friends should count on you to respect their values.

Feelings Of Jealousy And Possessiveness

> No friend of mine can be a friend of his!

> You don't own me.

Feeling jealous or possessive isn't a sign of friendship or love. Most people need a little space to be by themselves sometimes. You can be a loyal friend without holding on too tightly.

Using Others

> I'm having a tough time in history. Help me out. Tell me the answers to the quiz.

People abuse friendship when they use others for personal gain. Exploiting, or using others, doesn't build caring relationships or true friendships. Unlike networking, where people help each other, users are looking out for themselves. Using others does not show respect for yourself or respect for others.

Sometimes people let themselves be used. Being used may make them feel needed. But nobody should do all the giving. Eventually it catches up and comes out in anger and resentment.

When a friend insists that you do a favor that goes against your values, be on the lookout. You are better off with someone who is caring and unlikely to take unfair advantage of the friendship.

When You Hurt a Friend

"Don't take it that way—I meant to be nice!" You've probably hurt your friends occasionally without meaning to. You said something one way, and your friend took it another way. This kind of hurt usually doesn't cause lasting damage if you clear it up right away. But if you don't confront it, you may lose your friendship.

Often it's hard to undo the damage when you hurt someone. How do you get a relationship back on track?

- Admit your mistake. Don't run away.
- Apologize with sincerity. Admitting your mistakes shows you care.
- Allow enough time for the relationship to get back on track. The healing process is often slow.
- Watch out—don't make the same mistake again.

Power Plays

Power is a part of many human relationships. Someone usually assumes power and tries to control the relationship or make the decisions if nobody else does.

> "I'm tired of sitting around trying to decide what to do every Friday evening. So I've rented a video for us to watch," Bev announced.

Her friend Sharon could agree. If so, there wouldn't be a power conflict. But, she could take Bev's power away by refusing to watch the video. Or, she could insist on doing something else, challenging Bev's power.

Friendships often break up when people can't figure out how to deal with power. The best relationships come from cooperation and from sharing power. Learning to deal with power plays in constructive ways and to cooperate builds friendships.

Mistreating Others

> You wouldn't have any friends if it weren't for me. You know that people put up with you just because they like being with me. You wouldn't be anybody without me.

Sometimes people mistreat their friends. They are critical of their friends' clothes, the way they dance, or the traditions of their families, for example. The person who is always critical unconsciously is trying to look good by putting others down.

People also mistreat others as a way to break off relationships. They become hostile or snobbish instead of being honest about the need to break things off. Sometimes people are critical of their friends in order to get into a new group. At other times, people lie or blame their friends because they feel angry.

How To End Friendships

When you have a true friend, it's hard to imagine that you won't be friends forever. But friendships change. You change. You develop new interests, goals, and experiences. Your friend changes. Sometimes you just drift apart.

At other times you are separated from a friend because of things you can't control, such as a family move or school schedules that don't match. Or friendships fade because you don't take enough time for each other.

But there are times when one friend wants out and the other doesn't. If you are in a destructive friendship, you need to get out. When you want to end a friendship, you can try either of two approaches. The first is to be direct, but kind:

- Explain why you need to end the friendship. Focus on how you feel, not on the other person. For example, "When we're together, I feel like a big zero because my ideas

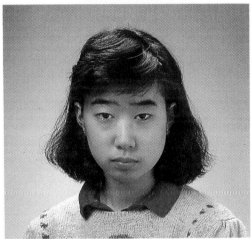

People enjoy being around others who are "up" rather than those who are "down."

aren't important." Or "I don't want to hang around people who take drugs because I don't want to get involved myself."
- Be honest about the reasons rather than telling a lie or mistreating someone who's been your friend.

The other way to end a friendship is to ease out of it quietly. Find other activities that gradually take more and more of your time.

Caring People Make The Best Friends

The friends you want to keep are caring people who share values with you. You can overcome barriers to friendship through caring. When you genuinely care about yourself and others, it shows. You make people feel valued. They like being with you. Some ways to show that you care are

- **be friendly.** Don't be afraid to smile or say "Hi!" to other people at school, even if you don't know them really well. People in some places have to be cautious about speaking to people they don't know. But if you see people regularly at school, you can usually feel comfortable speaking to them. Don't always wait for your friends to greet you first. It is hard for people to know you care if you appear unfriendly or cold.
- **take an interest in others.** Ask them casual questions to start conversation. Listen to what they have to say. Avoid too much talk about yourself. Showing interest in others is what friendship is about.
- **be sensitive.** Notice how others feel. Treat them with kindness. Caring people don't destroy others with lies, gossip, or ridicule. When others seem down, ask, "Want to talk about it?" If they don't, respect their feelings. They may not want to talk, but they will know that you care.
- **participate in school and community activities.** Volunteer to help. People usually welcome a willing worker. Active people show they care.
- **enjoy yourself and others.** You are unique and special—one of a kind! People would be lucky to have someone like you for a friend. When you believe it about yourself, others will believe it, too. People who care about themselves and others usually know how to have fun. They can be serious when it is appropriate, too.
- **be positive.** Look for the good in people and in situations. When things go wrong, look for the humor. If there isn't anything to laugh about, think about what you can learn from the situation. Don't get bogged down by being negative all the time.
- **be willing to give and take.** Caring people are not narrow-minded or overly rigid.

When you care, you will find that friendships naturally develop. You are busy doing things. You aren't sitting around waiting to be discovered. You can also make a point of going places where you will meet people. School sports events, dances, and clubs are good places to meet others. If you are active and visible, chances are you'll soon have friends.

Join in where you will meet the kind of people you'd like to spend time with. If you like music, for example, you might join a choir to be with others who share your interest.

Life is so much better when it's shared with good friends.

Be aware of who you are, and try to imagine how others see you. You don't have to be the all-time best-looking person in school.

The people who are attracted to you because of the things you have are likely to end up being users. True friends will be attracted to you because you seem like an interesting person they'd like to know.

Face it—no matter how hard you try, you may never be accepted by some people. The fact is, you aren't going to accept everybody, either. But, if people snub you or put you down when you try to be friendly, it can hurt. You have to ask yourself if you really want that kind of friendship. Even if you get into a group that puts you down, you'll always be the outsider. The fact is, not everybody is caring toward others. If you have to quit caring in order to be "in," being "in" costs too much!

Life would be dull without friendship. Everybody needs friends. Friends help you feel like you belong—not just to your family, but in the world. Friends help you to give and receive care beyond your family circle.

SECTION REVIEW
STOP AND REFLECT

1. Describe barriers you've run into in trying to make friends.

2. Tell about a friendship you've had or known about that hasn't worked. Why do you think it didn't work?

3. Name some ways you can be a more caring friend.

CHAPTER 13 REVIEW

Putting Your Values To Work

STRENGTHENING YOUR VALUES

Reread *A Teen Speaks* **on page 275. Then answer the following questions.**

1. What are some values Mary seems to have?
2. What might be some reasons that Mary felt left out in grade school?
3. What might her expectations have been in grade school?
4. How can expectations affect friendships?
5. Do you think a physical handicap is a problem in developing friendships? Explain your answer.
6. Do you think race is a problem in developing friendships. If Diedra and Mary weren't both the same race, do you think this friendship would have developed? Explain your answer.
7. What are some ways to make new friends whose values are the same as yours or similar to yours?

INTERPRETING KNOWLEDGE

1. Make a collage, using photographs or magazine pictures, depicting close friendships.
2. Write and present a skit about one or more barriers to friendship.
3. Write a report on a book, story, or poem that includes a great friendship.

SHARPENING YOUR THINKING SKILLS

1. In what ways can a true friendship help you learn more about yourself?
2. Give some examples of friends in television programs or in the movies. What values do they share?
3. What are some words you would use to describe a true friend? Place a star beside each of the words that also describes you.
4. What is the difference between an acquaintance and a friend? Give an example in real life of an acquaintance becoming a friend.
5. What are some of the most important personal needs that your friends help you to fulfill?
6. Explain how some friendships can be destructive or abusive of others.
7. What kinds of friends do you predict you will have as an adult?

APPLYING KNOWLEDGE

1. Individually or in a small group, make a list of questions to ask about friendships. Use your questions to interview several teachers, older relatives, or other adults about what being a friend means to them. Report your findings to the class.
2. Select a poem, song, television show, movie, or magazine advertisement that illustrates prejudice or stereotyping. Share and discuss your selection with the class.

CHAPTER 13 REVIEW

Putting Your Values To Work

PRACTICING DECISION-MAKING SKILLS

Read about each situation. Then answer the questions.

Situation A: Melanie is a very good student, and she has just completed a tough science assignment. Rob, her next-door neighbor and good friend, went away with his family over the weekend to visit his uncle. Late Sunday night, Rob knocks on Melanie's door and asks if he can copy her answers to the science assignment.

1. What should Melanie do? Explain your answer using the decision-making process.
2. What values would be represented in this decision?
3. Would your decision be different if Rob had been called unexpectedly to his uncle's because of an emergency?

Situation B: Your best friend got a haircut and thinks he or she looks great. You think the haircut is really bad, so you don't volunteer your opinion. Later in the day, your friend asks what you think.

4. What are your choices?
5. What are the probable consequences of each choice?
6. What would you do?
7. What values would be represented by your decision?

Situation C: Kim and Steve are high school juniors. For over a year, they've been seen as a couple. Recently Steve has begun to feel that Kim is too possessive and makes too many demands on his time. When he goes out with the guys, she gets mad. Steve thinks it's time to end the relationship and tells her so. For two weeks they don't see each other except at school. Then Steve asks Kim's best friend, Corinna, to go to the movies on Saturday night.

8. What challenge is represented in Corinna's decision?
9. What choices does Corinna have?
10. What are the possible consequences of each choice?
11. What should Corinna do? What values are represented by this decision?
12. Do you think it is okay for Steve to go out with the guys occasionally? Is it okay on Saturday nights? When is it not okay? Explain your answers.
13. Do you think it is okay for Steve's girlfriend to go out with the girls sometimes? On Saturday nights? When is it not okay? Compare your answers to those for the previous question. Are there differences? Why or why not?

Situation D: Yvonne and Brian have been friends since they were small children. They live in the same neighborhood and share a lot of common interests. Recently, however, Yvonne has begun to think of Brian differently. She really wants him to ask her out. So far, she's been very careful to hide her feelings from Brian, but she is beginning to feel awkward around him.

14. What advice would you give Yvonne? What advice would you give Brian?
15. What options are available to Yvonne?
16. What are the possible consequences of each of the options?

CHAPTER 14

Understanding Your Sexual Identity

A TEEN SPEAKS

It's easy to be confused about yourself these days. Sometimes I wonder who I really am—whether I'm "Daddy's little girl" or the toughest player on my basketball team. In some ways I'm the perfect lady. I like to get all dressed up. I'm a good cook. I'm really great with kids.

At other times I'm a real tomboy—maybe that comes from living on a farm. Talk about a tough basketball player! When our girls' team plays the boys, we keep them running. You let a guy just try to get past me to make a basket. Another thing—I love to drive our tractor and help with the plowing. I'm as good at it as my brother is. And I can do my share of chores. Some people say that's men's work, but I like doing it.

So which person am I—the perfect lady or the tomboy? And why do some people make me feel like I can't be both?

Beth Ann

SECTION 1

SEXUAL IDENTITY

everybody is born a sexual human being. This means that you are either male or female, with sexual characteristics, thoughts, and feelings. Everybody has both *masculine,* or manlike, and *feminine,* or womanlike, characteristics, thoughts, and feelings.

Picture a proud father cuddling his newborn baby daughter. Even though the father is considered very masculine, his gentle touch is a trait that society often labels "feminine."

The feminine side of men is an important side of their personality. And women have a masculine side, too, that is just as important.

What Is Sexual Identity?

Sometimes people try to hide or deny that they have both masculine and feminine parts of their personality. Unless you accept and develop both, you can't be a complete person. Imagine a world in which men never showed tenderness or sympathy. And what would life be like if women never showed strength?

Sexual identity is how you see yourself as a sexual human being. People's sexual identity is reflected in many ways. It shows in how they talk, walk, look, and dress. It also influences how they think and how they relate to others. Your sexual identity is a normal part of your total personality.

Where Does Sexual Identity Come From?

Ever since I was a little child, my parents wanted me to try new things. They gave me all kinds of toys. My favorites were the old pots and pans. I also loved a soft, cuddly doll. As I got older, my parents taught me how to do household chores as well as simple car repairs. We did athletic things together, like swimming, touch football, and soccer. My dad wanted me to be a "real boy." But he also helped me not to be embarrassed about the gentle side of myself.

Sexual identity comes from two major sources—heredity and environment. These two work together to shape a person's sexual identity.

Heredity And Environment— "Why I'm me."

Heredity means traits that are inherited from your parents. **Environment** is everything else that surrounds you and affects the way you develop or behave. A child inherits some characteristics from the parents' genes. These include the child's sex, eye and hair color, and even some abilities and disabilities. These are determined before birth. Environment has to do with your surroundings and experiences. The way a family treats a child and the opportunities the child has are environmental influences, for example.

It's hard to judge which affects children more, heredity or environment. People think Lin was born very feminine because her mother is an extremely attractive and clothes-conscious woman. But her parents also spend a lot of time encouraging Lin to develop her feminine traits. They bought her pretty dresses and gave her dolls to play with. Both heredity and environment are responsible for Lin's sexual identity.

Everybody has both feminine and masculine traits. It's important for you to accept both sides of your personality.

Our biological makeup as male or female is inherited. These differences are what cause men and women to be attracted to each other. But, many characteristics that people think are inherited actually come about because of environment. For example, you may hear someone say, "Martin is all boy. He got that from his grandfather." Although Martin likes to play rough-and-tumble games, this trait was not inherited. It was learned. Martin may be copying his grandfather's behavior.

Woman to Woman

In American society today, more and more women are getting the career opportunities that men have always had. The difference in male-female roles has narrowed as more women have achieved success in the workplace. At the same time, men have assumed more responsibility for housework and child rearing.

In many parts of the world, women continue a more traditional role, with home and family as the center of their lives. Culture defines behavior and determines opportunity.

- Many Japanese women defer to their husbands although they still feel in control of their lives. Traditionally, Japanese women walk behind their spouses.
- Modesty is essential to the reputation of many Islamic women. They keep their arms, legs, and even faces covered to avoid glances from men.
- In some rural African areas, women bear responsibility for both home and income while men enjoy their leisure.
- Illiteracy, or the inability to read, is higher among women worldwide than among men. Seventy percent of illiterate people are women. They haven't had equal chances to learn.

Paid or unpaid, women contribute to the world's assets.

Statistics show that
- women do two-thirds of the world's work, yet they receive only one-tenth of the world's wages and own far less of the world's assets.
- rural women produce over half of the food in Third World areas and 80 percent of the food in Africa.

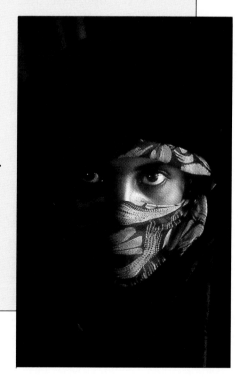

Roles change over the years, but basic values remain the same.

People learn a lot about their gender—their maleness or femaleness—from the people around them. Many of these lessons have been passed down from generation to generation—all the way back to the beginnings of recorded history.

Traditionally men have been expected to be in charge—to fight, hunt, and to do heavy labor. Think of the cave man, ready to fight ferocious animals to keep his family safe. Adventure stories provide a similar picture. Many classic tales concern the trials men face in bringing food and supplies to their families.

Traditionally, men were taught to be aggressive, strong, tough, and smart. Men were supposed to be brave and not cry when they were sad or hurt.

At an early age, boys learn what men are supposed to be like. They mimic the behaviors they see on television, and in adults around them. By age 4, most boys already show more aggressiveness and competitiveness than girls. They are also more likely to show independence and later understand that they can "go after what they want in life."

For centuries women have spent their time close to home because they're the ones to bear the children. Instead of being in charge, they have been expected to be helpers for their families or communities. In most societies, being a well-informed woman with creative ideas was generally discouraged. Instead, women were expected to be gentle, sensitive, and emotional.

Little girls have been encouraged to be like the women before them. By age 4, little girls often show signs of being more dependent and cooperative than most boys.

Parents, aunts, uncles, grandparents, and friends are an important part of a child's environment. Their words and actions help form the child's sexual identity. "Good for you, Susie. What a sweet girl." "Keep on trying, Pete. We know you can make it—you're a tough guy!" They are letting children know what is expected of them as males and females. This is the beginning stage of developing a sexual identity.

Today many parents question whether the past values of child rearing are really best for today and for the future. Some people think that boys and men should be encouraged to value and to play a more active role in household activities and in child rearing. They think girls and women should be encouraged to be more aggressive and to go after what they want in life outside the home—including the career and the income they most desire.

Many of today's magazines and books influence your sexual identity. So do television shows, movies and other people. The way toys, children's stories, magazines and books, television and movies present men and women is changing. But there are still plenty of television programs and advertisements featuring beautiful dependent females and "macho" males. A lot of the messages, whether spoken or unspoken, tell you how people think females and males are supposed to look and act. Of course, it may be hard to measure up to the extreme physical beauty or the enormous physical strength that you see on the page or on the screen. It's especially difficult when your body is going through changes that you don't always understand or expect. But accepting and developing your sexual identity is a way of valuing yourself.

SECTION REVIEW
STOP AND REFLECT

1. Describe the most masculine person you can think of. List his masculine characteristics.

2. What are some feminine qualities he shows?

3. Describe the most feminine person you can think of. List her feminine characteristics.

4. What are some of the masculine qualities she shows?

5. Notice all the TV commercials you see in one evening. Make a list of the commercials that appeal to sexual identity or contain sexual stereotypes.

6. What masculine or feminine characteristics do you think you have inherited?

7. Which of your masculine or feminine characteristics do you think are a result of your environment?

SECTION 2

ADOLESCENT SEXUALITY

Some of the most dramatic changes in your developing sexual identity happen during adolescence. Not only does the shape of your body change, but your thoughts and feelings are affected, and your values are tested. This is part of the normal development of your sexual identity.

Physical Changes

> My body is growing so fast. Mom is always telling me, "Amy, stand up straight!" But I'm afraid I'll be the tallest person in my class. It isn't any fun going to parties where I'm taller than every boy there.

It's easy to understand why a person's appearance is very important to his or her sexual identity. This is especially true during adolescence, when your body is undergoing changes that can undermine how you feel about yourself.

> Rod has begun to look at himself in the mirror more and more often. Because of visible changes in his body, he wants to see how he looks and how he is changing.

Teens often use the mirror to check for pimples. Like most teens, you're probably preoccupied with how you look. You may assume that everyone else notices your every imperfection.

You may also find yourself checking to see how your body compares with those of your classmates.

Making comparisons is a part of growing up. For instance, Jonetta compares her increased bustline with that of her best friend, Kay. Lester compares his height to that of the other guys at school. He wonders if he'll ever grow as tall and as muscular as his older brother Hank. Lester hates being short.

As you compare yourself to others, you are also testing your values. Do you

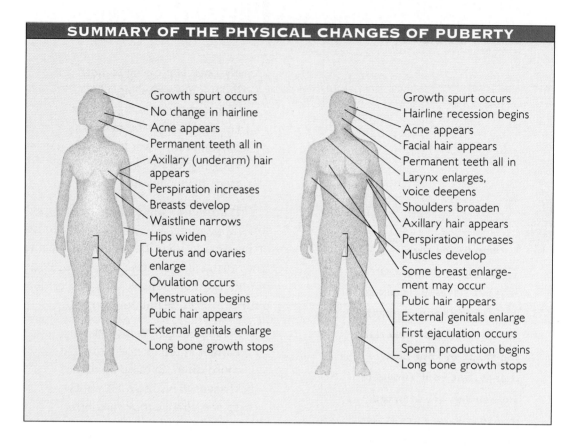

SUMMARY OF THE PHYSICAL CHANGES OF PUBERTY

- Growth spurt occurs
- No change in hairline
- Acne appears
- Permanent teeth all in
- Axillary (underarm) hair appears
- Perspiration increases
- Breasts develop
- Waistline narrows
- Hips widen
- Uterus and ovaries enlarge
- Ovulation occurs
- Menstruation begins
- Pubic hair appears
- External genitals enlarge
- Long bone growth stops

- Growth spurt occurs
- Hairline recession begins
- Acne appears
- Facial hair appears
- Permanent teeth all in
- Larynx enlarges, voice deepens
- Shoulders broaden
- Axillary hair appears
- Perspiration increases
- Muscles develop
- Some breast enlargement may occur
- Pubic hair appears
- External genitals enlarge
- First ejaculation occurs
- Sperm production begins
- Long bone growth stops

value yourself even if you aren't like everybody else?

Being a certain age does not mean that your body will grow in a particular way. **Growth spurts,** those times when your body begins to make big changes, start at different ages. Boys usually begin their growth spurt around 13½ years of age. But it can come as early as 11, as late as 15—or even later.

Girls usually begin their growth spurt around 11½ years. But it is normal for a girl's growth spurt to start at 9 or at 13. Whether you grow early or late, your final height and development probably will be similar to that of one or both of your parents.

Changes Girls Experience — "Is this me?"

When Cara is with her girlfriends, they often talk about how much easier life is for boys than for girls. They feel that boys have fewer physical changes to adjust to. This may be because many physical changes that happen to adolescent boys are harder to see than those that girls go through. However, the truth is that both boys and girls have a lot to adjust to during adolescence.

Girls may first notice a change in their breasts as early as 8 years of age. Then it is not uncommon for a year or more to pass before any additional changes in breast size take place. Afterwards, increases in breast size and shape usually continue until a girl reaches 17 or 18 years old. **Pubic hair,** which grows in the pelvic area, and **axillary hair,** which grows in the armpit, usually start appearing when a girl is 11 or 12 years old.

Hormones, or body chemicals, increase their production during puberty. **Puberty** is the beginning of adolescence. Hormones bring about a number of changes in the female. For example, the reproductive organs—the ovaries, fallopian tubes, uterus (or the womb), and the vagina—begin to grow larger. Fat deposits settle on the hips, in the breasts, and on the upper back and arms. Girls may notice a weight gain during this time. The heart and the lungs also enlarge to meet the demands of a growing body.

Menstruation is a monthly discharge of bloody fluid released from the uterus. A girl's menstrual period usually begins after a girl has had a big growth in height. Some girls have their first menstrual cycle as early as 8 or 9 years of age. Others don't have a period until as late as 17 or more. If a female is involved in a very strenuous training program, such as sports or dance, her period may start even later.

> I was kind of surprised when my period started. I was only 10 years old and didn't think it would happen to me. Maybe a year before, my mother had told me about menstruation.

Teen Pregnancy

Menstrual periods occur approximately every 28 days. However, periods may not be regular during the first few years. There may be months when a girl doesn't have a period. But, once menstruation has begun, a girl may become pregnant when she has sexual intercourse, even if she is not having regular periods. In fact, it is possible to become pregnant even before the beginning of menstruation.

If you think you are pregnant, you should see a doctor or go to a clinic immediately. Good prenatal care, care before the baby is born, helps prevent birth defects. It helps keep the mother healthy, too.

> But all I heard was that if you don't have a period, you're pregnant. I don't think she told me anything that isn't true. It's just that I picked up on this one thing. When several

> months passed and I hadn't had another period, I just knew I was pregnant. But I didn't even have a boyfriend. I didn't have all the facts. I was imagining terrible things! I was scared. I felt stupid, and didn't have enough nerve to ask questions.

Menstruation is part of a female's being able to have a baby. But you have to have sexual intercourse to have a baby. In the female's body, an egg matures once a month and breaks loose from one of the ovaries in the lower abdomen. The egg then goes into one of the fallopian tubes. If the female has had sexual intercourse around the time of **ovulation,** when the egg leaves the ovary, a sperm from a male may fertilize the egg. If this happens, the fertilized egg goes to the uterus and attaches itself to the uterine wall and becomes an embryo or the beginning of a baby. This is called **conception.**

If the egg is not fertilized, it breaks apart as it passes from the ovary through the fallopian tube. This takes several days. About 14 days after the egg leaves the ovary, a menstrual period begins. The thick, rich supply of blood and tissue that had built up in the uterus to support the life of a baby is not needed. As a result, the uterus sheds this blood as part of the menstrual cycle. The average menstrual period, or shedding of the blood, lasts approximately five days. However, it may take only three days or as long as seven or more.

It is important for women to follow good health habits during their menstrual periods; to exercise regularly and bathe or shower daily; to eat a balanced diet with plenty of fiber, iron, and water; and to make sure they get enough sleep and rest.

During your menstrual period, you may worry that you're bleeding too much or too long or that you're passing blood clots. If you have concerns like these, tell your parents, school nurse, or doctor. Practicing good health habits and seeing a doctor at recommended times are very important.

Changes Boys Experience — "I'm so awkward!"

> Avery and his friends are in the seventh grade and are starting to grow facial hair. They can't wait to begin to shave. Avery's brother is 17 and shaves every day. He says Avery is too young to shave.
>
> Avery is beginning to get interested in girls. Unfortunately, most of them are taller than he is. They seem more interested in dating older guys. It is a frustrating time for him. Even his voice, sometimes squeaky and at other times deep, gives him problems.

Physical changes, such as a cracking voice, can be embarrassing for a male. Family members and friends may kid a boy when they hear his voice crack or change. Or he may be in for teasing when they notice a dark line of hair above his upper lip. Many males will agree that going through puberty with all these changes isn't easy.

> I feel so awkward. It doesn't matter what I do to my hair, any haircut makes my ears look huge. And I feel really stupid around girls—a lot of my friends do, too. Another thing—I don't understand why I wake up with wet underwear sometimes. It's embarrassing. I wonder if I'm normal.

Nocturnal emissions, or wet dreams, begin to occur during adolescence, usually around 13 years of age. "Nocturnal" means "during the night," and "emissions" are things that are "ejected" or "sent forth." Nocturnal emissions, the **ejaculation** or emission of sperm from the penis during sleep, are nothing to be embarrassed about. Instead, they are a natural part of growing up. Wet dreams are only a response to an increased production of hormones.

Boys also worry about inappropriate erections. During adolescence many boys find themselves embarrassed to be having erections many times a day in all kinds of situations. The sight of a pretty girl, the brush of a girl passing in the hall, even just thoughts about girls—all of these may cause an erection. This, too, is a normal expression of the increased level of hormonal activity in a teen boy's body.

During adolescence, most boys notice an increase in the size of their tes-

Going through puberty can be difficult. Just remember, it happens to everyone.

ticles and penis. Even so, they may worry about the size of their penis. A boy can feel self-conscious if his penis is smaller than that of other boys. In reality, penis size has nothing to do with the ability to experience or give sexual satisfaction.

Some adolescent boys may feel a hard lump or swelling in one or both of their breasts. This swelling happens to over 60 percent of all adolescent males. In some boys, the swelling is more noticeable than in others. There may be soreness, with lumps under one or both breasts. Boys with this condition often worry because they think they are going to grow breasts and turn into girls. If this is happening to you, you may be questioning your masculinity. However, the swelling is entirely normal, and you won't turn into a girl! Within about a year or a year and a half, it will go away.

> ### Talking Helps
>
> Some males find it helpful to talk about their feelings during puberty—especially when they are worried about some of the changes in their bodies. Others worry about their size or their height. Sometimes growth takes place so fast that you feel weak and worn out, even when you haven't been doing much. If you have any of these concerns, you may decide to see a doctor who can evaluate your growth and the changes that are taking place. Others who could help are your parents, a school counselor, a school nurse, or a teacher you trust. Don't hesitate to ask for the information and support you need.

One of the final stages of puberty for males is increased facial and body hair. This often happens after a male is 16 or 17 years old. The need to shave may shift from once a week to about three times a week and, finally, every day for most males.

Psychological Changes

Many teens feel that adolescence is a lot like being on a roller coaster. Emotions and feelings can change rapidly during this time. There may be days when it seems as if everyone else has it all together while your life is going to pieces. There are times you act as if you don't care about the things you value most.

Sometimes you feel bad about the way you act toward your family, your friends, and your teachers. Perhaps you don't even know why you get angry at them. At other times, you're not sure if you said or did the right thing in class or at a party. It just seems that your feelings, rather than your brain, are running the show. Even though friendships and family love and support are very important, you may be afraid to let others know what you feel or what you need.

> Being a teen is tough. Like, asking a girl for a date is scary. I don't know what to say. I worry that she might turn me down. I worry about a lot of stuff—like making the baseball team and what people will say if I don't. I guess you could say I'm shy. But I have to admit it—I've done some dumb things to get attention.

Every human being has specific needs, physical and psychological. Your behavior is based on meeting these needs, like doing "dumb things" to get attention. There is a reason or need behind everything you do.

You may be more aware of some of these needs than of others. For instance, it's fairly easy to listen to your stomach. It tells you that you need food and water every day. And it is easy enough to know when your body needs sleep or relief from pain.

Maybe you haven't given as much thought to some of your other needs. Unless these needs are met, your sexual identity may not develop. Your self-confidence may not be as strong, and at times, you may feel unwanted and unsure of yourself. When these needs are met, you feel valued. You feel self-worth. And you are able to care about yourself and others.

Love And Support

Two needs that are very important to adolescents are love and support from family and friends. Teens need someone to say, "I care about you." They need someone to love them when it seems as if no one else cares. Especially when a girl's body is doing strange things, or a boy's voice is cracking, each needs love and support from people who matter to them. Even if their faces are covered with acne, they need to be understood and loved. Many teens feel so much need for love that when they start dating they want to get serious right away. If you're like most teens, however, you don't want to get serious too soon.

Respect

These are some remarks that show respect: "I've got time for you." "What you think and feel is important to me." "I admire how you are adapting to all the changes going on within you." People may tell you in so many words, or they may show respect in the way they treat you. Being respected by family members, friends, school staff, and other people helps you feel good about yourself.

Opportunities To Succeed

Most teens need help in identifying tasks they are good at. They need opportunities to succeed at those tasks. And they need recognition.

> Sure, I get good grades. But who cares? All my dad wants is for me to make the baseball team. I can tell him I've got all As, and he'll ask me about my batting average. I can tell him I got the lead in the school play, and he'll ask why I didn't make that big play at practice. It doesn't seem to matter how good I am at anything except sports. And kids at this school? You're nobody unless you are a big jock!

When people don't feel valued for what they do well, they sometimes look for other ways to get the recognition they need.

Everyone needs love, support, respect, and acceptance to become an emotionally healthy individual.

Acceptance By Both Sexes

Teens need acceptance by both sexes. They need male friends and female friends. Being accepted by both males and females can help reassure you that you are okay.

Before children are old enough to go to elementary school, everybody plays together. Who plays with whom has nothing to do with their gender.

When children go to school, this begins to change. Boys start spending more time with other boys. Girls spend time with other girls. In first and second grade, both boys and girls are likely to be invited to a birthday party. But by about 8, including the opposite sex in any kind of planned get-together becomes less common.

> Our club is for boys only. We don't allow any girls in here.

> I want a party for my birthday. I want all my best friends—and no boys!

This is a natural part of declaring independence from adults and developing sexual identity. For most 8-year-olds, there isn't much attraction to the opposite sex. Boys won't have any more to do with girls than they have to. And girls may dream about being brides or being mothers, but they don't want to have anything to do with the boys their own age.

That all ends. Sexual attraction begins, and it can make friendships awkward. But sexual attraction is your body's way of letting you know that you are physically capable of reproducing. Your body will be ready before you are emotionally ready for the responsibility of intimacy and parenthood.

Sometimes people get the idea that any relationship with the opposite sex must involve having sex. That's not true. It is important to develop friendships with both sexes based on common interests and shared values. You need friends like this all your life. Having a good friend of the opposite sex to talk with and to have fun with helps you develop your skills of caring and sharing. So does dating. Even when you are really serious about someone you are dating you don't need to have sex to show love, caring, and commitment. In fact, waiting to have sex until after you are married is a way to show how much you care about yourself and your partner.

Privacy

Do you sometimes find it hard to get away from others? Is your home often like Grand Central Station, with nowhere to go to daydream or to be alone with your thoughts? Most people need time alone to think. Also, they need opportunities to talk privately with their friends. This is normal. Being alone or with your friends is one way to help develop your own identity. You need time to think about the changes your body is going through. Sometimes it is helpful to discuss your feelings about these changes with friends you respect and trust.

SECTION REVIEW
STOP AND REFLECT

1. What do you think are some of the things that boys and girls your age worry about most as their bodies show physical changes?

2. Do you think boys or girls go through the most changes during puberty? Who has the toughest time? Why?

3. List ten things you thought about a lot when you were a child. List ten things you think about a lot now.

4. What things are alike on your two lists? Why?

5. What things on your two lists are different? Why?

6. Describe the biggest change you have experienced in your feelings since you were a child.

SECTION 3

SEXUAL ROLES THROUGHOUT LIFE

Morris saw himself as a "real man." He liked bragging to the guys about how far he could go with his girlfriend Betty. However, Betty ended up pregnant. Morris and Betty weren't ready to get married, but they valued family. And they wanted to be responsible for their baby. Their parents couldn't support them. Both Betty and Morris had to drop out of school to get jobs when they got married. Most of their money went to pay rent for their small apartment and to the doctor who delivered their baby. Neither Betty nor Morris has had a chance to finish high school. Both dream of finishing school someday, but they don't know if it will be possible.

Roles At Home And At School

Your sexuality affects the way you see yourself and the roles you play at home and at school. Just because you want to be very masculine doesn't mean you have to get your girlfriend pregnant to prove it. Just because you want to prove your femininity doesn't mean you have to let a guy talk you into having sex or become pregnant.

Sexual identity affects your life now and in the future. It affects the roles you play, the jobs you select, your dating life, and who you'll eventually choose to marry—or whether you will choose to marry at all.

Role responsibilities are not the same in every family. In some families, men and women have clearly defined responsibilities for decision making and chores. In other families, these responsibilities depend upon who is available to do them. Family roles depend on what works for the family. As long as family members are comfortable with their roles, the family operates smoothly.

Job Selection

Jimmy wants to be a long-distance trucker someday. He looks forward to driving a big 18-wheeler across the country. Although he's short and not very muscular, he's willing to work hard and do his best.

Li is a heavy-set six-footer who dreams of being a set designer in New York. He loves the theater and wants to be part of the "acting scene." He enjoys doing creative things such as designing costumes for the school play or serving as the makeup artist for the actors.

Both Jimmy and Li see themselves as masculine. However, their sexual identities are quite different. Jimmy's sexual identity has a lot to do with his controlling a piece of machinery, such as a big truck, or being in a job that is held by many more males than females. On the other hand, Li is interested in a job that suits his creative side. He doesn't see the role of a set designer as being either masculine or feminine. It is simply a job that would allow him to be creative and do tasks that he enjoys.

Your sexual identity probably will play a part in the jobs you select. Unfortunately, some people's ideas about sex roles keeps them from becoming all that they are capable of. By thinking that females are weak and emotional and males are tough and practical, people sometimes fail to consider women and girls for certain jobs. Males are left out of various jobs as well.

In the real world of work, you need to consider all your personal qualities. What are your strengths and weaknesses? What are your interests? To put yourself in a certain role, just because of your sex, is unfair to yourself. And it is even more unfair when others put you in that role. That's why discrimination on the basis of sex for most jobs is against the law.

Whether a person is male or female has little or nothing to do with their qualifications for the job.

Dating

Dating is one way in which sexual identity affects your life. Most teens want to date, either informally in groups or more formally in pairs. But finding someone you want to go out with and making it happen are not so simple.

> It used to be so much easier when I was little. I could play with any of the other kids—it didn't matter whether they were boys or girls. But now there are all these unwritten rules to follow. I want to have friends. I'd really like to get to know that girl named Jasmine who's in my science class, but I don't know what I'm supposed to do.

Getting to know someone of the opposite sex can be confusing and awkward. These are some of the feelings you have as a result of your natural growth and development. Try to remember that being yourself is the best approach to attracting others. Relationships built on honesty have the best chance of going well. It's almost impossible to enjoy and keep up a relationship based on deceit.

> Preston told Josie that he came from a family with a lot of money. He dressed nicely and always had cash to pay for their dates. Josie didn't think money was everything. But she did enjoy having nice things. She really enjoyed the presents Preston gave her.

Being yourself is the best approach on a date.

> Then one day Josie went shopping with her mom and saw Preston working in the grocery store as a bag boy. At first, Josie wondered why he had to work if his family was wealthy. Then, she realized that he hadn't been telling her the truth. Josie wondered if Preston was doing it for her. She decided to talk with him about their relationship.

For their relationship to last, Josie needs to tell Preston that she values him more than the things he buys for her. If the relationship is to be based on trust, Preston needs to be honest about who he really is.

Friends Can Help — "Will you introduce me?"

Going up to a person you like and speaking to that person is difficult. Using the phone isn't much easier. You may fear that you will be turned down, and you probably aren't sure how well you could take that.

Friends can help you "break the ice." They can involve you and the person you are attracted to in group activities so you'll have a chance to get to know each other. Or your friends can introduce you to the person you're interested in.

You can take the lead yourself. While you are with the group, you might tease the person or smile or joke with the person. Or you can call attention to yourself in a quiet way. Being in a group of friends gives you courage.

Sometimes people don't date because nobody asks them out. Or nobody acts interested, so they are afraid to risk asking someone out. Other people don't date because they just aren't interested. When you aren't dating and everybody else is, you may worry that you're not normal. It's hard to understand why nobody seems to want to go out with you.

It may be hard to believe now, but a lot of people don't begin dating until late in high school or in college. And they are really nice, "normal" people.

Sometimes people get pressured into dating just because everybody else is. But most people who date are dating for fun, not because anybody is pressuring them.

Teens Suggest Ways to Attract Others

Here are some suggestions teens have for getting other people interested in you:

- Express an interest in what the other person is doing.
- Be a good friend and act in a responsible manner. If you make a promise, follow through. Let others know they can count on you.
- Avoid situations where you may be tempted to do things that are in conflict with your values just to attract someone. Be yourself.
- Don't talk about your faults, but don't brag all the time either. What you say reflects how you see yourself. Choose words that are positive and that show you care about you!
- Try to look for the best in other people, too. Say good things about them, and never make unflattering or uncomplimentary remarks behind someone's back.

Caring Relationships — "I like us being together!"

A caring relationship that can withstand the pressures of everyday life involves a learning process. You need to learn about another person — what he or she enjoys, what bothers him or her, what he or she is good at, what his or her limitations are. You also need to learn about yourself and where you're headed in life.

People who get along because they have similar values and because their personalities blend well are said to be **compatible**. Compatibility in interests, values, and life goals is critical to success in any lasting relationship. Dating offers an opportunity for two people to discover whether or not they are compatible. Dating a number of people over a relatively long period of time lets you compare how compatible you are with different individuals.

Attraction And Fun Aren't Enough — "Values matter!"

Alan and Janice really fell for each other. Best of all, they enjoyed being together a lot. They loved the same music. They both got along well with their parents. They both had plans beyond high school. And they both loved riding horseback. They seemed perfect for each other.

As they got to know each other better, they found out their values about family, religion, and politics were very different. Janice planned to combine a family with a career in business. Alan expected his wife to stay home and take care of the kids. They also had very different values and religious beliefs. And their political views clashed.

Neither was willing to change. It became obvious that a long-term commitment wouldn't work.

Sexuality plays a big part in the persons you are most attracted to. But for a relationship to last, there must be more than physical attraction and good times. If there are too many negatives or if there aren't shared values and goals in the relationship, making a careful decision will be important. Breaking up may be the best thing.

Group activities help teens to get to know each other in a comfortable atmosphere.

Mate Selection

Most people would say that Sylvia is really together. She knows her strengths and weaknesses. She is cute and she enjoys the female side of herself. However, she doesn't mind getting her hands dirty or doing hard chores for her parents whenever they need her. Sylvia has a good idea of what she wants to do with her life when she finishes school. She's been dating for several years and has met a number of guys she enjoys going out with. Someday she wants to be married to somebody who feels good about himself. She wants somebody strong. But she also wants him to be gentle and kind and not afraid to show how he feels.

When you get ready to choose a mate, your sexual identity will probably play an important role in the process. Will your mate have to be handsome—or beautiful? Do you want someone who is willing to do the tasks that need to be done, regardless of whether other people consider them "woman's work" or "man's work"? Will you select someone who is willing to show the "feminine" as well as the "masculine" side of himself or herself?

As you think about the kind of mate you want in the future, ask yourself these questions: What kind of person do I need to be to attract the kind of person I want as a mate? What changes will I need to make? Am I willing to make those changes? You can't change other people to suit you. But you can grow and change in ways that attract the kind of person you want to spend time with.

The choice of a mate is a lifelong decision. Now is a good time to get to know more about your own sexual identity and the traits and values you feel are important to you. Then you will be ready for the kind of loving, caring relationship that can build a strong family.

What do Boys Like in Girls? What do Girls Like in Boys?

In one study teens were interviewed to find out what they liked in each other. The boys said they liked girls who
- are not too aggressive,
- talk and listen,
- don't show off,
- have nice manners,
- dress well,
- look good.

The girls who were interviewed said they liked boys who
- are nice looking,
- are clean and neat,
- act friendly,
- are cooperative,
- don't brag about themselves a lot.

Breaking Up

When you break up with someone, use "I statements." State what you feel and what you intend to do. Don't stay in a negative relationship because you hate the bad feelings that go with breaking up.

Deal with the loss by getting involved in new activities. It might be a great time for volunteering. There are always people who can use the help you offer. Sitting around moping won't do you any good. It won't help anybody else, either.

This is also true if someone breaks up with you. When somebody you love wants out, it's hard to take. Feelings of loss, disappointment, and rejection are overwhelming. But you can't make a relationship work by yourself. And you can't make somebody love you.

No matter how attractive a person is to you, there has to be love, caring, trust, respect, and responsibility shared between you. Otherwise, the relationship won't last.

Feeling Good About Who You Are

For many teens, puberty is a time of comparison and self-doubt. "Am I pretty enough?" "Do girls think I look good?" These are typical concerns. Perhaps you wish you could change the way you look. Maybe you wish you could date more. Maybe you wish you didn't feel so much pressure to date so you could do other things.

Boys are often afraid that, if they give in to feelings of tenderness or display their emotions, they are being too feminine. They are afraid to hug or touch another guy for fear people will think they are homosexual. Girls are often afraid that, if they act in aggressive ways, they are being too masculine. These concerns are normal.

Many teens are dissatisfied with their personalities. Some wish they could move to a new school and try out different personalities. Others would like to break out of molds as quiet or shy people and become more outgoing.

Maybe you have some personal qualities that don't please you. Do you think you're too shy—or too loud? Teens are often said to have the lowest self-confidence of any age group. They wonder if they'll ever turn out to "be somebody." This may be because they don't know many of their strengths.

Everyone should feel, "I am a good and worthwhile person." What you think of yourself directly affects how you act. It also affects your sexual identity.

During adolescence you'll do a lot of comparing. Don't be too hard on yourself.

If you don't like yourself, you may not try to do your best or be your best. You may be afraid to show your masculine or feminine side, or to act on your values of caring or family. You might not feel that other people will like you. If you don't like yourself, you'll probably find it hard to like others. Before you can care about others, you must first care about yourself.

Everybody has problems and limitations. Some problems can be solved; others cannot. You need to recognize the difference.

If you are troubled about your own sexual feelings, talk with your parents about your concerns. You can also talk to a school counselor, social worker, or member of the clergy. Talking with a professional makes more sense than bottling up your feelings.

You will continue to develop and change during your entire lifetime. Getting a job, dating, finding someone you love, making a commitment, getting married, and becoming a parent, are all normal parts of your sexuality. They are all ways you have of showing what you value most.

SECTION REVIEW
STOP AND REFLECT

1. What kind of job do you hope to have someday? How does it relate to the masculine and feminine parts of your personality?

2. How could you reassure someone who is anxious about dating?

3. What kind of person would you most like to go out with? List the qualities that you find most attractive.

4. What do you think are the most common fears teens you know have about their sexual identity?

5. If you could change your personality for a day, what would it be like?

CHAPTER 14 REVIEW

Putting Your Values To Work

STRENGTHENING YOUR VALUES

Reread *A Teen Speaks* **on page 297. Then answer the following questions.**

1. What conflict is Beth Ann experiencing?
2. What things does Beth Ann enjoy in her life?
3. How are those things similar to what you care about and value?
4. What differences are there between what you value and what Beth Ann values?
5. If you were asked to describe yourself, what characteristics would you mention? What things do you like to do? What type of person would you like to be?
6. Beth Ann describes herself as both a perfect lady and a tomboy. Have you ever felt caught between different parts of your personality? If so, how do you think you can balance these parts of yourself?
7. Suppose you are a close friend of Beth Ann's. What advice might you give her about how to become all she wants to be?

INTERPRETING KNOWLEDGE

1. Write an essay titled "Who Am I?" Describe yourself in various roles in your life. You might include a quote or the lyrics of a song that has special meaning to you.
2. Write the lyrics to a rap or other kind of song about a significant conflict in your own life or in the life of a peer.

SHARPENING YOUR THINKING SKILLS

1. Where does a person's sexual identity come from? Give some examples of influences on your own sexual identity.
2. What types of psychological changes do teens experience? What feelings have you had that may be different from your feelings in past years?
3. In your opinion, what important psychological needs do teens have? How can these needs be met in your life?
4. Turn to page 315, reread the suggestions for attracting others. Do you agree with what these teens have suggested? Do you have other suggestions?
5. What qualities do you like in someone of the opposite sex? Are they the same qualities you like in your friends of the same sex?
6. How would you define a caring and loving relationship? Describe the type of relationship you would like to have with a boyfriend or girlfriend.

APPLYING KNOWLEDGE

1. Choose at least two areas of your life — physical, social, spiritual, emotional or academic. In each of these areas, pick one thing that you can work on to improve. Set some specific goals and keep a daily log of your efforts. For example, in order to improve your physical health, you might want to run or do aerobics several times a week.

CHAPTER 14 REVIEW

Putting Your Values To Work

PRACTICING DECISION-MAKING SKILLS

Read about each situation. Then answer the questions.

Situation A: You have met someone whom you think you could really care about. You would like to get to know this person better to see if anything could develop between the two of you. You are in several classes together and have started to hang out with the same group of friends.

1. What would you want to know about the person to help you decide if a relationship with him or her is right for you?
2. How might you find these things out?
3. What qualities would you absolutely want to have in someone you date?
4. What qualities do you consider attractive but not absolutely necessary?
5. What things aren't important to you at all?
6. What qualities would you dislike in a boyfriend or girlfriend but feel you could live with and work on?
7. What would definitely make you decide not even to go out with a person?
8. What can you do to see if these qualities are present in the person you like? Name some specific things you would look for.

Situation B: Your best friend is very upset and comes to you for advice and support. He has been dating someone for several months. His girlfriend has been pressuring him to go further sexually than he wants to right now.

9. What do you say to this friend?
10. What are his options in the situation?
11. What might be the consequences of the various actions he could take?

Situation C: You are thinking about developing a talent and pursuing an interest you have had for a long time. You are very excited about it, but there is one major problem. What you are interested in is usually done by someone of the opposite sex, and you know you will face a lot of teasing and questions about your decision. You feel strongly about pursuing this interest, but you also want the approval of your friends and family.

12. Describe an activity that is generally seen as something that only girls do or only boys do.
13. Suppose that this is the activity you are interested in. What decision will you have to make? What issues are involved in your decision?
14. What possible consequences of your decision might you face?
15. Suppose you are criticized for what you are doing. How might you deal with that criticism?
16. Would the disapproval of friends or family influence your decision? If so, in what way? If not, why not?

CHAPTER 15

Making Sexual Decisions

A TEEN SPEAKS

I'd heard so much about it. There's so much on TV and movies—people go out and then they have sex. It's just what happens. And, you know, the girls whisper on the street, in the locker room and everything. And the boys make dirty jokes.

I'd heard it was good. It was supposed to be some wonderful thing, and it wasn't. I didn't feel dirty; I didn't feel very good. I just felt, "Well, it's over. Big deal." It happened under the bleachers at the football field. Real romantic! Even the backseat of a car would have been better than that.

I wasn't really turned on. I mean, I liked the guy. I still do care a lot about him. We're very good friends now. But you know, he just seemed to need it so badly.

Most guys think that if you're going to have a girlfriend, then sex goes along with that. I bought the whole thing. But I don't think that way anymore. You can have a boyfriend, but you don't have to have sex. That's not necessary. But a lot of guys think so. Girls do too.

SECTION 1

SAYING *YES* AND *NO* TO SEX

Sooner or later, once you start dating, you think about how far you should go on a date. At first, holding hands seems like a big thrill. Then there's kissing and hugging—more thrills. After that, things can get heavy. How far you let your *sexual attraction*, or desire to be physically close, take you is something to think about ahead of time. Once you're underneath the bleachers at the football field, it is a little late to start wondering how to say *NO*. When things have gone so far, your body may be saying *YES, YES!* Your head may be saying *NO*, but the message could have a hard time getting through.

Sexual Attraction

Sexual decisions will reflect your values. Sexual decisions are also influenced by
- your understanding of sex,
- your awareness of the possible consequences of having sex,
- peer pressure

Teens who think about sex a lot are often afraid they are abnormal. Other teens aren't at all interested in sex. They too, are afraid they're abnormal.

Many teens are somewhere in between. Sometimes they think about sex, and sometimes they don't. All of these ways of feeling are normal. And it is normal to worry that you aren't normal!

Teens develop at different rates, sexually as well as physically, emotionally, and mentally. Sooner or later, your body will change so much that you'll find it hard to ignore sexual attraction. Your sex hormones will cause your body to react to sexual stimulation whether or not you intend it to happen. That is, you'll see a picture, be watching a passionate movie, or just see somebody you've liked a lot, and it hits! Your face gets hot and flushed. Your heart beats faster. You get a fluttery feeling inside. It is automatic. That's normal sexual stimulation. You find that somebody in particular can really turn you on. That's normal sexual response.

What you do about sexual feelings should be based on your values, not on your body's automatic response to sexual stimulation.

Declaring Independence — "I'm old enough!"

For some teens, having sex is a way of declaring independence from their family. Teens who have sex may be unconsciously saying to themselves, "I'm old enough to decide what I will do, and you can't do anything about it!" And that's really true! You are old enough to make decisions about what you will do with your body. Other people can't do anything about it unless you let them (or unless they are abusive and force you).

Reality — "It's not the movies!"

Many teens are already disillusioned about sex. They've tried it because they thought it was the ultimate experience. But for them, sex didn't turn out to be as good as the media or some of their peers build it up to be.

> I guess I expected too much. It didn't hurt or anything, but I just thought "Is this all there is to it?"

What you forget, of course, is that the people who are in the movies and on television are getting paid to act like they are being carried away.

Making love can be a wonderful experience. But it's more wonderful when you discover it in your own way with somebody you care about and are

Remember your goals — even when your body reacts to sexual stimulation.

willing to commit yourself to. That's why a lot of teens decide to wait to have sex until they have married or until they have found someone they are willing to make a long-term commitment to. Then sex is something very special. And it isn't just acting and it isn't using someone or being used.

Other teens are disillusioned about sex because it takes over a relationship. They may know individuals who had sex, then found that the relationships changed drastically.

People who have sex, then get dumped often feel used and convinced others are laughing at them.

All he wants to do is have sex all the time. We used to do a lot of fun stuff together. I can't even get him to talk to me now.

Sex can get boring when it has nothing to do with caring or it becomes the only thing your partner wants to do. Then relationships that were special can start to go bad. That's another reason many teens decide to wait until they are older to have a sexual relationship.

Some teens are disillusioned about sex because a relationship ends. Teens who have sex and get dumped can feel used and abandoned.

It was great. But then I wished I hadn't done it. I felt like I'd given her everything. Then she split. So what happened to all that love and caring she talked about so much?

Sometimes people want to be loved so desperately that they look for love by having sex. But it doesn't work. Then sexual intimacy isn't connected to emotional and spiritual closeness. Having sex because you want to feel loved can leave you feeling isolated and used. People who have sex without love, caring, and commitment are missing sex at its best.

Differing Values — "Everybody isn't doing it."

Sexual decisions aren't easy — especially the decision to wait. The sex drive is one of the body's most powerful drives. As a human being, it is part of your biological nature to have sex so that you'll have children. Then human beings will continue to survive. And sex can be no more than that — for some people it is only for having children.

> I don't think sex is dirty or anything like that, but sex is for having children. That's the reason I don't believe in using birth control. If you don't want to have children, don't have sex.

For others, sex is just another way of feeling good — sex for fun, no ties, no commitments.

> Guys like doing it. And so do girls — don't let anybody tell you they don't want it as much as guys. It's fun. So why all the hype about waiting and love? If it feels good, do it.

Or sex can be a way of expressing a deep sense of caring and intimacy.

> I take a lot of grief from my friends because I'm a virgin. They keep telling me what I'm missing or that they're going to find the right guy for me. Believe me, it has to be the right guy — somebody ready to be committed to me and to our relationship and that can't happen — not when I'm only 15. It's not like I haven't had my chances. But I'm not ready. They'd never admit it, but I think my friends really envy me for sticking to what I believe.

Sex shouldn't be either for family or for pleasure. It should be for both. Sex isn't just what you see in the movies and what some classmates brag about.

Saying Yes to Sexuality

Your body may be ready for sex before you are emotionally, intellectually, or socially ready for sex or ready for the consequences of sex.

Saying *YES* to your sexuality is appropriate. It is a way of beginning to accept responsibility for who you are. Saying *YES* to an experience you aren't ready for isn't appropriate. Pressuring somebody else into saying *YES* isn't appropriate either.

But having sex before you are ready can leave you feeling dominated by somebody else's wishes. It can leave you worried, guilty, and disappointed.

But how do you know when you're ready? Many people believe the answer to this question is "when you are ready for marriage".

It's possible to have a very satisfying relationship without participating in risky sexual activity.

Appropriate Sexual Behavior— "Not now!"

You've been getting ready for your sexual role all your life. It is a part of who you are. Sexual curiosity and discovery began when you were young. Children do a lot of sexual exploring, and that's appropriate. But an older child who is forcing a younger child to participate in inappropriate sexual activity like fondling, touching genitals, or sexual intercourse is wrong because it is exploittive, or uses another person. Having sex with a young child is also illegal. It can cause the other person to have emotional, social, and psychological problems.

When you see a brother or sister doing something inappropriate, it is a good idea to say, "That isn't appropriate." Then you should talk about it to a parent or another trusted adult.

Sexual curiosity is rediscovered as puberty comes along. Sexual attraction and desire begin. Then invitations to sexual play are inappropriate and likely to be symptoms of serious mental health problems or a complete lack of moral and ethical values.

Teens who feel a lot of shame, guilt, or fear about sex may be reacting to unhappy, early experiences. Someone may have said, "Shame on you," or "That's dirty," to their natural curiosity and urges. Or someone may have exploited them.

Myths About Sexual Behavior — "Shame, shame!"

A lot of teens grow up knowing little about their own bodies and feeling guilty when they try to find out. They begin to say *NO* to all sexual feelings. They feel guilty about **masturbation,** the touching of one's genitals for pleasure. They feel guilty about sexual thoughts or fantasies. They hear a lot of myths about sex, and they don't always know what to believe. These are some examples:

- **Myth:** Masturbation can make you go blind. There are a lot of myths about masturbation. You've probably heard some of them, too. Masturbating doesn't make people go blind. People have different ideas about masturbation which reflect their values. Some point out that masturbating can help relieve sexual tension. It can provide a safe way to experiment with sexual responses. It can help control natural urges. It can release general stress and tension.

  Others feel that masturbation can be used as an escape from emotions such as anxiety or depression. They are concerned that it may lessen feelings of self-control.

  Still others believe that sex is meant to be shared between two people. They think masturbation makes sex too selfish. In fact, some religions declare masturbation a sin for this reason.

  Masturbation is private. That isn't a myth. Masturbation in public is inappropriate sexual behavior. Forcing another person to masturbate is inappropriate, too.

- **Myth:** Having sexual fantasies means a person is sexually preoccupied or perverted. That just isn't true. Fantasy is a way of helping you think about sexual options. It is also a way of trying out in your mind things you couldn't really do — or might not want to really do. Obsessive fantasies — fantasies that occur all the time and that begin to replace reality — can be unhealthy. If you are preoccupied with a fantasy, it is a good idea to talk with a professional like the school counselor or nurse. Then you can find out what the fantasy is trying to help you resolve in your life.

  Obsessive fantasies aren't anything to be ashamed of, no matter how wild they seem. They are a way your unconscious mind has of telling you that there is something you need to work on. Forcing somebody to act out your sexual fantasies is inappropriate sexual behavior.

- **Myth:** Planning takes the romance out of relationships. Some people think that planning to have sex is wrong. That's their excuse for not using a form of contraception. But if you get swept away, it can't be helped — after all, you didn't mean for it to happen. That's a myth. Planning doesn't ruin romance. Planning doesn't mean you'll have sex either. It means that you decide ahead of time. You admit your interest in sex and plan so that your sexual behavior matches up with your values and goals. Planning means you don't get caught off guard and make a poor choice that you have to live with the rest of your life.

When you begin to accept your own sexuality and feel good about it, you are much more capable of making wise sexual decisions. You aren't at war with yourself because you have sexual feelings or because you haven't started to have them yet. You're able to relax and enjoy being a teen.

Being accurately informed about contraception will help you make sexual decisions later.

SECTION REVIEW
STOP AND REFLECT

1. Where did you get your ideas about sexual behavior? Who or what has influenced you most?

2. What values do you have that influence your sexual behavior? List as many of these values as you can.

3. Which of your ideas about sexual behavior are based on myths? Describe one of the ideas you have now or you had when you were younger.

4. What are some of the ways people shame little children about their sexual curiosity? What should they do instead to help children learn appropriate sexual behavior?

5. What could you tell a younger brother or sister about how to say *NO* to someone who asks them to behave in a sexual way that is inappropriate?

SECTION 2

THINKING ABOUT SEXUAL OPTIONS

Sometimes sexual decisions take you by surprise. You don't have time to stop and say, "Well, let's see, what are the options here?"

Thinking about appropriate sexual behavior is a way of getting clear about the options. Finding out what your family values and what they think is appropriate sexual behavior is another way of getting clear about the options. Teens who talk to their parents about sexual values and behavior are less likely to make sexual decisions that hurt themselves and others. Thinking in advance about how you will act on a date is another way of discovering your options, too.

Knowing What You Want

Thinking and planning ahead can save you from some awkward moments. When you've thought about your options, you aren't so likely to make a choice before you are ready for it.

Helen and Kyle had been good friends since grade school. Things had always been casual— they'd never even held hands.

One night after a movie, when Kyle walked Helen to the door, he said, "Hey, don't I even get a good-night kiss?"

Helen hadn't ever thought of Kyle that way. And he didn't wait for her to think about it. He started to kiss her. She turned her head because she wasn't sure what she wanted. Kyle's kiss landed on her ear. They both felt really dumb. Kyle left in a hurry, and Helen ran inside.

Helen and Kyle hadn't thought about their relationship. If Helen had been thinking ahead, she'd have realized that all boy-girl relationships can become something more than "just being friends." She wasn't prepared for the situation. A good-night kiss didn't have to be such a big deal. It didn't have to leave them both feeling silly.

Kyle didn't plan too well either. Even if the idea of kissing Helen good-night was something that just happened,

When you find yourself really embarrassed, take time to analyze the situation. Then talk to the other person, admitting your feelings.

he should have given her time to think. He was probably nervous. So he ended up acting pushy when he didn't mean to.

Of course the situation didn't have to end there. They could have just laughed about it. Now either Helen or Kyle could give the other a call and say, "Hey, I ended up feeling really dumb! Maybe we should talk. We've always been such good friends. Do we want something more?" Thinking ahead can help you act in ways that show what you value most.

Knowing Who You Want To Be — "It's just not me!"

What do you value in a relationship? What kind of person do you want to be? What kind of people do you want to be with? These are good questions to think about ahead of time. You aren't always reacting to the pressure other people may put on you. Sometimes you react to your own inner pressure — the things you want to try and aren't sure about.

There are also things you are pretty sure you shouldn't try yet. But you're tempted to try anyway. Thinking ahead helps you to understand who you want to be and to recognize the actions that help you be that person.

> Joe likes to think of himself as a caring person. He wants to get married someday. He already knows he wants to marry somebody who will value the same things he values. He wants somebody he can trust. And he says he wants to marry somebody who is willing to wait for sex and who will be true to him.
>
> Joe has been dating Ruthie for a few weeks. He asked her out because she has the looks he likes. None of his friends particularly like her. But, as Joe puts it, "Who needs personality when you look like that?"

One afternoon Joe was getting some stuff out of his locker. "Hey, macho man," Ruthie yelled. "So when do we get out of here and have some real fun?"

Joe felt himself blush and begin to sweat. One of the things he liked was that Ruthie let him get by with a lot more than any other girl. Even though he didn't like her treating him that way in front of other people. Joe didn't want to end the relationship just yet. He was having too much fun. So he didn't say anything.

Joe hasn't done enough thinking about the kind of person he wants to be. He is using Ruthie to experiment. He likes the idea of taking sexual liberties with her in private. But he doesn't want her going public. Joe imagines that he can learn a lot from Ruthie. But he isn't thinking about the possible consequences for himself or for Ruthie. He doesn't realize that he is being used just as much as he is using her.

Joe doesn't seem to know that you can't spend all your time acting like one kind of person and still expect to turn out like somebody else—the somebody you want to be. If you use people, you become a user. If you want to end up married to somebody you can respect, you should treat the people you date with respect.

The opposite of using behaviors are caring behaviors. They show that you value yourself and other people.

Standing Up For Your Values — "It's who I am!"

Iliana has thought a lot about the kind of person she wants to be. She's thought about the kind of sexual behavior she will accept from herself and others.

Iliana's really fallen for Jorge. They've dated some. She hopes he'll ask her to go steady. Even though she's really attracted to him, she wants to take one step at a time and not rush things.

But one night Jorge suddenly reached over and pulled her close. Iliana liked it a lot, but she wasn't sure what Jorge had in mind—they'd never made out or anything like that.

Iliana didn't wait. She pulled away and took Jorge by the shoulders. "Wait a minute, Jorge. I don't know what you're thinking, but I'm not real comfortable with this. I care about you a lot. But I think we should talk."

Jorge looked confused for a minute. Iliana held her breath. She was afraid she'd hurt Jorge and ruined her chances with him. But she didn't back down. Then he said, "Don't get all worked up, Iliana. I just want to

If you don't show respect for yourself and others now, chances are you won't do so as an adult either.

hold you. You're so pretty. I wouldn't do anything to hurt you. I'm not that kind of guy. And I don't want to rush you."

Iliana showed the values of respect, responsibility, and caring for Jorge and for herself. Even though Jorge surprised her by doing something unexpected, she was able to act in a way that was fair to both of them. She stopped before things reached a point where it would be hard to say *NO* or would be embarrassing to either of them. She told Jorge what she needed. She gave him a chance to tell her what he had in mind.

Jorge showed his caring attitude by accepting Iliana's feelings. He told her what he wanted. Their caring behaviors show what they value in another person and in a relationship. Iliana took a risk by stopping things. She ended up finding out that Jorge was the kind of person she really did want to go steady with. He wasn't just out to use her. Now she knew she could trust him—and herself as well.

As you think about your own sexual desires and how you will deal with them, you can begin to set some guidelines for yourself that are based on your values. It's never too late.

I've been sexually active, if that's what you want to call it, since I was in sixth grade.

How Far Do I Go?

Caring, respect, trust, responsibility, and family are values that build solid relationships—the kind that later become the foundation of your own family. They are good guidelines for deciding about your own sexual behavior, too. Ask yourself these questions. Do my actions show that

- I truly care about what happens to my partner as well as to myself?
- I am willing to allow negative consequences to happen to my partner—like getting pregnant or getting a sexually transmitted disease?
- I respect myself enough to set limits? Do I respect my partner enough to honor the limits he or she sets?
- I am becoming responsible? Am I able to set goals for the future and act in ways that will allow me to meet them?
- I listen to my head instead of being ruled by my hormones?
- I am someone to trust? Can my date relax with me without having to be on guard? Can my family trust me to do what I say I'm doing? When I say *NO* do others know that I mean it?
- I love and am committed to family—the family I am a member of and the one I will have in the future?

I've tried it with a lot of guys. It isn't any big deal. It makes me feel wanted, you know. I have something somebody really wants. So don't talk to me about sex like it's some wonderful thing I'm saving. What's to save?

Sex isn't special unless people make it special. Even if you've been sexually active, you can choose to wait. Even if you've been a user or let yourself be used, you can become caring. You start by knowing the kind of person you want to be. Then you plan ahead. Knowing what you want makes a difference when you must make choices quickly. And when you start saying *NO* to using or being used after you've been saying *YES*, you need to have a plan. If somebody has a problem with your plan, it is that person's problem! That's when it's time to end a relationship.

When you begin to date, ask yourself these questions. "Can my date relax with me? Can my family trust me to do what I say I'm doing?"

What If You're In Love — "This is it!"

You'll probably be in love several times before you're old enough and feel mature enough to settle down, get married, and start a family. The first time you fall in love it's hard to believe it won't last forever. But feelings do change.

Teens who decide to wait for sexual intercourse until they're married do so because of their values. They want to keep sex special for themselves and the person they marry. They don't want to feel the false sense of closeness that sex without commitment brings. They think it makes sex less than it can be. They want to enjoy their teen years as a time of fun and growing. They believe that getting sexually involved with somebody — even somebody they love — can keep the relationship from developing. They don't want to move attention from the relationship and focus it on sex. They believe that loving someone enough to wait is a good test of how much they are in love.

Deciding to wait isn't an easy choice. But it is still an ideal worth striving for. To save sex for marriage shows the value placed on marriage and family.

Knowing Your Sexual Messages — "But I said *NO!*"

One of the ways you can plan ahead is by finding out what sexual messages you are sending. If you are telling people *NO* and acting and looking like you mean *YES*, don't get upset if you're not taken seriously. Sexual messages are signals you give to another person about your sexual values. You send sexual messages through the clothes you wear, your body language, and how you act.

> Sasha's mom was upset. Sasha had just come home from the mall with a new sweater. It was a super color and looked fantastic with her jeans. Her mom didn't think so. "I won't let you out of the house wearing a sweater like that. You're too young to put yourself on display. You look like you're advertising your body!"

Sasha's mom was concerned about the sexual message she believed Sasha would send with the new sweater. She thought it was too tight and attracted too much attention to Sasha's breasts.

Clothes that call attention to sexual areas of your body can be like an invitation to have sexual contact — even if that's not what you had in mind. A lot depends on the culture you live in. For example, in some cultures Sasha's sweater would be okay. It would be expected that her father and brothers

"I'm uncomfortable when you keep asking me to go all the way."

"I understand you want more from our relationship."

"I can't continue going out with you if I feel pressured."

Your parents may object to some clothing choices, such as swimsuits, tight sweaters or pants, because they know these choices send misleading messages to those who don't know you.

would protect her from anyone who might get the wrong impression.

In our society, where people of so many cultures live and work in the same neighborhoods, it is a good idea not to assume too much. Sasha shouldn't assume that nobody will get the wrong impression. If her mom is that upset over the sweater, it's a pretty good clue that the messages Sasha would be sending are not okay messages in their culture.

On the other hand, guys who see a girl like Sasha in a tight sweater or tight jeans shouldn't assume they are being invited to touch. Before you decide what the sexual message is, you need to know the person and think about the total picture.

Actions are part of the picture. The way you carry your body and the way you act around other people can send many different messages:
- Come get me.
- I'm available.
- Look, but don't touch.
- Let's get acquainted.
- I just want to be friends.
- Keep your distance.

Rape

Rape is forced sexual intercourse. It is not an act of passion. It has nothing to do with sexual desire. Rape is an act of violence. People who rape are striking out at other people. The victim may be a woman, a man, or a child.

Somebody who forces a date to have sexual intercourse is raping that person. Forced sexual activity of any kind on a date isn't just a difference of opinion about how far to go. It is abusive, violent behavior. A date rape can cause the victim to feel confused, guilty, and ashamed.

Rape is never the victim's fault. No one causes a rape to occur by what he or she says, does, or wears.

You can protect yourself against rape.
- It is safer to go places with other people.
- If you go alone, tell someone your plans.
- Walk briskly, with purpose.
- Stay in well-lighted, populated areas.
- Have your keys ready when you are going to your home or car.
- Lock all doors and windows in a car and at home. Always check inside your car before getting in.
- Don't open the door to strangers.
- If you have car trouble, stay inside the locked car with windows up. If anyone wants to help, have that person help by getting the police.
- If you're being followed or feel threatened, go to a public place. Run, yell "fire," scream—make all the noise you can.
- Try to keep calm and think clearly.
- Don't be alone with someone who makes you uneasy.
- Learn self-defense.
- Don't use drugs or alcohol—they make you vulnerable.

You can protect yourself against date rape.
- Know who you are dating. If you don't know your date that well, go with another couple. Stick to public places.
- Don't let your date make all the plans.
- Don't spend a lot of time alone or in isolated places.

What you actually say also sends sexual messages, even if you're just kidding around. If you're always talking about making out or making comments like "I need some loving," don't be surprised if your date makes a move, or others think you're available. What you talk about and the kind of language you use make a difference in how other people understand your messages.

So what messages do you send by your actions, your dress, and your words? Your messages are part of your reputation. They help establish the kind of person you are. They show what you value. If what you say you value and what your messages say aren't the same, change your messages.

Knowing self-defense techniques is a good idea for people of all ages.

SECTION REVIEW
STOP AND REFLECT

1. What do you think is appropriate sexual behavior for people your age? Make a list of the sexual behaviors you think are appropriate for a first date. For a second date. For steady daters. What values are related to each behavior?

2. Describe a situation you can think of (real or imagined) when a date is moving too fast. Describe what is happening. Tell how you (or the person you thought of) could slow things down in a caring way.

3. Without using names, tell about someone whose sexual messages aren't the same as the sexual values he or she talks about having. In what ways is that person sending mixed signals?

SECTION 3

FACING THE CONSEQUENCES

Some of the consequences of sexual decisions are ones people don't want to talk about. You always figure they won't happen to you. That's one of the things about being a teen. It's normal to feel that bad things won't happen to you. But feeling invincible doesn't keep bad things away. When you make unwise choices, feeling like nothing bad will happen won't save you. That's a myth.

Saying *Yes* To Consequences

When you say *YES* to sexual intercourse, you may be saying *YES* to more than you want. You may also be saying *YES* to
- getting pregnant,
- getting somebody pregnant,
- deciding about abortion,
- having a baby with low birth weight, mental retardation, or birth defects,
- getting trapped on welfare,
- getting STDs.

These are the big consequences everybody likes to pretend won't happen. Many teens would rather believe myths and lies. But many more teens have started looking at their options ahead of time.

Reputation — "I'll always respect you."

Some teens imagine that having sex won't affect the way they see themselves or the way other people see them. Or, they figure that if it does affect them, it will make them feel good and be looked up to. And for some, that may happen — for a while. Then the consequences start to catch up.

Boys may suffer because they're supposed to be sexually experienced. If they haven't experimented with sex, they probably don't want to admit it to anyone. A lot of guys try to cover by bragging about all their sexual activity. They suppose that by lying they can keep the pressure off. It may work for a while.

But the truth is, you feel better about yourself when you're honest and live up to your values. And you feel better about yourself when you're tough enough to control your impulses. You can cave in, like you think everybody else is doing, in order to look like a macho guy. But, when you think about it, being strong and responsible is a lot harder than acting macho. It takes a lot of strength, character, and downright toughness to do what you think is right when everybody seems to be doing the opposite. Being responsible is hard. But it leads to better feelings about yourself in the long run.

What about now, the short run? Did you know that more than half of the male teens under 17 have chosen to say *NO* to sexual intercourse? A lot of guys are talking big in order to feel big. How you end up feeling about yourself 5, 10, or 20 years from now is a lot more important than how a few other teens feel about you now.

Girls get pressured, too. They get a lot of mixed messages from society. Women of today are supposed to be sexually liberated—as free as guys. But girls who go too far too soon end up with bad reputations. Guys talk about how much they want and need sex. Then they often don't want to have anything to do in public with the girls who give it to them. A lot of girls get used.

Girls hear that everybody is doing it, too. In reality, over 60 percent of female teens under 17 have chosen to wait. And a lot of girls who end up giving in soon decide to say *NO* and stick to their decision.

Saying *NO* isn't any easier for girls than it is for guys. A lot of guys expect the girl to set the limits—it may not be fair, but it is still true. That puts pressure on a girl to be strong, especially when her boyfriend tells her how much he needs her. The truth is, a guy who talks about needing you really means that he wants to use you. What he actually needs is some courage to be responsible. And what you need is another boyfriend—one who will respect you and care enough about you not to pressure.

For girls, the chances of getting pregnant are always there. Not only do you risk your reputation, but you risk your future when you decide not to wait.

Pregnancy — "It won't happen to us."

"I won't get pregnant." This is just a myth. But a lot of people believe it. Half of the teenage women between 15 and 19 who get pregnant believe this myth. They think they can't get pregnant. They simply don't know enough about their own bodies to know when they are most likely to get pregnant. They don't use birth control. And the guys they have sex with don't know or care enough to use birth control.

> I never realized I was taking such a big chance. I mean, I didn't get pregnant the first time. So I didn't think I was going to get pregnant the second or the third or the fourth or the fifth.

Some teens think that a girl has to be having regular periods to get pregnant. That's another myth. As soon as a girl starts having periods, she can get pregnant—even if she doesn't have them more than a few times a year.

Some people pretend that if they do get pregnant, they'll have a wonderful life together as teenage parents because they're so much in love. That's another myth. Almost 90 percent of the teenage guys who get girls pregnant abandon them.

> Just a couple of months ago I was a kid. Now I'm a woman. I ended up getting involved with this guy. The next thing I know, I'm sitting here staring out this window, and trying to get some direction for my life. I'm pregnant. And I'm alone.

Teens who do get married most often end up divorced. That's because being a teen parent is very hard, even when you care a lot and try. Teen parents have to give up a lot of freedom and a lot of the fun of just being a kid.

> I love the baby and everything, but it was just hard to accept the fact that I wouldn't be able to go out whenever I wanted.

One moment of irresponsibility can dramatically change your life forever.

Not only does the teenage mother find herself spending all her time with a baby, but she's likely to be taking care of babies for some time to come. Most teen mothers who have a baby before they are 17 have another by the time they are 19. That's why so many teen mothers become trapped on welfare.

Birth Control — "It doesn't ruin romance!"

A lot of teens think that using birth control ruins romance. That's a myth. Birth control isn't nearly as hard on romance as an unwanted pregnancy. And it's a lot more romantic than getting AIDS or any other sexually transmitted disease.

People have strong differences of opinion about the use of **contraceptives,** products designed to prevent pregnancy. These opinions may be related to religious values. You need to know what your family values and decide what you value. If you believe that use of contraceptives is wrong, then your sexual behavior needs to line up with your values. Only one method of birth control is 100 percent safe, and that's not having sexual intercourse.

Some teens are too embarrassed to find out about birth control or to buy contraceptives. They don't want anyone to know that they are sexually active. But imagine how embarrassing it would be to tell people that you've gotten somebody pregnant—or that you are pregnant. Thinking you can keep others in the dark about your sexual activity is another myth.

> There were rumors going around—untrue rumors—that I didn't know who the father was, that it happened at a party. I guess kids do that because they don't know what it's like to be pregnant at 15.

Teen Pregnancy Is Risky Business

There are serious risks to being pregnant as a teen—both to the mother and to the baby. Think about these:

- Babies born to teen mothers are more likely to be premature or have low birth weight.
- Babies born to teen mothers are twice as likely to die in infancy as other babies.
- Babies born to teen mothers are more likely to be mentally retarded or have birth defects than other babies.
- Teen mothers have more complications during pregnancy than other mothers. These can include miscarriage, toxemia, hemorrhaging (or excess bleeding), and a higher risk of death than other mothers.

Other teens don't use birth control because it bothers their partner. He or she doesn't like using birth control. If the relationship is difficult, it is just one more thing to argue about. So they take their chances. They think that by not using birth control they'll make the relationship better. That's another myth.

STDs And AIDS – "It's not my problem!"

STDs, or sexually transmitted diseases, include any disease that is spread through sexual contact. Sexual diseases can be spread through oral, genital, or anal intercourse.

School is one place to learn the facts about STDs. What are other ways to learn the facts?

> I can't believe it happened to me. I guess no one does. You think you know someone; you think he's all right. You love someone; you trust him. He was the only one for me, so I figured I was the only one for him, too. Believe me, the last thing I was thinking about was catching some disease. But I sure should have thought of it. I was lucky, too. I got it caught and taken care of before it made me real sick and before it could do anything bad to my baby.
> Now that's something I don't understand—how could you let someone catch something from you that would hurt the baby?

> Don't let anyone tell you you're stupid to wait to have sex. Waiting has got to be better than all the terrible things that can happen. I'm going to take care of myself now, and I think if people care about each other they have to make sure they are honest with each other. It's not just love, it's responsibility. You don't have any right to make someone else sick.

Anyone who has sexual contact with another person is at risk of getting STDs. About 2,000 teens become infected with an STD every day. Often they don't have any symptoms, so they infect their partner without even suspecting it.

There's No Such Thing As Safe Sex

Having sex is not caring or loving when you're risking your welfare and that of your partner. If you do have sex, use safer sex practices:
- Limit your sex partners. Every time you have sex with someone you are at risk of getting STDs that person has picked up from other partners.
- Use a latex condom correctly. AIDS is carried in semen, so other forms of birth control won't protect you. This also means that without a condom you can get AIDS from oral, vaginal, or anal intercourse.
- Don't use oil-based lubricants like Vaseline, Crisco, or baby oil. They weaken a condom.

You can't catch AIDS by talking to or being around someone with the disease.

People who are not sexually active will not get STDs. Choosing not to be sexually active is a responsible health decision.

AIDS stands for acquired immunodeficiency syndrome. It is caused by a virus (HIV) that attacks the body's immune system. This means that the body can't defend itself. Other germs—bacteria or virus—can enter the body and cause infections. The body's immune system can't do anything about these infections. So even a simple cold spreads and becomes life-threatening.

AIDS is unlike other STDs, which can be treated or cured. Scientists haven't found a cure for AIDS. A person who has AIDS will die. And it is spreading at an alarming rate, especially among teens.

The HIV lives in blood, semen, vaginal secretions, tears, and saliva. It gets spread from person to person through blood, semen, and vaginal secretions. You get AIDS by having sex with an infected person, by sharing needles, or from contaminated blood. It is also spread from a pregnant, infected mother to her baby. And that baby is doomed to mental retardation or to death.

Aids Isn't the Only STD

These are common STDs:

- Chlamydia is the most common STD. It attacks the male and female reproductive organs. It is spread by sexual contact with someone who is infected. And it can be spread to a baby at birth if the mother is infected.

  Untreated, chlamydia can cause scar tissue to build up in men and women. The scar tissue may prevent them from having children.

- Gonorrhea is highly contagious. It attacks mucous membranes of the penis, vagina, rectum, or throat. It is spread through sexual contact.

  Gonorrhea increases a pregnant woman's chances of having premature labor and a stillbirth or a baby with a serious eye infection.

- Syphilis is one of the most dangerous STDs. It is passed from person to person during sexual intercourse. Once it enters the body, it can be life-threatening to both men and women. A pregnant woman is four times as likely to have a miscarriage and twice as likely to have a stillborn child. A baby born to a woman with syphilis is likely to be born with syphilis.

  Untreated, syphilis can cause heart disease, blindness, paralysis, and insanity.

- Herpes simplex virus (herpes II)—causes blisterlike genital sores. No cure exists. Herpes is transmitted through sexual intercourse. There is no sure way of knowing when the disease is contagious.

  Pregnant women with herpes II run a higher risk of miscarriage or premature birth. There is a high death rate among babies born to infected mothers. Babies also have a higher risk of brain damage if they pass through the birth canal when the infection is active.

  Treatment for all STDs is an important personal responsibility. An STD won't just go away. If you have an STD, you need to be treated. And you should tell everyone with whom you've had sexual contact. They might not like it, but they have a right to know.

  Using condoms and contraceptive foam helps protect against STDs. But the best protection is not to have sexual intercourse.

AIDS Prevention: A Worldwide Effort

Acquired immunodeficiency syndrome (AIDS) is a global problem that has spread from continent to continent.

From the mid-1970s to 1981, AIDS spread around the world unnoticed. In 1981, only a few thousand cases were reported. By 1988, 350,000 cases were estimated worldwide, double from 1987. By 1993, at least one million new AIDS cases are projected by the United Nations' World Health Organization (WHO).

No disease in history has evoked such a worldwide response or global solidarity. With WHO's initiative, almost all of the world's five billion people now know about the disease.

Much more needs to be done to fight AIDS. People need help to follow healthful behavior; then AIDS won't spread further. We need to avoid discriminating against those who already have AIDS so they can openly seek help.

Only a worldwide effort, involving every person, will stop the spread of AIDS.

People infected with the HIV virus may go for years without having a single symptom. Meanwhile they carry the virus and can give it to everybody they have sex with.

One reason teens are in so much danger of getting AIDS is because many of them experiment with drugs. When a person is high, he or she isn't likely to use good judgment about saying *NO* to sex or about using safer sex practices.

When people inject drugs into their veins, they get blood on the needle. HIV grows in blood. So if one person who uses a needle is HIV-infected, all the others who use that needle will shoot both drugs and AIDS into their bodies. This happens even when they can't see the blood on the needle.

That's another reason why it's important to stay off drugs or get off drugs. Even if you don't do drugs, you can get AIDS if you have a sexual relationship with someone who uses or has used injected drugs.

Before 1985, people who received blood transfusions could get blood contaminated by HIV. Now all blood used in transfusions is screened for signs of HIV. You can't get AIDS by giving blood because only new, sterile needles are used. But you could get AIDS from doing a blood brother or sister rite.

A pregnant woman can have HIV without knowing it. She can give AIDS to her baby. The virus passes from her bloodstream to her growing baby. Babies born with AIDS never get a chance at life. Most die within two years after they are born.

STDs, including AIDS, can ruin your life. That's why it is smart to learn how to protect yourself from getting them. The best and most responsible protection is not to have sexual intercourse as a teen. But if you do, avoid behaviors that put you at high risk of getting AIDS or other STDs.

Saying *Yes* To Your Future

There aren't any charts or guidebooks that can tell you who's the best person for you or what will happen for sure if you get involved in a relationship and have sex. Those things depend on you.

Saying *NO* to inappropriate sex is a way of caring about yourself. It's also a way of being in charge of your life. Showing caring behaviors toward the people you meet and spend time with is a way of being in charge of your life, too.

There are many ways to show that you care besides having sex. In fact, saying *NO* shows you care enough to be responsible and wait.

SECTION REVIEW
STOP AND REFLECT

1. What myths about pregnancy do some of the people you know believe?

2. Tell why it would be hard for you to be a parent while you are a teen.

3. Why do you think some teens who are sexually active refuse to use birth control?

4. What does your family believe about birth control? How are their beliefs related to values they have?

5. Why do most people who know about AIDS say the best way to prevent AIDS is not to have sex?

6. Is it dangerous to go to school with someone who has AIDS? Explain your answer.

CHAPTER 15 REVIEW

Putting Your Values To Work

STRENGTHENING YOUR VALUES

Reread *A Teen Speaks* **on page 323. Then answer the following questions.**

1. How does Lisette feel about her first sexual experience?
2. What things influence the views people have of sex? What values might have influenced Lisette's decision? What are some particularly strong influences on you?
3. What things might Lisette have considered before she decided to go ahead? What options did she have?
4. What might have been the consequences of each possible course of action?
5. Whom might Lisette have talked to before making a decision?
6. Suppose Lisette made the decision to wait to have sex. How might Lisette have communicated her needs to her boyfriend?

INTERPRETING KNOWLEDGE

1. Find at least five examples of positive male-female relationships in television, movies, books, music, or art. Why do you think they are positive? What do your selections say about your own values?
2. Read at least two magazines geared to teen readers. How do these magazines treat the issue of sex? What values are present? How do the values these magazines display compare with your own beliefs and values?

SHARPENING YOUR THINKING SKILLS

1. List three factors that may influence an individual's decision about sex. How important is each of these factors? How does each factor influence the decisions you make?
2. When one person says *NO* to something sexually, the other person must always respect this decision. There are no exceptions. List at least 10 effective ways of saying *NO* and getting out of a difficult situation. If you say *NO* to someone, how might that person deal with the situation in a caring, responsible manner? Be specific.
3. Define **sexually transmitted diseases.** How can they be spread? How can people protect themselves from them? Describe three STDs mentioned in this chapter. How does a person catch these diseases? What are the symptoms? Why are immediate detection and treatment so important? Where can you go to get more information about STDs?

APPLYING KNOWLEDGE

1. What additional questions do you have about sex? Where can you go to get accurate information? Research at least three questions you have. Write a summary of your findings.
2. Write a brief skit about two teens, one of whom is pressuring the other sexually. Write several endings in which the teens deal with this situation in a responsible and caring way.

CHAPTER 15 REVIEW

Putting Your Values To Work

PRACTICING DECISION-MAKING SKILLS

Read about each situation. Then answer the questions.

Situation A: One of your closest friends has just started dating for the first time. You have known this friend since childhood. The two of you have often discussed the people you like and the kind of relationship you hope to have in the future. Now you want to provide your friend with the best support you can, since you know she will be facing many important issues.

1. What issues do you think you and your friend should discuss now?
2. What decisions might she face in the future?
3. What types of questions will you raise with her?
4. Are there any issues that you feel should not be discussed? If so, what are they?
5. Suppose you discover that your values are different from those of your friend. How might this affect the advice and support you give?

Situation B: Suppose you do not like the girl your best friend is dating. After carefully examining your feelings and motives, you still honestly believe that she is not right for your friend and that a relationship with her could prove to be very destructive.

6. What options do you have?
7. What might be the consequences of each different action you could take?

Situation C: You have been dating someone you really care a lot about. You and your boyfriend or girlfriend have a relationship in which you can talk about your feelings honestly and openly. Now you both need to talk about the physical part of your relationship. Before you do, however, you want to be sure you know what you really want and value.

8. What issues must you think about as you make any decision?
9. Whom can you talk to about these issues to get support and/or guidance?
10. After you have thought these issues through on your own, what are your options about what you will say to your boyfriend or girlfriend?

Situation D: Suppose he or she reacts in a less than supportive way. You discover that the good communication you thought the two of you had does not exist after all.

11. What options do you have now?
12. What might be the consequences of each different action you could take?

Situation E: You accidentally overhear a conversation in which two people are discussing you in a relationship—your values and your behavior towards your boyfriend or girlfriend.

13. What would you want them to say about you?
14. What steps could you take in your life to ensure that this positive picture of you will really develop?
15. What things that they might say would really bother you? What can you do to prevent this picture from developing?

UNIT 4
Caring For Your Family

CHAPTER 16 UNDERSTANDING FAMILIES
CHAPTER 17 FINDING HELP FOR TROUBLED FAMILIES
CHAPTER 18 BUILDING A STRONG FAMILY FOR TOMORROW

CHAPTER 16

Understanding Families

A TEEN SPEAKS

Don't get me wrong. I love my family. But sometimes they really embarrass me. Take my mother. Mom is really pretty, but she puts on this makeup. You can see where it stops between her face and her neck. My friends have even noticed it. She puts on too much blush and wears this bright red lipstick all the time, even when it doesn't match what she is wearing. She says that's the way my stepdad likes it.

My friends say she's okay. But they think she looks really weird. She makes a big emotional deal of it when anybody comes over. She has to ask everybody how they are and how their families are. My friends don't seem to mind so much, but I just hate it when she does that.

I can bring anybody home with me I want to. But that's another problem. Our house is always a mess. We eat in the kitchen because the dining room table is always piled up with papers and junk. Mom hates to clean. She'll start to clean up and see a magazine article she's wanted to read. Next thing you know, she's off in another world. The mess doesn't seem to bother her.

I've been trying to help clean up so the place won't be such a wreck when my friends come in, but my little brother and sister are real slobs. They sit in front of the TV all the time, playing with junk and leaving it everywhere. One day some of my friends were over and we were going to watch a movie. One of them sat down in the chair and there was this big crunch—he looked under the cushion and here was all this garbage: grape seeds, used tissues, pieces of paper—even a soda can. I could have died right then.

Mike

SECTION 1

ACCEPTING YOUR FAMILY

have you ever wished you could trade your family in for a new one? Sometimes when you look at your family, you see how different they are from other families you know. This can make you feel glad you have the family you do. Or it can make you wish things were better for you than they seem to be.

Appreciating Family

Nearly all people can think of things they'd like to change about their families. For example, some people wish that

- their little brother would stay out of their stuff—while others wish they had a little brother.
- their big sister wouldn't boss them around—while others wish their sisters would pay attention to them.
- their family wasn't so large—while others wish they weren't the only child.
- their dad wouldn't laugh so loud at his own jokes—while others wish they had a dad.
- their mom wouldn't wear so much makeup—while others wish their mom would take an interest in her appearance.
- their grandparents didn't have to live with them—while others wish they could see their grandparents more often.

Even though most people wish they could change at least some things about their family, people are usually very protective of family members. Have you noticed how you feel when somebody else tells you how to change your family? Usually when outsiders start criticizing your family, you get huffy. Or, if somebody picks on a family member or tries to hurt someone in your family, you are ready to fight.

Some of the things that may irritate you most about family members are things you probably didn't even notice a

If you compare different families you'll see many similarities and many differences.

few years ago. That is part of growing up. As a teen, you are moving away from the time in which you were totally dependent on your family. You are looking toward the time when you will be on your own. The time between dependence and freedom can be awkward. But, it can also be a time to look at family life and think about the kind of family you want to establish.

What Is A Family?

If you want to be technical about it, a **family** is a group of two or more people who live together and/or are related by blood or marriage. Families are usually made up of people who are related to each other. But they don't have to be. Sometimes people who don't have families band together and take care of each other. These people become a family. Some of the kinds of families you might see in your community are

- one parent and children,
- two parents and children,
- grandparent and grandchildren,
- parents, grandparents, and children,
- parent, other adult (for example an uncle, aunt, or a friend), and children.

Sometimes a single parent with children lives with another single parent who has children. And sometimes two single parents get married and combine their families into one big family.

The thing that makes a group of people into a family is the fact that they
- **care about and for each other,**
- **respect each other,**
- **accept responsibility for themselves and the family.**
- **make a commitment to stay together over time.**

Groups of children who have no parents and are reared in an orphanage often become a family to each other. In some countries, where families have been split up by war, children band together and take care of each other. Wanting and needing family is a basic human need.

All families experience problems and joys. All families have their ups and downs. No family is perfect. Family members have to work together to make life better for one another.

> A young mother put it this way: "A family is the nest we have for our children. You teach a child there's a place where he belongs. You teach a child what's right and what's wrong. You teach a child what the world is all about. In a family, there's caring. You don't walk away. It's not a job. It's your life on the line."

It's true, your life is on the line in your family. And the story of every person is the story of growing from complete dependence on family to freedom and **interdependence,** choosing to depend on each other. While there are many kinds of families all over the world, every family is alike in this way.

Dependence — "Who'll take care of me?"

Life begins with total dependence on others. The greatest dependence comes in the early stages of life, before a baby is born. After a baby is conceived, the baby receives everything—food, and oxygen—from its mother.

A pregnant woman's choices about food, exercise, and rest directly affect her growing baby. If the woman chooses to take drugs, use alcohol, or smoke, she is passing these harmful substances on to her baby. If a mother is addicted, her baby is addicted.

A mother's care for herself is reflected in the health of the unborn baby.

Even after birth, a baby cannot live without care. And right from the beginning, the stage is set for each baby to learn about values. A baby who cries and is comforted learns to trust others. A baby who is fed when hungry learns about care. And an angry or frightened baby who is tenderly handled learns about respect.

Who you are begins with who your family is. You may love it or hate it, but you are part of your family. Even if you run away from it, your former family is still a part of you. The experiences you had as a young child, your ideas about yourself, and your values are all tied in to what you have learned from your family.

Your Family is a Place to Grow

A healthy family is a place to grow, to experiment, and even to make mistakes. Family members are there to help you learn. They trust you and you trust them. It's safe to talk about things that concern you.

You learn about rules in the family. At first the rules are restrictive. **Restrictive** rules set limits:

- "No, no! We don't put fingers in the fan."
- "No, you can't walk to school by yourself."

As you grow older, rules become prescriptive. **Prescriptive** rules tell you what you should do:

- "Pick up your things."
- "We expect you to be in by 11 o'clock."

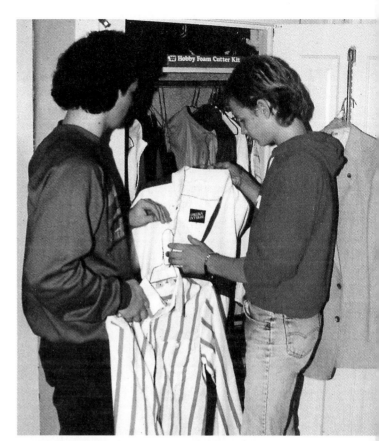

Asking permission before borrowing a shirt from your brother is just good common sense.

When you are ready for independence, rules should be internal. **Internal** rules are within you.

- "As soon as I finish cleaning the garage, I'll be over."
- "I'm going to study for my test tonight, and if I feel prepared I'll go out tomorrow night. Otherwise, I'll study tomorrow night, too."

In a healthy family, you learn about **intimacy,** a feeling of closeness and familiarity. Your family sees you at your best and your worst. They accept you for who you are.

A Global Family Album

Everywhere in the world, families are the basic unit of human living. Everywhere families nurture their young, preparing them for adult life. In this way, families are alike. Family structure differs among cultures.

Most Americans live in nuclear families with one or both parents and children. In other societies, people live in joint family systems.

- In rural India, most married couples live with the groom's family. Their marriage success depends on how well both the family and their village accept them and how well the couple adjusts to both.
- Although the practice isn't common, Islamic culture allows a man to have four wives. The Koran places a condition, however. The husband must treat all the wives equally and justly.
- Older people are shown much more respect in Asia than in America. Adults are expected to take care of their aged parents, so they often live together. Grandparents, in turn, share the work of raising the young and offer counsel, guiding the family through life.
- When couples marry in the Philippines, they choose two "sponsors," an older man and an older woman. The sponsors don't live with them, but they do advise them during their marriage. Sponsors take part in the marriage ceremony.

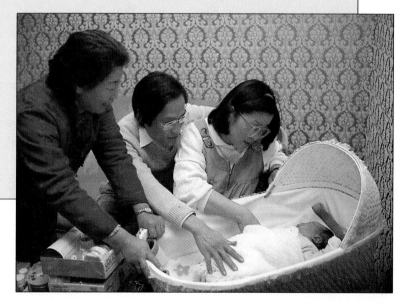

Your Family Teaches You to Get Along

You learn how to get along in life at home. No family can always meet every member's needs for physical and emotional support all of the time. So you have to learn how to share, take turns, and delay gratification. To **delay gratification** means to wait for something you want. Most needs are met, most of the time.

But, when the growing child's needs aren't met by the family, the child keeps looking for ways to have them met. The way children learn to get along in the family is the way they usually try to get along in the world.

> Gillian's family was always busy. When she was a toddler, she spent the day with a neighbor who took care of several preschool kids. When her grandma came home from work at the end of the day, Gillian wanted to play. But Grandma had to take care of the family. She didn't have time to play. So she put Gillian in the playpen and turned on the television. Her older brothers and sisters always had homework to do or friends to play with after school. They didn't have time for Gillian either.
>
> Gillian learned that if she made a fuss she could get her grandma to notice her. And she learned that a sure way to get her brothers and sisters to notice her was to pick a fight. They didn't ignore her then! That's the way Gillian has been all her life. She seems to keep looking for attention and raising a fuss to get it.

When a family is **unhealthy,** it is hostile and uncaring. When parents want to keep complete control, children have a tough time gaining freedom. Often they stay dependent on their parents. They are afraid to try to make it on their own. Or they begin to wage an emotional war on the family in order to break free. They rebel and defy parents, sometimes all their lives. They want to be free, but they can't seem to get free.

When a family is **healthy,** it is loving and caring. Most of a child's needs are met. Children begin to outgrow their dependence on parents. They gradually assume more and more responsibility for themselves and are ready for adulthood when it comes.

Independence — "When will I be on my own?"

The older you get, the more you can do for yourself. As a teen, you may be completely capable of taking care of your own physical needs. If you had a job, you could take care of yourself. You could cook your own meals, or learn how. You have your own ideas about clothes, so you don't need somebody to tell you how to dress. And you have your own friends.

But the nice thing about family is getting to enjoy being a teen without having to take on adult responsibility. Most teens are still dependent on family to provide food, shelter, and clothing.

Besides, have you noticed how much you need family to take care of you at certain times:
- when you are discouraged,
- when you are sick,
- when you are lonely,
- when things aren't going right?

And have you noticed how much you need family to share important moments:
- when something wonderful happens,
- at special times like birthdays or holidays,
- when you need to be yourself?

You can get ready for independence by doing things like
- getting a part-time job,
- taking responsibility for spending or saving the money you earn,
- spending time apart from your family (for example, going to summer camp or visiting grandparents for a week).

You won't be truly independent until you stop behaving in ways that are childish. In healthy families, leaving childhood is a natural process. There are growing pains, like the times you think nobody understands you or the times your parents think you are impossible. But growing toward independence moves along.

In unhealthy families, the path is rocky. Sometimes children never outgrow being childish. They become adults who are still childish.

An important step on the road to independence is learning how to raise questions with your parents
- without attacking or being defiant,
- without being sarcastic,
- without complaining,
- without feeling guilty.

Parents are People, Too

A bunch of us were talking to Mrs. Mason after class. We were complaining about our parents. It got started because Ralph's mom let his little brother start staying up past 9:00 o'clock on weeknights. She hadn't let Ralph do that until he was in junior high. Ralph was saying how unfair that was. And the rest of us jumped in. We all have gripes about our folks. Then Mrs. Mason said, "Hey, give your parents a break! They're just people, too. Sometimes people make mistakes, even parents. Did you ever think they might be doing the best job they know how to do?"

That got me thinking about my folks. I guess I've never really thought of them as "just people." I've always thought of them as my parents—like they were supposed to have all the right answers and not make any mistakes.

You don't have to be 16 for all part-time jobs.

It is hard to see your parents as people. When you were young, your parents or the adults who took care of you were bigger than life. In your eyes, they seemed all-powerful. They provided for your needs. They judged your actions as right or wrong. They handed out rewards and punishment. They were always right. They always told the truth. They always knew what to do.

As a child, you had a hard time seeing that your parents could make mistakes. You may not have noticed that they didn't always mean what they said. And you didn't realize they didn't always know what to do. This makes it hard to see them in ordinary terms. You may see them as more perfect than they really are—then be terribly disappointed when you see them make mistakes. You may imagine that they expect more of you than they actually do.

You take a big step toward independence when you begin to see your parents, or the adults who care for you, as people. At some point along the way, you may learn to appreciate them as friends—not pals, but special friends. When you start to talk to your parents and find out what they think about and what really interests them, you are moving toward being independent from them.

Families Change

Being independent doesn't mean that you give up your family. It doesn't mean that you'll ever reach the point where you can't benefit from their advice. It doesn't mean you'll ever stop needing their love. But it does mean a new relationship with family members.

The fact is, you may fuss with your brothers and sisters a lot, but you probably also share a lot. As you become more independent, you will find that sharing is easier. And you'll learn to value your brothers and sisters as persons.

You may get frustrated with your family, but you probably don't stop loving them. Your family can embarrass you, but you care about it. As you become more independent, you will be able to value your family even more. Your family helps you to learn that true freedom requires responsibility. The more you pitch in and help, the more you become involved in family decisions, the more ready you are for independence. It is give and take that makes a family work.

To be truly independent, you must outgrow your childhood dependence on your parents, or adults who take care of you, and on your family. You have to be able to shift your loyalty to someone else to start your own family. And you must be prepared to take on the role of a parent. If this goes well, you will be more than independent, you will be an interdependent adult.

Interdependence — "We take care of each other!"

Becoming independent of your family is not the same as throwing them away. A person who is truly free is interdependent. When you are interdependent, you can live away from your family, but love them. You can cherish the values they have taught you, but stand up for the values you have claimed for yourself.

The key to interdependence is the freedom, desire, and ability to take responsibility for
- your own thoughts,
- your own feelings,
- your decisions.

Then you can share with others because you have something to share. You may want to ask your family for advice. But you won't let them dictate what you should do. You will be able to make your own decisions. You may always respect your family's values, but you will act mainly on the basis of those values you have taken for yourself, many of which will probably be those of your family.

SECTION REVIEW
STOP AND REFLECT

1. List some things you would change about your family if you could.

2. Why would you want to make these changes?

3. What kind of family do you live in? Who are the members?

4. In what ways are you still dependent on your family?

5. How are you becoming more independent?

6. What are some ways you might accept your parent, or the adults who care for you, as people?

7. Do you know of a family that is a good example of what you feel a family should be? Describe that family.

8. Ask your parent the questions above about their families when they were growing up.

SECTION 2

STRENGTHENING YOUR FAMILY

Some people think that the idea of family as two caring companions is on its way out. They say that the idea of two people getting married and sticking together no matter what is outdated.

Family as a group of people who are "stuck" with each other probably ought to be outdated. But family as two people who are caring companions, who respect each other, and who are loyal to each other, will never be outdated. It represents an ideal that is totally in keeping with the best of American values.

Recognizing Healthy Families

Why do some families seem to have so many problems? Why are some couples still together after 20 years of being married while others divorce? Why do some single parents seem to be able to provide a caring home while others seem unable to get along without a partner? Why do some kids seem to like their families while others can't wait to escape?

Loving, caring families don't just happen. There isn't one magic formula for building a strong family, either. There are many different kinds of healthy families. But there are some things that healthy families seem to have in common. Healthy families

- **care about family.** They stick together even when times are tough.
- **communicate with each other.** They listen and work to understand each other.
- **support each other.** Each member of the family is important for who he or she is, not for what he or she looks like or does.
- **trust each other.** And they keep promises they make to each other.
- **show respect for each other.** They are honest about their feelings, but they don't dump their worst behavior at home.
- **allow differences of opinion.** They disagree without attacking each other.
- **take time for each other and for the family as a whole.** They share cele-

It's okay to show your friends that you like your brother. What values do you think are being reinforced here?

brations and traditions, and they find ways to have fun together.
- **include everyone in important family decisions.** They discuss issues that affect the whole family, like finding a new house or apartment or deciding where to go on vacation.
- **pitch in and share responsibilities.**

Loyal Families — "I stand up for my family!"

Families show their loyalty by sticking up for each other.

John is on the high school baseball team. His little sister, Lorie, is always following him around. Sometimes John gets impatient with her. But once, after a big win, the team was planning to go get burgers and celebrate. Lorie ran after John. Mike, one of John's friends, commented, "I don't know how you stand it having a brat like her for a sister!"

John turned to Mike, "She's really not so bad." Then he walked over to Lorie and gave her a big hug, in front of everyone.

Caring Families — "How can I help?"

How do you show you care for other members of your family? Do you give somebody an unexpected hug once in a while or help out around the house without being told? Here are some ways to show you care and to strengthen your family:

- **Show appreciation for things done well.** "Thank you" works at home, too!
- **Encourage others,** even your parents. "That must have been a hard job — you really stuck with it."
- **Show concern when a family member has a problem** and look for ways to help. "I see something's bothering you. Can I help?"
- **Stop criticizing** in a hurtful way. If you must criticize, be tactful and constructive. Don't nag.
- **Try to put yourself in the other person's shoes** and see things from his or her point of view. "Are you feeling that I get to go first all the time? That must seem unfair!"
- **Start conversations** about something the other person is interested in. "You've been working on that cross-stitch sampler for a long time, Mom." "I see you've got quite a paper-clip chain started, Delvin."
- **Listen** to what the other person says about his or her interests and follow up. "How come you like cross-stitch so much?" "So do you have to find the paper clips for the chain, or can you go out and buy a box of them?"

Caring family members aren't afraid to show affection for each other. An unexpected hug can make someone feel really special.

A loving, caring family believes that family is important. They stick together, come what may. This is **loyalty,** being faithful to someone. It shows commitment. When you are loyal, you don't betray trust.

Loyalty means doing what is best for the family. It means not gossiping about family members. It means facing tough issues. It means not hiding problems — even big ones such as addiction, abuse, and teen pregnancy.

- **Do things to strengthen the other person's interest or skill.** "The lady at the craft shop said a lot of people like to use these needles for cross-stitch, so I got you a package." "My friends and I found these paper clips for you to add to your chain."
- **Do things the other person's way** part of the time. "Okay, Dee, if you want to watch cartoons first thing, I'll watch them with you."
- **Keep your promises,** and don't make promises you can't keep. "Hey, Justin—I promised we'd go to the park after dinner, so let's go!" "No, I won't promise to watch videos with you when we get back because I have a lot of homework. I don't know if I'll have time."

Growing Families — "Look what I can do!"

In healthy families, everybody belongs. All the family members know they are in it together. They help each other to grow from dependence to independence. They practice the kind of interdependence within the family unit that will help the children be capable of choosing interdependence when they are adults.

When all family members help out with the chores, the family unit is strengthened.

When Families Need Help

Since individuals who make up the family aren't always perfect, it is impossible to have a perfect family. It is not, however, impossible to have a loyal, loving, healthy, caring, respectful family. Building a strong family takes work by every member of the family. In a healthy family, one member's problem is the family's problem.

When you have a problem try to talk things over with your parents. Choose a quiet time that's convenient to everyone.

Fighting — "Let's fight fair!"

Even healthy families get into fights sometimes. In unhealthy families, there may be constant fighting, bickering, yelling, and hitting. Families need help with fighting when the family members have developed a pattern of negative conflict.

Sometimes we do things that make conflict even worse. Treating others as if they had no sense at all or going on a rampage just makes matters worse. You can learn to fight fairly.

Using the approaches that make conflict positive works at home, too. Some things that help keep a fight fair are

- choosing the best time and place to discuss the problem.
- focusing on the problem, not on hurting other people.
- not exaggerating. Don't threaten.
- being as honest as possible.
- getting your feelings out in the open, but not dumping them all over everybody.
- giving good reasons for what you think—using "I statements."
- trying to see the other person's side, even if you think that person is wrong.
- agreeing to a solution everybody can live with.
- not giving up if the first solution doesn't work. Try something else.

Professional counseling can help families solve serious problems. Sometimes the hardest part is taking the first step — walking in the door.

Not every parent will be open to fair fighting. And not every parent who is open will be open all the time. Sometimes a parent will feel very strongly about a particular matter and announce, "There is no discussion necessary."

What do you do then? You can talk to another adult family member or to an adult outside the family. Get advice about helping your parent see and appreciate how you feel about the issue. Your school's guidance counselor, a clergy member, or a teacher can help.

It is very important that you talk with someone. Talking with people who have more experience than you do really helps. There is always more than one side in a conflict. It is as important for you to see your parents' side as it is for your parents to see your side of things. Talking it out with someone will help you do that.

Trouble — "Who can help my family?"

No family totally escapes troubled times. The next chapter deals with some of the kinds of trouble that can strike any family and discusses how to get help.

Your family will surely go through troubled times. All families do. Your family may be — or become — unhealthy, too. When there are problems in your family, every family member suffers. To help the whole family, at least one family member has to admit that there is a problem and has to start getting help. That family member can be you.

Getting help for your family member will enable you to become the independent young person and interdependent adult you are meant to be. Getting help shows you how much you value family, how much you care about family members, and how responsible you are.

Getting Help

A good place to look for help for serious problems is the **extended family**, other relatives who may not live with you. The extended family knows and loves your family. It is often easier to confide in them than in someone outside.

But sometimes even extended family members act to hide the problem rather than helping with it. This is because deep family problems often get passed on from generation to generation. That is, the problems you are experiencing in your family may be the same problems that were passed on to them by a family that didn't get help.

This is why you sometimes have to go outside the family to break the pattern of an unhealthy family. Chapter 7 gives examples of where to go for help when a family is in trouble. Chapter 9 on recovery also makes some valuable suggestions about how to get help.

Families Are Worth Fighting For

Fighting to get help for your family will make you feel better about yourself and about your family. Fighting to help a friend get help for his or her family will make you feel good about yourself, too.

Helping your family to be more healthy today will help you with your own family in the future. It will help with relationships outside the family, too. When you work at being a strong, healthy family member, you are becoming a person who is ready to face the future. It is a way of growing up caring!

SECTION REVIEW
STOP AND REFLECT

1. Tell how your family shows loyalty for various family members.

2. What do you think the best family would be like? How would members treat each other?

3. Think of some ways you could show caring for each member of your family.

4. Who could you go to if your family were in trouble? List the names of some people who would help.

CHAPTER 16 REVIEW

Putting Your Values To Work

STRENGTHENING YOUR VALUES

Reread *A Teen Speaks* **on page 355. Then answer the following questions.**

1. What do you think Mike values? His mother? His stepfather?
2. What values does Mike seem to be lacking? His mother? His stepfather?
3. If you were one of Mike's friends, how would you describe Mike's family?
4. What differences seem to exist between the things Mike cares about and the things his family cares about?
5. How do you think Mike wants his family to behave? What does this say about Mike's values?
6. What might Mike do to improve the situation? What might his family do?
7. If someone were to criticize Mike's family, how do you think he would respond?
8. If Mike had a chance to change his family, do you think he would? Why or why not?
9. What are some of the good qualities of Mike's family?

INTERPRETING KNOWLEDGE

1. Create a collage that describes your family. Share it with a family member.
2. Write a poem or song that describes the special qualities of your family.
3. Think about a song that reminds you of your family. What song did you choose? Why?

SHARPENING YOUR THINKING SKILLS

1. Give examples of "restrictive," "prescriptive," and "internal" rules that you follow.
2. In what ways are your family members interdependent?
3. Compare the person you are now to the person you were five years ago. In what ways were you more dependent on your family then?
4. In what ways is your family different from any other?
5. In what ways has your family taught you to be independent?
6. Name three characteristics of healthy families.
7. Name three characteristics of unhealthy families.
8. Name five things you could do to help keep a fight "fair."

APPLYING KNOWLEDGE

1. Think about a family television show that you watch regularly. How is the family in the show like real families you know? How is it different?
2. Plan an activity to share with another family member or with your entire family. After completing the activity, present a short oral report to the class. Explain why you chose the activity you did, what you learned about the other person or persons while doing the activity, and what you enjoyed most about the shared time.

CHAPTER 16 REVIEW

Putting Your Values To Work

PRACTICING DECISION-MAKING SKILLS

Read about each situation. Then answer the questions.

Situation A: Carol's parents are divorced; Carol and her two younger sisters live with their mother. In order to support the family, Carol's mother works very long hours. As the oldest, Carol is expected to take care of her younger sisters while her mother is at work. Sometimes Carol feels like a second mother.

One of Carol's sisters, Jean, begins to have a hard time in school. She doesn't do her work and is rude to the teachers. She has even skipped school a couple of times, and her grades are getting worse and worse. The principal knows Carol's mother works long hours, so she discusses Jean's situation with Carol and asks Carol to relay the discussion to her mother. Carol promises to do this. However, when her mother gets home, Carol sees how tired she is and how many problems she has to deal with already. Carol isn't sure how to handle this.

1. If you were in Carol's place, what would you have done or said?
2. Is it Carol's responsibility to deal with this situation? Why or why not?
3. What do you think Carol is feeling?
4. What are some actions Carol could take?

Situation B: Same as Situation A. This time, Carol happens to overhear an argument between Jean and one of her teachers. Jean starts to cry but continues to lash back at the teacher. Finally, the teacher says, "Well, I'm going to tell the principal and your mother about your attitude." Jean just shrugs her shoulders and walks away. Carol goes up to the teacher and says, "I know she isn't the best student, but you didn't have to talk to her the way you did." The teacher says to Carol, "Why are you sticking up for her? I know you're her sister, but you really don't know the whole situation. I'll handle this with your mother."

5. What decision needs to be made?
6. What are the main challenges facing Carol?
7. What are the possible actions Carol could take?
8. What are the possible consequences of each action?
9. What decision should Carol make?
10. How might Carol rehearse this decision?
11. If Carol were to take the action you think best, what do you think the result of her actions might be?

Situation C: Same as Situation B, but this time Carol catches up with Jean. Jean says that the teacher accused her of cheating on a test. "Honest, I didn't look at anyone's paper! The teacher just assumed I did because I got a B. Now she's going to tell Mom and she'll get upset with me. But I didn't do it, I didn't!" While she is listening to all of this, Carol is remembering that Jean has gotten nothing but very low grades all year. And yet, Jean sounds as though she is telling the truth. Carol feels stuck.

12. What should Carol do now?
13. What choices are open to Carol?
14. Carol is faced with being loyal to her sister and trying to be objective at the same time. What advice would you give her?

CHAPTER 17

Help For Troubled Families

A TEEN SPEAKS

The whole family was sitting around before dinner one night. Suddenly, Mom said, "Oh, something's burning! There goes the dinner!" But it wasn't the dinner.

Dad opened the back door where the smoke seemed to be coming from. When he did, there was something like an explosion. I'll never forget how he looked when he said, "The house is on fire. We can't get out this way."

Mom and Dad led us out and away from the house. Dad wanted to go back in and get some of our things, but Mom wouldn't let him. She said the most important things we had were our lives and they weren't worth the risk. It was a good thing he didn't go back. The whole house was gone in such a short time. We lost everything.

At first I guess we were kind of in shock. Neighbors came and helped. I remember somebody wrapping a blanket around me. It was cold outside and none of us had on our coats. Then we went to stay with Grandma.

People were really good to us. They gave us a shower with a lot of furniture and stuff for the house when we get one. We got a lot of old clothes, too.

I don't mean to sound ungrateful, because I do appreciate their caring. But sometimes I just wish I could have a new shirt or something that somebody in this town hasn't worn before.

It's been almost a year now, and we're still staying with Grandma. We'll be here until we can get another house. I love Grandma, but miss my things so much. And now that we're with her, it's like we're a different family.

SECTION 1

ONE FOR ALL AND ALL FOR ONE

a family is like a living thing. It has its own personality. No two families are alike. Each family has its own way of doing things and its own network of relationships between members. Each family has its own history and its own future.

There is something about a family that can't be explained by looking at its members alone. A family is more than a collection of people who live in the same space. This is why when trouble comes to a family — trouble like a home burning to the ground, a sudden illness, or death — it changes the family. When trouble comes to a family member, it changes the whole family, too.

Families Face Trouble Together

When one family member has a problem, the whole family has a problem. To understand what is happening to a family member who has a problem like getting along with friends or like drug abuse, you have to understand how the family works.

> Nick's sister has anorexia, an eating disorder. Nick thinks that the problem must come from the group of kids she hangs out with. Their idea of beauty is to look like a slim fashion model. They have teased Nick's sister about being too fat. Now that she has anorexia, it must be their fault.

In reality, his sister is trying to deal with her feelings in inappropriate and destructive ways. For her to get well, the whole family will need to work together.

> Jeb is failing at school. He says he hates school and the work is boring.

It may be that things are dull at school, and Jeb hates it. But the way he has chosen to deal with the problem comes from what he has learned about dealing with problems in the family. For Jeb to become more responsible about school, the whole family will need to work together.

The point is, any problem with a family member relates to the whole family. This is true in every family—even healthy families where there is a lot of love, care, and support for members. The whole family has to take the grief when one member gets into trouble. The whole family bears the burden when one member has a problem. The opposite is true as well. When one member achieves success—like making an A on a term paper, the whole family shares his or her pride.

Families In Balance— "It's under control!"

Every family faces stress and difficulty. That is part of life. Family members learn how to work together so the normal stress of living doesn't get out of hand. The way they work together keeps some kind of order or balance in the family.

Suppose you have a really rotten day—it can happen to anyone. When you get home, you're in a terrible mood. Your little brother starts getting after you to play with him, and you blow up. He starts crying. You have caused a disruption in the family order. Your older sister says, "What got you in such a foul mood? Better go cool off!" Your grandmother picks up your little brother and comforts him. She tells him to leave you alone for a while. Later you get yourself together and apologize to everyone. Meanwhile the family has dealt with your stress. The balance and order among family members has been kept in a healthy way.

But sometimes, a family can work together to keep the balance in ways that are unhealthy.

> Rob's Aunt Ella lives with his family. She is known for her terrible temper and bad moods. His mom makes excuses for her, "That's just Aunt Ella. She's always been that way." The whole family is organized around Aunt Ella's moods. Rob's parents know that if dinner is late, Aunt Ella will create a fuss. So, no matter what, meals are served on time, and there are no excuses for being late. Aunt Ella gets nervous and upset when there is too much activity in the living room. So nobody gets to use the living room, except to

How can the trouble of one family member affect the other family members?

sit. Aunt Ella controls the family with her bad temper. All the other family members cooperate to keep Aunt Ella happy so they won't have to deal with her anger when she is crossed.

Family balance is maintained. But to keep that balance, everybody has to tiptoe around Aunt Ella. Other family members can't get angry and act out. Everyone has to hold back feelings and keep behavior in check for fear of Aunt Ella's reactions. This makes the whole family unhealthy.

All families have a way of keeping things in balance. It may be a healthy balance in which everybody has an opportunity to express feelings. Or it may be an unhealthy balance, where everything is organized around one "difficult" member. But a balance is achieved. When things happen to destroy the balance, a family can be in stress or in serious trouble.

Families Out Of Balance — "This is out of control!"

Some of the things that can happen to destroy family balance and create unusual stress and trouble are
- disaster or emergency conditions,
- change,
- depression,
- tragedy,
- violence,

Any of these can happen to any family. Healthy families usually handle crises better than unhealthy families. But even the healthiest of families may need support when family balance is destroyed.

You can't always keep bad things from happening. And you can't expect somebody to carry all the burdens for you. Knowing how to get help is a way of caring for your family.

Where to Find Help

When the balance in your family is destroyed, all the members need help. By asking the right people for help, you can be the one who starts you family on its way back to a healthy, balanced life.

These are the people who are most likely to be able to help you:
- **your parents.** Your parents may be able to help you understand and deal with a family problem. They may know which social agencies and professionals can help the family. However, if your parents are too involved in the problem themselves, they may not be ready or willing to discuss it with you.
- **members of your extended family.** Your grandparents, aunts, uncles, and other relatives may be able to offer help and advice.
- **a member of the clergy.** Most members of the clergy have been trained to help people understand and deal with family problems.
- **a school counselor or trusted teacher.** There may be an adult at school with whom you feel comfortable talking. A counselor or teacher will

Children In Global Crises

In the United States, family life may be disrupted by personal conflict, money problems, substance abuse, and work stress, among others. In many countries, family life is destroyed by war, political upheaval, and social turmoil. The greatest tragedy is what happens to children, who are the future of each nation and of the world.

Children have become the weapons of war. In Cambodia in the 1970s, young teenage boys were separated from their parents and taught to oppose traditional family life. They became soldiers in an army that killed millions of fellow citizens.

Children have also become the victims of war. They have been maimed by booby-trapped toys as acts of terrorism, while parents were left unharmed. In the early 1980s, thousands of Guatemalan children were killed in civil unrest.

What kind of adults will the survivors of these wars become?

also be able to contact social agencies or professionals who can help your family.
- **your family doctor or school nurse.** A doctor or nurse can be especially helpful in dealing with problems that involve illness or injury. A doctor or nurse who cannot help you directly will probably be able to put you in touch with agencies that can help.
- **a youth worker.** A local community organization, such as the YMCA, the YMHA, or the Red Cross, may have a youth worker. This person's job is to work with young people as individuals and in groups. A youth worker can help you identify and understand your family's problem and can help you contact social agencies.

A guidance counselor or a trusted teacher may be able to help you with a family problem. If not, she can refer you to the proper professional.

The Code of Silence

In order to help yourself and the rest of your family, you have to admit it when there is a problem. The hardest step is telling yourself the truth. You have to be able to admit, "My family is in trouble. I am in trouble."

Families in trouble often follow an unwritten code of silence. Family members tell nobody—not even each other. So they don't get help. Through secrecy and silence, they keep family problems going.

When one person in a family sees or experiences a problem, that person has to find the strength to admit it, own up to feelings, and confront the problem.

This is true whether the problem you are confronting is ordinary, as when your feelings are being ignored, or extraordinary, as when a parent is abusive or a brother or sister is in deep trouble. You may have to say, "It really did happen. They decided my grandpa should have my room without even asking me. Now I have to share my little brother's room." Or you may have to admit, "It really did happen. My mom really did whip me with a wooden coat hanger." Or you might have to say, "It really did happen. My sister stole all of the grocery money to buy booze."

Then you need to admit your feelings. They've very real. If you don't acknowledge them you may not be able to confront the problem. You have to say, "My feelings count too! I need to tell my family how I feel. I love my grandpa but its not fair for them to give him my room without even talking with me about it." Or you admit, "It really did hurt. Hitting me with a wooden coat realize that everyone needs to have a part in making decisions that affect them.

Or, you'll help your mom avoid dealing with her problem by pretending she didn't whip you with a wooden coat hanger. And chances are she'll do it again to you or one of your brothers and sisters.

Or you'll help your sister hide her alcohol dependency.

Common Family Problems — "What can I do?"

Some family problems are easy to recognize. If a hurricane destroys your house, it's easy to see that your family's life will be changed. Other problems aren't so easy to see. If your older sister moves out or your grandfather moves in, everyone can recognize that life at home will be different. But they may not be prepared for the strong feelings that family members will have as a result. Changes like these are bound to throw your family out of balance—for a while.

Some of the problems that families commonly run into include:
- **decision making**—who decides what,
- **responsibilities at home**—who does what,
- **space and privacy**—when people need to be alone,
- **relationships**—in and outside the family,
- **conflict**—how differences are handled,
- **resources**—who gets what.

Silence doesn't solve anything. Talking can be tough, but it helps.

hanger was wrong." Or, "My sister is hurting me. She is hurting the family."

To take action in solving big or little problems you need to tell yourself, "I have to get help so I can heal. I have to quit hiding from the problem. My family needs help so it can heal."

This is a tough step. In most families everyone has been taught to act caring. So when your feelings make you seem uncaring, like not wanting your grandpa, it is hard to speak up.

You are expected to be polite to your parents and to respect them. You imagine you shouldn't jump in and tell your parents they have a problem, or you have a problem with a decision they've made. But if you don't help, you will keep that problem going. You'll harbor bad feelings toward your grandpa and your parents. And your parents may not

Many healthy families settle their problems by holding regular family meetings where all members have a chance to have their say and hear what other family members think. All members have some input on decisions, even though a parent may have the final say.

Disaster or Emergency Conditions

A family member may become sick with leukemia or AIDS. Several family members may be in an automobile accident together. A flood, a tornado, or an earthquake may strike. Any number of life-threatening events can place a family under serious stress.

When there is an emergency, a family often moves into high gear and begins working very efficiently to take care of the crisis. The family members create a new, emergency balance. Sometimes they don't let themselves express feelings of anger, sorrow, fear, or guilt because they are too busy dealing with the emergency situation. Often they don't get enough rest because they are giving every minute to keeping the emergency balance going. Then, all of a sudden, a family member simply can't cope. One member breaks down under the stress and gives up or begins to let out a flood of feelings.

> Taylor's dad was in a terrible accident. He was in a coma for several weeks. Everybody in the family pitched in to help. Milly and Don, who were in high school, began doing all of the cooking and cleaning at home. Taylor helped with the younger children. Their mom spent all her waking hours at the hospital with Dad. Taylor could hardly stand to see his dad so helpless. And it was hard to see his mom so worried.
>
> One afternoon when he was at the hospital, Taylor asked, "Mom, what if Dad doesn't get well?" Without warning, his mother turned on him. "How dare you say such a thing! You're just waiting for him to die so you won't have to do so much." She began to cry uncontrollably.

Taylor's mother couldn't cope at that particular moment. She didn't mean any of the terrible things she said to him. But she was exhausted from her long hours and worry. And she dumped her fears and worries on Taylor.

Everyone in Taylor's family needs help. The family emergency is wearing them down. Watching someone you love suffer for a long time can be shattering. Sometimes it creates resentment toward the sick or injured family member. It's important that these family members talk about their feelings—to a trusted friend or among themselves.

A disaster is often damaging to many people at the same time. Families are deprived of the ordinary materials and information they need to carry on their daily lives. A disaster spares nobody. People are overwhelmed emotionally, and often show signs of shock—lack of emotion, slowed reaction and movement. When families face disaster,

A family move can create all kinds of problems as well as adventures. What might a teen do to make a move more positive?

they need support services from agencies such as the Red Cross.

Families need immediate support in dealing with the aftereffects of the disaster, such as cleaning up a house damaged by flood waters or finding a new place when a home is destroyed by a hurricane. But families also need long-term support in dealing with the emotional effects of a disaster. Relief agencies are usually on hand when a disaster strikes a community. They provide short-term assistance and usually help people find ways to get long-term assistance.

Change

Changes can keep life from becoming routine and dull. Every family changes as members grow. For instance, family membership can change if an older brother or sister leaves home or grandparents move in. When family membership changes, new rules of behavior have to be worked out. These range from who uses the bathroom when to who makes choices about what money is spent for.

Sometimes changes are planned—for example, the family may decide to move to a new apartment. Others are unexpected—a parent can get transferred to another city or lose a job.

Moving means leaving behind much of what you know. It means adjusting to a new place, new school, new neighborhood, and finding new friends. The loss of a job means a loss of income. Making a move or finding a job can be a challenge and fun for the whole family. But these changes can also create intense and painful feelings within the family.

Howard's family had always been solid. Both his parents had jobs, and everybody helped at home. Sometimes it was hard to spend time doing housework when he wanted to be out. But Howard knew that he was needed. He knew that both his parents enjoyed their work. Since both parents worked, the family could enjoy extras, like taking vacation trips.

All that ended when his mother lost her job. Now she can't find work, and the family paycheck has been cut in half. All of a sudden there is no money for extras. Sometimes there is hardly enough to buy groceries.

Howard's mom cries a lot now. She blames herself for losing the job. She keeps looking, but nobody needs her. She says she's been replaced by machines.

Howard's family is in trouble. People are often embarrassed to admit that they don't have enough money to get by. They don't always know where to turn to get new training when the work they do is no longer needed.

When a family is having trouble dealing with any disaster, emergency conditions, or change, the whole family is thrown out of balance. If you and your family are trying to adjust to a move, a lack of money, or a change in family membership, it is important to get help. Your school counselor or a member of the clergy are good resources for dealing with any troubling changes. They may not be the people who end up giving you the most help, but they can help you find the people who will.

SECTION REVIEW
STOP AND REFLECT

1. What keeps your family in balance?

2. Describe a way that family balance has been threatened in your family. How did family members deal with it?

3. What trusted adult would you probably talk to about a family problem? Why do you think that person would be able and willing to help you?

4. Tell about a disaster or an emergency experienced by a family you know.

5. What kind of help did the family get?

6. What other resources might they have used?

7. Describe a change experienced by a family you know. Tell how the family coped with the change.

SECTION 2

TOGETHER IN TROUBLE

healthy families can become stronger as they face troubled times. But they may need support to do so. Unhealthy families can be seriously harmed by trouble. They need support to deal with trouble and to develop into a healthy family.

Facing Additional Trouble

Many different kinds of trouble can throw a family out of balance. Some problems are especially difficult because they can be hard to admit. Remember, though, that ignoring a problem will not solve the problem—it will only get worse. When your family has to face trouble that involves depression, loss, or violence, you can help. You can admit that there is a problem, and you can find help for your family even if nobody else will.

Depression

Everybody feels depressed some of the time. And people can go through mild depression from time to time. That is, they find it hard to get their work done because they feel so discouraged. They feel tired, irritable, sad, anxious, and full of doubts.

When you or a member of your family is depressed, the whole family suffers. If depression goes on for more than a week or two, it becomes serious.

> Becki's uncle has lived with her family since she can remember. He has always been moody, but lately he just wants to sleep all the time. He falls asleep in front of the television. He won't come to the table for

dinner. He seems distracted and won't carry on a conversation. When he does talk, he is always putting himself down.

Becki talked to her dad about it. He told her, "Never mind Uncle Fred. He's just in one of his moods. Just be as cheerful as you can. It will make him feel better."

But Uncle Fred is not getting better.

When somebody is depressed, being cheerful doesn't help. Leaving that person alone doesn't help. Neither does denying that there is a problem.

Uncle Fred's problem affects the whole family. He needs the family to help him get well. He will have to start by working on the depression. When depressed people begin to work on the problems that have made them depressed, they can be hard to live with. They are often angry and hurtful to family members, even though they don't mean to be. They seem to be getting worse, not better. The family may need to have some family counseling in order to cope with Uncle Fred.

To get help, Becki has already done one important thing: talk with her dad. But, since he refuses to see the problem, there are some other alternatives. Becki could
- talk with the family doctor. Doctors know about depression.
- talk with the school nurse. The nurse will be able to help Becki take steps to get help. A person not specially trained can't help with

Your school nurse is trained to help you with personal or family problems.

depression. That is why it is so important to talk with a doctor, nurse, social worker, or psychologist.
- talk with a member of the clergy.

Loss

Some of the important losses that families may experience are through suicide, divorce, and death. All three are tragic because they bring an end to the family as the family members have known it. But families can learn how to overcome such tragedies.

Suicide

People commit suicide when they feel completely hopeless. Most are suffering from extreme depression. Unless you are an expert, you probably can't tell when a depressed person is likely to try to commit suicide. That is why when someone you know is depressed, it is critical to get help for that person.

It is also important for the whole family to get help. People who are in danger of suicide need the love and support of family members.

> Maxine's older sister Karen had been depressed for months. She was unhappy. She no longer did anything with her friends. She was tired all the time. Maxine used to admire the way Karen looked. But Karen didn't seem to care at all anymore. Sometimes she didn't even comb her hair.
>
> Then, all of a sudden, Karen seemed to become more like herself. She was almost cheerful. Maxine was relieved until she stumbled across her sister's open journal. She didn't mean to snoop, but something caught her eye and frightened her. The journal read, "I can see myself dying without anybody to care. Each day is worse than the last. What is there to look forward to? What is the use?"
>
> Just as Maxine turned away from the journal, Karen came in. "What have you been doing?" she asked. Maxine couldn't lie to her. When she told her what she'd read, Karen said, "Oh well, I was in a bad mood. But I was just being dramatic. I'm really okay now."

Suicide

A family needs support when a member of the family is thinking about committing suicide or when a family member has actually committed suicide.

- If someone you know threatens suicide, take the threat seriously.
- If there is no help, take the suicidal person to the emergency room of a hospital. The doctors and nurses there will see that the individual gets the right kind of help.
- Don't hesitate to call for emergency help if the person will not respond to your talking. Even if you're not sure how serious the threat is, get help.
- Check the yellow pages under "Suicide"—There may be a local crisis hot-line.

Maxine's sister seems to be better, but that does not mean she is out of danger. In fact, people who decide to commit suicide often feel a sense of relief and begin to act more like themselves. They often say that they are okay. But they are still in great danger.

Many schools have support groups for students whose parents are going through divorce. Can you think of ways a support group might be helpful?

Maxine needs to see that her sister and the family get help. She can help by talking with her parents about what she has seen in the journal.

The family can help by

- **listening.** They should talk with Karen and tell her that they love her and are worried about her. They should ask her to tell how she is feeling. And they should try to understand what she says without getting upset or acting horrified.
- **asking.** They should ask Karen if she plans to kill herself. A person who is thinking about suicide is often relieved to have a chance to talk about it. They should ask her to wait to do anything until the family can meet with a doctor, member of the clergy, or therapist.
- **taking the problem seriously.** Even if Karen is very low-key and says she is okay, they must take her seriously.
- **getting help.** They should do something specific. They might call the doctor, a member of the clergy, or a therapist so that Karen knows something is being done to help her.

Sometimes the code of silence in a family can make it difficult to get help for a family member. If Maxine's parent's don't want to go outside the family for help, Maxine needs to find the courage to do it herself.

If someone in the family commits suicide, family members have to come to grips with the tragic loss. Often they blame themselves. They think about what they might have done to keep the tragedy from happening.

A family recovering from a suicide needs to deal with the death of the family member. But the problem of suicide presents additional challenges. Support groups and counseling programs help many families who have been through this terrible experience.

Divorce

Divorcing parents and their children—young or old—have to deal with loss. They all face the loss of their family as they have known it. If one parent moves far away, the rest of the family must deal with the loss of that family member. Everyone must deal with the new economic arrangements—that is, property and income that was shared is divided up. Everyone has to adjust to new living arrangements.

Some family members who experience a divorce must also deal with anxiety, anger, and depression. Parents and children alike feel guilt and anxiety. Many need help in overcoming these feelings in a positive way that will lead to growth.

Divorcing parents are often too self-involved to think about getting help for their children. But if your family goes through a divorce, you need support. Talking to close friends about what is going on at home can help a lot. Talking with a member of the clergy, a school counselor, or a teacher can help, too. Adults will help you recognize that the divorce involves issues between your parents; it is not your fault.

Death

No family can escape death. Everybody who is alive will die sometime. When there is a death in the family, everyone experiences grief. People go through grief in very different ways. Family members need to make room for each other to mourn in their own ways.

Many people approach death by denying that it will happen. They don't want to believe that it can be true. With love and support, they come to recognize the reality of a death and are able to express their anger and frustration over it. This seems to occur before they are ready to make peace with the fact that death has happened or will happen soon.

> Jasmine's dad had a sudden heart attack and died. At first Jasmine couldn't believe that it had happened. But her grandparents, uncles and aunts, and cousins were all there to support the family. People from her church brought in food for them. There was so much going on, Jasmine hardly had time to think.
>
> A week after the funeral, everybody has gone home, and the house seems strange and empty. Jasmine's mother just sits and stares out the window. She seems unable to figure out what to do next.
>
> To make matters worse, Jasmine's little brother has become impossible. He could always be a pest, but now he is worse than ever. He is demanding. When he doesn't get his way, he has temper tantrums. Jasmine has found herself losing her temper with him, which hasn't helped at all. Without her dad there, everything seems to be falling apart.

Once all the activity surrounding a funeral quiets down, a family is left alone to face the loss. This may be when they need the most help. In addition to the emotional support and loving care that people need to overcome grief, they may have some practical needs.

Sometimes it is hard to figure out what to do next. There are hundreds of details that a parent must deal with when his or her partner dies. Flare-ups among family members are common because it is such an emotional time.

To get help, Jasmine can talk to a good friend, a member of the clergy, or a teacher. Sometimes it takes someone outside the family circle to provide help when a family is grieving.

Talking with a good friend can help, but this should not replace talking with an adult who can help Jasmine find support for her family. Finding the right support can make a great deal of difference in how a family deals with the crisis of death.

Support comes in different forms. For some, books are very helpful.

Violence

When a family member is the victim of a violent crime, the whole family is affected. Violent crimes occur when someone is attacked and beaten or raped.

Rape

Rape is an act in which someone is forced to have sex. It is above all things, a violent and hostile act.

When a family member is raped, it is a crisis for the whole family. Sometimes a rape victim becomes afraid, anxious, extremely angry, and very emotional. Or a rape victim may seem to be calm and subdued. In either case, family members need to recognize that there is a crisis. The whole family needs to help.

The victim may feel guilty. The victim may think he or she used poor judgment and caused the rape to happen. Weeks or months after the rape, a victim can begin to be afraid to be alone or have nightmares about violence.

> Peggy was coming home from play practice after school one evening. The elevator in her apartment building hadn't been working. So she wasn't too surprised when this guy who was hanging out in the lobby said, "You'd better use the stairs—the elevator is out."
>
> Peggy had already started up the stairs when she realized that the guy was following her. She started to run and he grabbed her. He put a knife to

her throat and told her to keep going or he would cut her. He took her to the roof of the building and raped her.

After he left, she kept sobbing. She felt so frightened and ashamed. Finally, when she got her strength, Peggy went downstairs to the apartment. Her brother Hank saw her come in and knew something was terribly wrong.

To help Peggy, Hank should
- **call 911 for emergency help.** If parents are not at home, Hank should call them after he has called 911. It is important to get medical treatment following a rape.
- **comfort Peggy,** but not let her take a shower or bath. She needs to go to a hospital as soon as an emergency unit arrives. It will be important for her to have a medical examination before she showers or has a bath.
- **call a rape crisis center** for family counseling. Sometimes family members become impatient and wonder why the victim can't let it go and stop thinking about the experience. Families need support in providing care for the victim and in dealing with their own feelings about the rape.

Parents usually take over and lead the family in a time of crisis. But knowing what to do in case parents can't be reached, or if parents are in too much stress to act, can help the family.

Assault

An **assault** is a violent attack. An assault victim has many of the same problems experienced by a rape victim, because the victim's life has been threatened. A person who has been beaten or hurt has to recover physically and also must deal with the emotional trauma caused by the attack. If someone in your family is assaulted, you should
- **call the police, 911 or the family doctor.** Medical recovery is always the first step, but the police must be notified about an assault.
- **call your parents or extended family members** for support.

No family wants to think about rape or assault happening to them. But terrible things do sometimes happen, and the family must be prepared to deal with them.

Violence Within the Family

Violent abuse is a real problem that happens in many families every year. Violent abuse can be physical. That is, parents or other family members may hurt children or teens by hitting them. Parents can also hurt children or teens by failing to provide adequate food, clothing, shelter, education, or health care.

Violent abuse can also be verbal. A parent may constantly berate, belittle, or put down a child or teen.

Violent abuse often happens like an explosion. Something happens inside the adult. It has very little, if anything, to do with what the child is doing at the moment, even though the child may think so. The parent explodes and strikes out.

After hurting the child, the abusing adult usually feels guilty and may be very loving and caring. But the abuse happens again later. A sickening cycle goes on with the family. Other family members may know about it.

Sexual abuse is a particularly frightening form of physical abuse. Any sexual activity that involves a child or young person and an adult is **sexual abuse**. The abuse may involve touching genitals or breasts. It may involve forcing the abused child to masturbate. Or the abuse may involve sexual intercourse.

Sexual abuse is a very difficult subject for most people to talk about. In fact, most people who are abused don't tell anyone. They usually feel as if they have done something wrong, dirty, or bad. They are left anxious and confused. They may be afraid that they will be blamed or hurt if they talk to anyone.

Sexual abuse can come from an older family member of the same sex or from someone of the opposite sex. An adult who sexually abuses anyone has very serious problems.

> Roland woke up early one morning and noticed his dad coming out of his younger sister's bedroom. He didn't think too much of it. He figured that Dad was helping her with one of her anxiety attacks. His sister had been acting really weird lately. She had started acting afraid of everything.
>
> Later Roland discovered something he didn't want to know. His sister was being sexually abused. It made him feel sick. He couldn't bear to think about it.

Sometimes members of a family will see that another member is being sexually abused. It is a terrible and shocking discovery. Usually, they try to pretend it isn't happening.

Ignoring the problem never works. The code of silence does not help anyone. Family members in serious trouble may not be able to help themselves.

To get help, Roland could talk to his mother. This may not work. The parent who is not involved in sexual or physical abuse may refuse to act. He or she may have become a silent partner in abuse. However, the parent who is not involved may want to act. Sometimes speaking to that parent is enough to push him or her into action. That parent may be able to recognize that there is a terrible family problem and that the whole family needs help.

If Roland's mother won't act, he shouldn't stop there. He could call Child Welfare Services and talk to a counselor about the problem. Child welfare workers are experienced in dealing with violence within families.

Or he could call the family doctor or see the school nurse. A person with medical training can see that families get the right kind of help.

The important thing is for Roland to get help. If you are the victim of violence in your family, or if you think another family member is, you have to ask for help. It is a way of being caring and responsible.

An abusing family member needs individual counseling or therapy. The entire family needs emotional support and counseling to help deal with the painful consequences of violence.

Keeping The Family Together

Violence is the kind of terrible problem that happens to some families. But some kinds of trouble will strike every family. Death and illness are two that no family escapes. And no family is safe from disasters such as fires, hurricanes, or earthquakes, which are random and life-threatening.

The best way to deal with any kind of trouble is to have a healthy family life. Most families can be healthy with some help. When family members work together to strengthen the family, they find that all of their relationships improve—even those outside the family.

SECTION REVIEW
STOP AND REFLECT

1. Describe someone you know who seems to be depressed a lot. (You don't have to use names.)

2. What could family members do to help the depressed member?

3. What would you say to a family member or friend who was talking about committing suicide?

4. Describe a loss that a family you know has experienced.

5. How did family members cope with the loss? What else might they have done?

6. Why do you think people have trouble talking about violence within the family?

This family, homeless due to an earthquake, found temporary shelter, food and assistance in an emergency center.

CHAPTER 17 REVIEW

Putting Your Values To Work

STRENGTHENING YOUR VALUES

Reread *A Teen Speaks* **on page 375. Then answer the following questions.**

1. What values do you think Craig's mother has?
2. What values do you think Craig's dad has?
3. During a fire or similar emergency, do you think your values would be more like the values of Craig's mother or more like those of his dad? Explain your answer.
4. What values did neighbors and other community members exhibit after the fire?
5. Have you ever helped an individual or a family after an accident or illness? If so, explain what you did and how you felt afterward.
6. If you or your family ever experienced a fire or serious illness, what would you want your friends and relatives to do to help out?
7. What did Craig mean by saying, "Now that we're with Grandma, it's like we're a different family"?

INTERPRETING KNOWLEDGE

1. Write and present a skit about a family in trouble. Afterwards, discuss with the class steps that the family could take to get help. Make a list of some of the sources of help in your community.
2. Present a dance or pantomime that expresses the feelings of family members in troubled families.

SHARPENING YOUR THINKING SKILLS

1. Describe some of the differences between a healthy family and an unhealthy one.
2. What are some of the factors that destroy family balance and create unusual stress and trouble?
3. List several resources that can help families become healthy again.
4. Why do families in trouble often follow an unwritten code of silence?
5. What are some family problems that are easy to recognize?
6. Name some changes that can cause a family to get out of balance.
7. In what ways does the depression of a family member affect the entire family?
8. What are some steps a family can take to help a family member who is considering suicide?
9. Explain why sexual abuse is a difficult subject for its victims to talk about.

APPLYING KNOWLEDGE

1. Look in newspapers and magazines for stories about troubled families. Select one family. Write a letter you would like to send to that family, suggesting ways to keep the family together and where to get help locally.
2. Locate and share with the class different kinds of music, poems, and essays about families in trouble.

CHAPTER 17 REVIEW

Putting Your Values To Work

PRACTICING DECISION-MAKING SKILLS

Read about each situation. Then answer the questions.

Situation A: Refer to *A Teen Speaks* on page 375. It's been over a year now since Craig and his family moved in with his grandmother. His father has recently been laid off from his job because of a slump in the economy. Although he has looked seriously for a job, he hasn't found one yet.

Lately, Craig's father has started to drink a lot, and his moods are beginning to affect the rest of the family. Last week, he lost his temper because dinner was late. He screamed at Craig's mother and punched a hole in the kitchen wall.

Craig's mother works from 8 to 5 Monday through Friday. She's worried about her husband and wonders if she should get a part-time job on the weekends.

1. What should Craig's mother do?
2. What are some of the challenges his mother must face?
3. What choices are available to her?
4. List some of the positive and negative consequences of each choice.
5. What do you think will happen after Craig's mother takes the action you suggest?

Situation B: Tim's mother and father are getting a divorce. His mother plans to stay in their house. His father is moving into a two-bedroom apartment across town.

Tim is devastated. He feels alone and scared. Although many of his friends have divorced parents, he never thought it would happen to him. He has always felt close to both his mom and dad, and now he wonders if he must choose sides. Should he continue living with his mom? Or should he ask to live with his dad? Neither of his parents has put any pressure on him so far.

6. What are some of Tim's values?
7. What would you advise him to do?
8. What advice would you give his parents?
9. Name some of the possible choices Tim has in choosing where to live.
10. What are some of the positive and negative consequences of each choice?

Situation C: Ruth's brother, Miguel, has AIDS. He's 25 and has taught for three years. Over the summer his symptoms have become quite severe, and he doesn't plan to return to teaching in the fall. Although he has had his own apartment since he left home to go to college, he has moved back in with his parents and Ruth.

Ruth loves her brother and is proud of him. She knows he was a very good teacher. Although Ruth's parents love Miguel, they are embarrassed and ashamed about his illness. They refuse to discuss it, and instead act as if nothing has happened.

Ruth is afraid that her brother may die soon. She would like her family to be close and to offer as much support to Miguel as possible.

11. What should Ruth do?
12. How do Ruth's values differ from those of her parents?

CHAPTER 18

Building A Strong Family For Tomorrow

A TEEN SPEAKS

The way I figure it, I'm not through being a kid yet. I don't want to even think about having my own family! If I ever get married—and that's a big if— it will be after I have been out on my own and had some fun. I'm not getting trapped until I'm ready to settle down.

Believe me, I know what I'm talking about. My folks got married when they were really young. They're always talking about how much they've had to struggle. And they fight all the time. Sometimes I get so tired of hearing them yell at each other. If I ever do get married and have my own family, we're not going to yell at each other. I can tell you that much.

I don't think we have a very happy family. I mean, you see these families on TV where everybody loves each other. They always work out their problems. The house is always clean. Everybody always ends up smiling. We're just not like that. It's not like I don't love my folks. I do love them. But we are definitely not a TV family.

If I could find somebody who really loved me, then maybe it would be different. Maybe I'd want to get married. I think if you really love each other enough, everything will work out. But I won't get married unless I find somebody who really loves me. It would have to be true love.

SECTION 1

TRUE LOVE

Is there such a thing as true love? How can you find it? And, if you find it, how can you keep it? In most fairy tales, the men and women face many problems and go through terrible trouble. But once they find true love, they get married and they live happily ever after. It sounds so simple—once the right person finds you or you find the right person. But, living happily ever after takes more than most people imagine.

Times Are Changing

We were thinking we'd learn to love, and then we would get married. We wanted to have a really good marriage—not a bad one, like so many are. We thought we'd be able to profit from everyone else's mistakes. You look out your window on the street where I live, and that's all you see: mistakes and mistakes and mistakes. And you see children who won't know how to avoid making more mistakes when they grow up. I know. I tried, and look where I got. I hope, someday, I'll find the right man and I'll be the right woman for him, and we'll get married and then, a few years later—not before—we'll start having kids.

Some people think marriage and family are a thing of the past. The fact is, nobody has a crystal ball to show exactly what marriage or family life will look like in the years to come. Instead people who study the future make guesses based on **trends,** the direction in which things seem to be going.

One important trend involves working women. More women work outside the home than ever before, and this trend is expected to continue. Of the women who get married, many will marry much later than before—some after they turn 30. Many married women will work outside the home. This means there will be even more changes in the way housekeeping chores and child care are worked out within families.

Many men already share or take major responsibility for housekeeping and child care. It is expected that in the future more older brothers and sisters will take on the responsibilities of housekeeping and caring for younger brothers and sisters.

There will still be some households in which there is little change. These families will continue with the traditional family pattern. The man will work outside the home, and the woman will manage the home.

Another trend that may affect families in the future involves new jobs and new ways to work. These may mean that parents can work at home, for example from computer terminals. It is possible that more children will be educated at home. The technology exists to take the classroom to the home via television and computer.

While some people think family life will be harder in the 2000s, others think families will have even more time to be together.

Valuing Marriage — "So why marry?"

Marriage has existed in almost every culture at every point in time. As far back as you can find records of human life, you can find records of marriage. Most religious and ethnic traditions value marriage. They provide special ceremonies for getting married. And they make it hard to end marriages.

For many couples, the wedding is a religious service. It takes on additional meaning for them as a time to take vows or make promises. Their vows are made not only to each other, but to their God and to other members of their religious community.

A lot of men take major responsibility today for housework and child rearing. What values are being reinforced here?

A religious wedding ceremony allows the couple to make their commitment public and before their God.

People get married for many reasons. Some of the most important reasons include
- love,
- companionship,
- shared interests,
- common goals,
- children,
- economic security,
- intimacy.

The strongest reason for marriage is probably companionship. Most people want to marry because they want someone to trust completely and openly. They want someone who will accept and love them for who they are. Marriage is a deep and satisfying friendship. Each partner values the other as much as himself or herself. They share joy and sorrow.

People go through a marriage ceremony because they want to make a public commitment to each other. Whether they have a religious ceremony, such as one performed by a priest or rabbi, or a civil ceremony, such as one performed by a judge or justice of the peace, their ceremony shows the world that they are making a serious commitment and a binding legal agreement. A **commitment** is a special kind of caring that includes a long-term promise. By taking marriage vows they are saying, "We are in this together. No easy outs. We are a couple, and we're going to support each other come what may."

Living Together — "Who needs a piece of paper?"

Some couples decide to live together without getting married. This arrangement may seem a lot easier than getting married. Living together doesn't require the same public vows. It doesn't require the same commitment as marriage. When you marry, you are promising to live and work with this person for the rest of your life. You are taking on legal obligations of support and protection.

Some couples who live together may be very committed to each other and the relationship. But living together without the formal, public commitment allows one or the other of the persons to walk away. There are more exits to the relationship and fewer protections.

> If half the people who get married end up divorced, why make it so hard? If this doesn't work, I'm out of here. No lawyers. No fighting. No big deal.

Public vows do not allow either partner just to walk away. Getting married does make it harder to end a relationship. Separation and divorce are not matters to be taken lightly. There are important moral and legal issues involved.

Many school organizations rely heavily on volunteers. These volunteers see a number of rewards for their service. Some have even met their spouses through volunteer activities.

Building Strong Marriages

Living happily ever after starts with facing some hard facts. The first is that getting married involves more than having a ceremony that will be remembered forever. Some teens spend a lot of time dreaming about their wedding day and don't do any thinking about being married.

A dreamy wedding ceremony can be wonderful—something worth planning for. But it's over in a very short time. Marriage involves a serious commitment to one other person. Once you get married, life changes forever. Living happily ever after takes more than sitting around looking at videos of the wedding day.

Fairy tales may be misleading about what a good marriage takes. But they have one thing right—true love begins with the right partner. Whom you choose to marry makes all the difference!

The Right Partner — "How will I know?"

A lot of people think that Mr. or Ms. Right will be moved by fate to find them. In most cases, it is not fate but circumstances that bring people together.

> We met at synagogue. We'd actually gone to the same junior high, but it was pretty big and we didn't really run into each other. But the synagogue isn't so big, and I guess we just started noticing each other. Then we found out that we have a lot of common interests besides our religion. And yes, we were really attracted to each other!

A lot of what happens to a couple gets carried along by romantic love and emotions.

> I know she's right for me. When we are together, I have this special feeling all over. She really takes my breath away.

> He is so fabulous. One look into those brown eyes and I just melt away.

Falling in Love

Being sexually attracted to another person and feeling the closeness of love is one of the most wonderful experiences you can have. It gives you special feelings. But those special feelings don't last forever. In fact, many teens have already experienced being in and out of love more than once.

Romantic attraction or "falling in love" gets people interested. A person

Being in love — with the right person — is great!

you didn't know well before is suddenly open to you. You are open to that person. The two of you are crazy about each other. The more intense this experience feels, the more some people take it as proof of true love. But the sudden feeling of emotional closeness you experience simply can't last forever. It isn't possible because the more you get to know somebody, the further away you get from falling in love. Sudden closeness is over and gone. If there is to be any closeness at all, it has to be developed.

At this point in a relationship, many couples begin to think about having sex. Having sex with another person brings you into physical closeness in a way as intense as the emotional closeness of falling in love. A lot of people confuse intense physical desire and sudden emotional closeness with true caring. They think that wanting each other sexually means they are in love.

Love can last. It requires a lot of caring as well as respect both for yourself and for others.

> Sonja has really fallen for A.J. She hasn't dated many other guys, but she doesn't care. All she wants is to be with A.J. She feels such a special closeness to him. It is like somebody had tied A.J. to her heart. She can feel the pull.
>
> Last week A.J. told her he wants to have sex. Sonja isn't sure. She is afraid she'll lose him if she doesn't.
>
> One thing is holding her back. Her sister Della Mae had sex with her boyfriend when she was just about Sonja's age. Della Mae got pregnant. Della Mae kept the baby, but her boyfriend split. Now Della Mae, Sonja, and the whole family take care of the baby. Della Mae had to quit school and get a job.
>
> Della Mae says, "Sonja, I know that look you and A.J. give each other. Just be careful. I'm not saying A.J. will do what my boyfriend did to me. But you think about it. Just be sure that what you decide to do will be as good later as it is now."

It is hard to resist the attraction of being crazy about somebody. And it is hard for Sonja to imagine that she and A.J. will ever feel anything but the closeness they now feel. But that kind of closeness doesn't last forever.

Being Loved

A lot of people think that finding true love is a matter of finding someone who will love them. They want to find someone who will make them feel a special way about themselves.

People often forget to ask how they will love somebody back. The true love it takes to form a strong marriage involves sacrifice. You have to be able to give. To really give, you have to be free enough of your own concerns to pay attention to what someone else needs. And you have to be able to take care of your own needs.

People who study marriages say they can predict which marriages are more likely to end in divorce than others. The age when people marry is a key factor. You've probably heard that most teenage marriages end in divorce.

People who marry at 16 or even 19 probably still haven't had enough time to develop self-understanding. When people marry so young, they are usually trying to get away from their problems. And, unfortunately, they take what they are running away from right into the marriage.

For a lasting marriage, there must be caring, responsibility, nurturing, and supporting—as well as true friendship. Few teens are mature enough to be the caring, responsible, nurturing, and supporting partner they need to be. Few teens have enough money or job skills to give marriage a strong economic base. They end up struggling to survive instead of working on the relationship.

When you've known somebody for a long time, you have less chance of presenting yourself as anything other than who you are. You can see them for who they are, too.

If you don't know someone really well, you don't know what that person values most. You haven't had time to establish good communication. It's true that gazing into someone's eyes may be a form of communication, but you can only gaze for so long without getting bored. The same goes for making out and even for having sex.

When a special feeling, emotional or physical, is all you are looking for, you'll want to leave as soon as that feeling dies. You could spend your whole life going from one experience of falling in love to the next without ever knowing what it is like to have a deep, satisfying, and lasting relationship.

In a strong marriage the couple shares the same values.

For a marriage to last, the couple has to be able to communicate about their wishes, hopes, dreams, and feelings—which will include anger as well as fondness. They have to be able to talk and listen to each other about what they value most in life.

Values are important in a marriage. If marriage partners don't share a common core of values, it will be difficult to stay married. When romantic attraction is gone, there has to be some reason to stick around. What keeps couples together is a deep satisfying friendship. They like and respect each other and enjoy being together. They like the idea of being responsible for each other's well-being. They don't have to have sex to have a good time, but when they do have sex it comes from real caring, respect, and affection as well as physical desire.

Goals are important, too. A couple that shares the same goals will be headed in the same direction.

Love is active. It is primarily giving, not receiving. When you love, you give without expecting to get something back. Your giving shows that you care—it is the joyful sharing of what you have.

Loving includes these values:

- **Care.** When you care, you are actively concerned for the life and growth of those you love.
- **Responsibility.** When you are responsible, you are ready and able to respond to the needs of another. You are ready and able to help with that person's physical, emotional, and spiritual needs.
- **Respect.** When you respect, you see a person as he or she really is. You want to help that person to grow. You look at what that person is, not what or whom you need the person to be.
- **Trust.** When you trust, you no longer doubt. You can be open and not be afraid of being used.
- **Knowledge.** When you know, you can enter into a relationship without being blinded. You know about the other person, and you learn more about yourself. You feel with the other person, and you learn more about your own feelings. That's why you don't get bored with the same person—even if you are married for 50 years!

SECTION REVIEW
STOP AND REFLECT

1. Describe the kind of family you hope to have in the future.

2. How will your family be like the family you've grown up in? How will it be different?

3. Think of a couple you know who seem to have a good marriage. Tell why you think they have stayed married.

4. In what ways will you need to grow up before you are ready for a family of your own?

5. What are the qualities you would look for in a marriage partner?

SECTION 2

PREPARING FOR MARRIAGE

The best preparation for marriage is becoming the caring, responsible person you are meant to be. As you grow and mature, you have more to offer. And you are more capable of drawing from a relationship.

Once marriage vows are taken, commitment involves legal, emotional, spiritual and practical responsibilities each partner has to the other. When there are children, these responsibilities are increased. Marriage isn't a simple matter of two people deciding they want to live together. It is a very complex agreement between two people that will involve them for years to come. There is no substitute for maturity in entering such a relationship.

Engagement

Suppose you find somebody to whom you are really attracted and who seems to have the maturity and values to build a marriage. You want to make the commitment of marriage—but how can you be sure? Even though you can't know for certain, it makes sense to be as prepared as you can. The way two people prepare for a marriage can make a great deal of difference.

An engagement is a good way to explore ways in which you and your partner are alike and unalike. Many of those discoveries will have taken place before the engagement. But this is a time to look seriously at differences and to decide how likely your marriage will be to last.

It is not impossible for people who are opposite to marry and stay married, but it is more difficult. It is dangerous to assume that differences between people just don't matter. It is also dangerous to assume you can change differences between you and your partner. You can't count on changing or reforming someone just because you love him or her.

> I know that Tara drinks a lot when she relaxes. I don't like it. But I know that when we're married I'll be able to help her stop. She just really needs someone to take care of her.

If you are engaged to a substance abuser, you're likely to be married to a substance abuser. If the person you love promises to quit, that person needs to quit before you get married.

If you're engaged to someone who occasionally gets angry and slaps you around, then you can expect to be slapped around when you are married. Use the period of engagement to put your differences on the table. If they are too big to overcome, cancel the wedding. Better to cancel the wedding now than end up having to cancel the marriage later.

Harmony in areas of religion, culture, education, and childhood experiences are essential to success of a marriage. **Compatibility,** or a sense of harmony and agreement, in these matters is like the glue that holds a marriage together:

- **Religion.** Religion is the basis of many of our most important customs, celebrations, and beliefs. People of different religions who marry must find ways to develop common customs, celebrations, and beliefs. This can be very difficult to do. It frequently means going against what one has been taught all of one's life. Religious differences can enrich a marriage, but very often, especially when there are children, these differences create serious problems.
- **Culture or ethnic background.** Culture forms the basis for many of our experiences. Whether you are African-American, Caucasian, Asian-American, Mexican-American or Spanish-American, part of how you feel about the world is based on ideas that have been handed down by your culture. Marriages between people from different cultures or ethnic groups can work. But a cou-

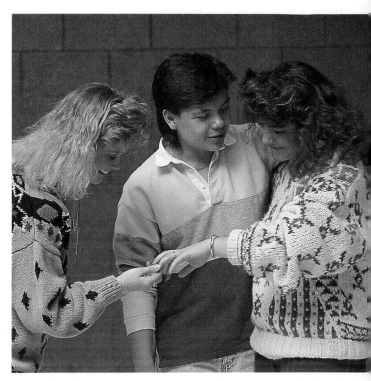

Don't think that an engagement means your partner will change.

ple will face many difficulties, including not being welcome in many social situations. During an engagement period, you need to face honestly whether or not you share enough in common to overcome cultural differences in your marriage.
- **Education.** Formal education is very important to some people. Others find on-the-job experience more critical. Your job, what you can afford, where you will live, and whom you will associate with are affected by your education. Two people who share similar goals and values about education can grow together.

Dating someone with a different background means you'll have additional problems. It's important to discuss early how you'll handle these problems.

- **Childhood experiences.** Couples who share similar backgrounds or childhood experiences are more likely to have similar ideas about how their own children should be reared. They also come into the marriage with similar points of view about life.

The couples who face tough issues together during their engagement will be more ready for marriage. True love is more than holding hands, making out, and enjoying good feelings. It is being together in love: working on problems, facing issues, giving each other room to fail, forgiving, planning a future, and growing together.

Living Happily Ever After

Once the vows are said and the ceremony is over, the work of marriage begins. Each partner comes to the marriage full of expectations of what married life will be like. Adjusting to married life is something every married couple goes through. It takes time and effort to bring expectations about marriage into line with reality.

Marriage is a task. It requires insight, understanding, and hard work to keep a marriage together. It takes sharing deeply, being responsible, loving, and trusting. Couples who get married hoping that they will live happily ever after just because they are married are in for some real surprises.

You can fall in love with someone but be unable to develop a satisfying relationship. Love won't make it happen. Love doesn't settle every problem. Things don't work out because you hope

they will. Marriage means a lot of togetherness:
- eating together,
- sleeping together,
- having sex together,
- managing money together,
- taking care of the home together,
- rearing a family together,
- developing friends together.

All that togetherness can make marriage partners feel crowded. They have to learn to deal with their conflicts. They also have to adjust to the wishes, talents, and personalities of two different people.

A Lasting Marriage — "We're in this to stay!"

A partner in a strong marriage thinks like this: "When adjustments are necessary to make this marriage work, I am willing to make adjustments in myself first. I can't change my partner, but I can work with my partner to figure out how we can both be satisfied in our relationship."

How much adjusting a couple needs to make will depend on their personalities. Your partner may not be perfect, but neither are you!

Marriage today can be very rewarding. But it requires work because relationships between partners are changing. Because married life is so close and intimate, husband and wife are bound to disagree on some things.

A lot depends on what people expect from marriage. Marriage is a relationship that grows and changes. To meet the changes, a couple must share a common set of values and goals for the marriage. They need to have in mind what they want from marriage. A good marriage takes

- **learning how to engage in positive conflict.** Getting close and being intimate require working through conflict. Having the person you love most of all disagree with you can be hard to take! But it is important to deal with conflict in marriage in positive ways. What matters most is that a couple understand why they quarrel and what to do about it.
- **building open, honest communication.** Without communication, you can't know another person deeply. In a strong marriage, partners talk about everything that affects their life together. They don't avoid certain subjects, like finances, sexual problems, or who does the chores. Instead, they recognize that working through problems will make family life run more smoothly.
- **recognizing the need for everyone to continue to grow and develop.** A healthy family supports and affirms all of the family members, not just the adults or those who are outstanding at what they do. Strong families view the future as a time that holds many opportunities for each individual. These families work together to bring each member's dreams closer to reality.

Before marriage, it's important to know what your expectations are.

Decisions In Marriage — "Who decides who decides?"

Once you are married, your decisions affect at least two people and later, when children come, more. Many decisions need to be made with your partner. The choices a couple makes early in their marriage set a pattern for how they will relate throughout married life. That's another reason why good communication skills are important in building a strong marriage.

"Where Will We Live?"

A couple should have made decisions about housing long before the wedding. But the decision about where to live comes up repeatedly in most marriages. The average American family moves every five years. Once a couple is married, they have to find out if they can live together in the space they have.

> When we first got married, we thought we could get by in this little studio apartment. It was one big room with a tiny bathroom no bigger than a closet off to one side. We soon found out that we couldn't stand the place. There wasn't anywhere you could go to be alone except the bathroom and sometimes you've just got to be by yourself!

Other questions about housing are important, too. How close to work does the couple want to live? How close to their parents should their home be? What kind of housing can they afford?

"Who's Going To Work?"

Job decisions have to be considered early in marriage. Usually, these decisions have to be considered again and again. Some kind of agreement is necessary in order to build a healthy family life. A couple has to consider questions like these:

- Will both partners work?
- Are two incomes needed?
- Do both need to work for self-fulfillment?

- Do their job schedules mesh so partners have some time together?
- What happens to their jobs when children come?

Modern couples have to come to agreement about the importance of each partner's job to the family.

> Everything was going along fine until she got this job offer. It meant that she'd have to transfer to California. And here we are in St. Louis. It's not like I can just pack up and go.
>
> My boss says if I stick with the company I can expect a great future.

Is one job more important than another? Answering questions like this can help a couple decide what they value about work. It can also help them establish guidelines for future family decisions.

"Can We Afford This?"

Some of the most important issues facing married couples concern finances. Many couples say they never have enough money. Sometimes what they mean is that they spend everything they have. They don't feel satisfied with how they use their money. At other times they truly do not have enough money to take care of all their needs.

> I'm not good with money. When I see something I want, I buy it if there is a penny in my pocket. That's just the way

It's also important to know your spouse's expectations.

> I am. And I don't like being nagged about it.

In a healthy marriage, both partners plan expenses, and both take some responsibility for handling the money.

"Who Will Do What?"

When a couple gets married, they must decide who does what every day. Just as an actor plays different roles at different times, each partner must take on different roles or jobs at different times of the day or different times during the marriage.

All couples have to work out their finances and decide how money will be handled. What values are important in these decisions?

> We don't measure who gives more, who works harder, who brings home more money. We accept that we both are giving everything we can give. Some days one can give more than the other. But it all comes out in the end.

Many couples look for new ways to divide household chores. They try to avoid thinking of some tasks as "woman's work" and others as "man's work."

"What About Sex?"

Married couples must make decisions about their sex lives. Some of these decisions concern health and pregnancy. Each married couple must answer these questions:

- What sexual practices are we comfortable with?
- Do we want children?
- If not, or if not right away, how will we prevent pregnancy?

Other decisions about sex concern trust and faithfulness. Because of the spread of AIDS and other sexually transmitted diseases, many couples are taking another look at the sexual "freedom" of the past. There is more concern with fidelity to one's partner. **Fidelity** means you will treat your partner in a special way that you treat no other person. It means being true to your wedding vows. It means not having sex with anyone else.

Fidelity is one important sign of how much you respect your partner. It builds a strong foundation of trust and a strong basis for future family life.

"What About Family?"

Parenting decisions last. Once you decide to have children and they are born, you're a parent for the rest of your life. So it makes sense to take time and prepare yourself. Knowing what you're getting into makes parenthood easier.

Rearing children is one of the greatest responsibilities any individual ever faces. There aren't any foolproof approaches to parenthood. What works with some children may not work with others. That's one reason you need to be comfortable with yourself before you have children. No matter how confident you are that you can be a good parent, you will find that you have to adjust and change as you rear a child.

> Well, when the baby finally arrived, I really couldn't believe it. Like, you know, you've always wanted it, and dreamed about how terrific it would be, and then you've got it. It's finally there. And it's kind of scary. When I took him home, I think I got slapped in the face. Everybody prepared you for, once the baby is there, how to care for it. But the actual thing about him being there and him tying you down, they didn't tell you that at all. When I take a shower, it's, "Hurry, hurry, he might wake up." And he cries a lot. He cries when he's hungry. He cries when he's wet. He cries when he's tired. Sometimes I just feel like crying too.

Parenting is most rewarding when it is planned and the couple has had time to prepare — mentally, socially, economically.

Before having children, both partners should ask these questions
- Am I ready to rear a child? How well do I handle myself?
- Do I understand what a child needs?
- Do I want a child for selfish reasons, such as to control my partner's behavior?
- Will I be good at setting limits and establishing discipline?

Parents have to be prepared to include children in all of their plans. A good way to be prepared for including your children is to get involved in your family now.

Different Families, Different Houses

How would you describe your home? Who uses what space? When? How? Does your home promote togetherness, privacy, or both? What values does your home show? Homes tell a lot about values, culture, families, and family interaction.

In China and East Africa, a family often has a one-room apartment and shares a kitchen and bathroom with several other families. This requires cooperation and socializing among families and family members. At the same time, limited space allows little chance for privacy.

In the Middle East, the sitting room is used mainly by men in the extended family. Women and children may use the room when the men are gone. Guests are entertained formally in another room with an outside door. This allows women and girls to avoid seeing male visitors. Women entertain separately.

A traditional Japanese home is open inside with paper-thin, sliding walls. The rooms are used by everyone in the family, so there is little privacy.

In many parts of Europe, homes are designed to ensure privacy with separate rooms and thick walls. Often socializing is done in public places, and the home is the private place of the family.

If your family doesn't have a family council, for example, you could start one. A **family council** is a family meeting where every member has a right to express his or her feelings. It would be good for your family now. It would also be good for you in the future, when you are dealing with children of your own.

Other Family Options

Some people never marry. Singles often choose to be alone because they believe they can be the best person and accomplish their goals in life by remaining single. Sometimes singles are alone because they have not found a partner with whom they are ready to share a life.

Family is an important value whether you marry or not. Some singles adopt children because they want to have the experience of parenting. Some singles have children, but choose not to get married for various reasons.

Living With True Love

Instead of thinking about marriage as living happily ever after, couples should realize that a marriage ceremony is the beginning. "Ever after" is like a long voyage for two people in a little boat. The two are setting out to sea. They don't know everything that will happen, but they know that you can't set out to sea without running into winds and rain. Sometimes there will be storms with high waves. And they might even be faced with a hurricane. They know they will need to work together to stay safe and secure.

This is the kind of relationship that builds a strong family. To live happily ever after, you have to be ready to get through the troubled times. You have to accept that not every minute will be pleasant. You have to talk to each other about how you feel and what you are thinking. You have to face conflict and deal with it in positive ways. This makes for something deeper than happiness—joy.

SECTION REVIEW
STOP AND REFLECT

1. Think about how your family handles decisions about money. What works well? What will you do differently when you have your own family?

2. How does your family divide up the work? Who works outside the home? Who does what work at home?

3. What do you think are the most important values to have in a marriage?

4. Why do you think family relationships are important for single people?

5. If you knew two people who were thinking about getting married, what could you tell them that would help them build a strong marriage? List the things you think are most important.

CHAPTER 18 REVIEW

Putting Your Values To Work

STRENGTHENING YOUR VALUES

Reread *A Teen Speaks* **on page 397. Then answer the following questions.**

1. What can you tell about Roger that indicates he may one day have a strong marriage?
2. What are some attitudes Roger might need to change before he can begin a strong marriage?
3. At this point, does Roger exhibit the caring, responsibility, and respect necessary for a strong marriage? Why or why not?
4. What additional caring, responsibility, and respect might be needed?
5. Roger refers to being married as "getting trapped." Do you agree or disagree? Why?
6. Roger says, "If you really love each other enough, everything will work out." Do you agree or disagree? Why?
7. Roger says he is not going to get "trapped" until after he has been out on his own and had some fun. Is it possible to have some fun after settling down? If so, how? If not, why not?

INTERPRETING KNOWLEDGE

1. In your journal write a letter to your parents or guardians, describing their relationship with you. Offer them your advice on how they could improve that relationship.

SHARPENING YOUR THINKING SKILLS

1. List four trends that may change the nature of marriages. Discuss what those changes may be.
2. Name four reasons for getting married. Explain why you think they are important.
3. What are some ways you might find the right partner for marriage? What can you do to help make sure you have found the right partner?
4. Before they are married, many couples say all they want is to be together all the time. When they get married, they find being together all the time creates problems. What areas of being together do you think you would find most difficult? Is it possible to be together too much? Why or why not?
5. Compare modern and traditional marriages. Which one would you choose? Why?
6. List three elements of a good marriage. Explain why they are important.
7. Why must you be comfortable with yourself and with your mate in order to raise a child?

APPLYING KNOWLEDGE

1. Talk with a married couple about the decisions they had to make when they first got married. Find out what decisions they continually have to make together. How do they make them? Do they think they make decisions fairly? Do you?

CHAPTER 18 REVIEW

Putting Your Values To Work

PRACTICING DECISION-MAKING SKILLS

Read about each situation. Then answer the questions.

Situation A: Jerry and Kate are very much in love. They have been dating only each other for over two years and have no desire to date anyone else. They have not, however, discussed marriage. Both seem to be afraid to make the commitment. Both are afraid that an offer of marriage will be turned down.

They live in separate apartments with friends. Kate's two roommates are moving out in a few months. This will leave Kate with an apartment she can't afford on her own. Jerry is having problems with his roommates and has decided to move. While talking over the situation, they realize that they can each save money if they move in together.

At an outing with Kate's family, they explain their plan to move in together. They are asked why they don't marry. Both get upset.

1. What is the challenge here?
2. What choices do Jerry and Kate have?
3. What are the possible consequences of each choice?
4. What do you think they should do?
5. Why, if they are so in love, are Jerry and Kate afraid of committing to each other?
6. How old do you think Jerry and Kate are? Do you think age is a factor?

Situation B: Jerry and Kate decide to get married. They and their families are very happy with the decision. The wedding is wonderful. The couple moves into Kate's apartment. They are ready to begin their married life.

As they spend so much time together, they discover that Jerry is a day person and Kate is a night person. Jerry goes to bed early and wakes up early with a smile on his face. Kate stays up late at night and sleeps as late as she can in the mornings.

Jerry is extremely neat, and Kate can't be bothered with daily cleaning. This drives Jerry crazy. He feels that he is always cleaning up her mess. She doesn't see what the big deal is.

7. Why do you think they have these problems even though they have been in love for nearly three years?
8. List three ways they might solve their problems. What are the consequences of each?
9. If you were Jerry or Kate, what would you do? Why?
10. How would you know whether or not your solution worked?

Situation C: The new problem with Jerry and Kate is money. While they both work and share expenses, Jerry assumes the responsibility for paying bills and managing their joint checking account. While balancing the checkbook for the first time, Jerry discovers that Kate spends $5 a week on flowers for the apartment. He is angry. He feels they cannot afford this expense and tells Kate so. She replies that she has been spending her own money on flowers and has no plans to stop.

11. What do you think they should do? Explain your answer.
12. What values are involved in making their decision?

UNIT 5

Caring For Your Community

CHAPTER 19   **PROTECTING THE NATURAL ENVIRONMENT**

CHAPTER 20   **BUILDING A CARING COMMUNITY**

CHAPTER 19

Protecting The Natural Environment

A TEEN SPEAKS

Last weekend I had to stay with my grandmother. She broke her hip in a fall six months ago. She can get around with a walker, but she still needs some help.

I'm not too crazy about staying at Grandma's house. She gets on my case pretty fast.

When I got off my bike in her yard, I grabbed a candy bar from my pocket. Before I reached her front steps, she was hollering, "Pick up that candy wrapper! It doesn't belong on the lawn."

"How can one little candy bar wrapper hurt the lawn?"

That got her started. She stayed with it all weekend.

Grandma told me to look at the fence around the school yard across the street. The wind had blown lots of paper cups, newspaper, candy wrappers, and other trash up against the fence. "That's what happens when everyone thinks, 'what harm can one candy wrapper do?'"

You know, she makes a lot of sense.

Tyrone

SECTION 1

UNDERSTANDING THE NATURAL ENVIRONMENT

Our environment includes everything around us—our family, our friends, our homes, our neighborhoods, our rivers, the air, the sun, our communities, the world. The *natural environment* includes those parts of the environment produced by nature, as well as our yards, our parks, the rivers, and manmade lakes.

People who study our natural environment are worried. Some of them think the time is coming when the earth may be nothing more than a speck of dust floating and spinning in space. They think that may happen unless we make caring about life—all life—an important value.

All Life Is Interrelated

All of life on the planet earth is woven together in many ways that support life. A threat to one kind of life is a threat to all the others.

Not everyone sees it that way, however. Some people see human beings as separate from the rest of the environment and superior to it. That is, they see the environment as something for people to use in whatever way they want, regardless of the consequences.

People who talk about the interrelatedness of all life on earth see human beings as a part of the environment. They believe that the health and well-being of human beings is related to the health and well-being of the rest of the environment.

Seeing human beings as part of nature requires people to respect and value the environment. It leads people to consider how any use of nature will affect life for other humans and other living things in the future.

When you value nature, you are concerned about your own quality of life. This means that you live in partnership with the environment and in harmony with the laws and principles of nature.

It has taken a long time to produce all the plants and animals on the earth. But now we have the ability to alter nature and destroy all living things. It doesn't have to turn out that way. We can work together to do our part to protect our environment.

A Look At The Environment

Our natural environment includes all living things and some nonliving things like rocks, mountains, and air. Each part of the environment is related to every other part. To affect one part of our environment is to affect the rest of it, too. People need to work for balance in the environment so that it remains life-supporting in the future.

> What's my environment? Okay, my environment is me + air + water + plants + animals + people I depend on. So, if you're blowing your cigarette smoke in my face, you're messing with my environment and I don't like it!

For a very long time, human impact on the environment went unrecognized.

The Sun — "Shine on me!"

All living things must have energy to live. Almost all living things get their energy in some way from the sun. Plants, especially those that grow on land, get their energy from the sun and provide food for many animals, including human beings. In the end, this means that human beings take energy from the sun in the form of food. Then humans convert that energy into another form of energy that enables us to work and play.

The Air — "I'm surrounded!"

Air is essential to life on earth. It provides oxygen, which all animals need. It also provides carbon dioxide, which all plants need. The air helps distribute water as rain. The air surrounding the earth is also a source of protection from the sun. It acts like an insulating blanket preventing ground-level temperatures

Sun, air, water, and soil are essential to healthy plant growth. Pollution can affect all or any part of our food chain.

from getting too hot during the day and too cold during the night. It also protects us from deadly ultraviolet radiation from the sun.

The air we breathe is made up of different gases. The most important are oxygen, nitrogen, and carbon dioxide. Most of the oxygen in the earth's atmosphere is produced by plants. Nitrogen has very little effect on animals, but it is important for plants, which in turn produce the oxygen that is essential for other forms of life. There is a very small amount of carbon dioxide in the air, but without it no life could exist. It helps plants produce oxygen. Air is also the source of carbon, which helps form nutrients that are important to both plants and animals.

Water — "I can't live without you!"

Seventy-two percent of the earth's surface is water. Only about 1 percent of the earth's water is found in rivers and lakes. It is this 1 percent that is vital to all land and water life.

With very few exceptions, plants and animals have to have lots of water. Some living things die almost immediately when they are deprived of water.

Plants get their food by absorbing moisture from the soil. The food we need is carried to every part of our bodies by blood, which is 90 percent water. Many of the body's waste products are eliminated from the body in water.

How Others Care for Their Environment

In the Soviet Union, over 20 million teenagers participate in an organization for the protection of the environment. Their clubs are found throughout the nation.

One of the purposes of the organization is to educate the public about the environment. Another purpose is to encourage conservation of natural resources. Members hold special activities during school vacations. These include "Day of the Birds," "Month of Saving Young Fish," and "Month of the Forest."

Students have organized round-table discussions with environmental experts and community leaders. They have written articles on environmental concerns and published them in local newspapers. In all these ways, Russian teenagers show that they care for the environment.

The Earth — "It's my home!"

The earth supplies important minerals that plants need in order to grow. Plants, in turn, provide much of the food that human beings and other animals need.

The earth provides fuels such as gas, oil, and coal. It also yields minerals such as iron, nickel, copper, gold, silver, and uranium, which have many different uses. It provides gems and jewels such as diamonds, emeralds, and onyx. Every resource from the earth can be used up. And when the resources have been used up, there will be no more.

Besides being essential to life, clean water provides for many of our recreational needs.

Plants

There is a tremendous variety in the kinds of plant life on earth. Some plants grow in the soil, while others grow in water. Plants are an important source of food for many animals, including human beings. Plants absorb minerals, which are passed along to humans in our food. And, plants are our most important source of oxygen, which is essential to life.

Plants also provide the raw material for houses, boats, paper, and many other necessary and useful products. In addition, plants are the source of many different medicines essential for the treatment of disease.

Animals

Animals range from the one-celled amoeba all the way to the human being. Some animals are so small you can't see them without a microscope. Others, like the elephant, are gigantic. Some feed on other animals. Others feed only on plants. Some animals, like human beings, eat both meat and plants.

SECTION REVIEW
STOP AND REFLECT

1. Describe the way someone with no respect for the environment would treat it. How would this behavior affect your environment?

2. How does your behavior affect the environment? What changes can you make in your behavior to show that you care about the environment?

3. What bodies of water are in your community? What would happen to your community if they became polluted?

4. What kinds of animals live in your community? What would happen if one or more of them could no longer live there?

5. If you could suddenly turn off all man-made noises in your environment, what natural sounds might you hear during the day? After dark?

SECTION 2

CARING FOR THE NATURAL ENVIRONMENT

Some people think they can do whatever they want to the environment. As a result, our air and water are polluted. Many plants and animals are near extinction. And life on earth is in danger. This abuse of the environment may mean that time is running out for us to save the earth for future generations of people.

Ways Our Environment Has Been Abused

Human beings have been polluting the environment for centuries. We have filled the air with smoke from our fires. We have dumped wastes into waterways or buried them in the ground.

We have also used the ores, fuels, and minerals that come from the earth as if there were a never-ending supply of them. When bothered by pests, we have sprayed them with chemicals and not stopped to ask if those same chemicals could create problems in our environment. Human beings have clearly not been very caring about the natural environment.

Air Pollution — "I can't breathe!"

Some of the most serious threats to the environment are from pollution. Air and water pollution threaten our natural environment.

People who live in America's large cities know that air is one of the most abused parts of our environment. Exhaust fumes from cars, trucks, buses, and industry have done a lot to pollute the air. So have smoke and gas from activities like burning coal for fuel.

People in large cities are directly affected by poor air quality. The U.S. spends billions of dollars each year fighting air pollution.

We don't have a limitless supply of air. When we use and abuse it, we can't go to the store and buy some more air. Even though the United States spends billions of dollars each year to fight air pollution, the problem is still very serious.

When petroleum, coal, natural gas, and wood are burned, carbon dioxide is released. Automobiles and industry pump carbon dioxide into the atmosphere. Some of it is absorbed by plants. But, when there is more carbon dioxide than can be absorbed, it begins to build up. A blanket of carbon dioxide insulation is formed around the earth. Just like a wool blanket, the carbon dioxide blanket tends to cause a warming of the earth's atmosphere. This could change the world's climate in ways that would harm or destroy living things.

Carbon monoxide is a colorless, odorless, tasteless gas. It is also very poisonous and, therefore, dangerous. It is in the exhaust fumes of every car. Persons who breathe too much carbon monoxide can die from a lack of oxygen very quickly.

Breathing in too much carbon monoxide all at once isn't our main cause for concern. Instead, people should worry about exposure to small amounts of carbon monoxide over a long period of time. Many of us get carbon monoxide from being around people who smoke. And smokers, of course, get more than nonsmokers.

Carbon monoxide is also harmful to plants. Even with a very small increase in the amount of carbon monoxide in the air, some plants age more quickly than usual. Others lose their leaves. Carbon monoxide interferes with the production of oxygen by plants.

Several different gases are formed in combination with nitrogen. These gases are among the most toxic substances in polluted air. You can see them in the atmosphere. On a day with very bad smog, the air looks reddish-brown due to the presence of nitrogen.

Some scientists think that some synthetic gases interact with the earth's ozone layer, causing chemical changes that are using up the ozone layer. This is a concern because the ozone layer shields us from high-energy radiation. If we use up the ozone layer, we may create very significant changes in the earth's weather. Furthermore, if more of the high-energy radiation reaches the earth, we may see an increase in cases of skin cancer. Air pollution is a deadly threat to the environment.

Water Pollution — "This water isn't safe."

Some cities haul their garbage out to sea and dump it, affecting plant life in the ocean. Some of that garbage is carried back to land and washed up on our beaches. Even more of the garbage is carried back to land in the flesh of contaminated seafood.

Businesses and individuals also dump wastes in rivers, and lakes. This dumping can disrupt the very delicate life cycle of the plants and animals in the water.

Every animal and plant that lives in water must have the right mix of carbon dioxide and oxygen. Consequently, we must work hard to maintain clean water.

Other Forms Of Pollution

Some forms of pollution affect both air and water. They show how closely connected all parts of the environment really are. Some substances that are harmful to the environment come from the earth itself.

Acid Rain

Pollutants that collect in the atmosphere stay there until it rains. Then they are showered back upon the earth. When the rain falls on forests, rivers, and lakes, it brings back a deadly cargo of pollutants called **acid rain**. When it mixes with the water in a river or lake, acid rain makes it very difficult for fish and other water life to survive.

Dumping garbage into the water disrupts the delicate cycle of plants and animals that live in the water.

Many hazardous wastes are so dangerous that people who work with them must wear protective clothing.

Hazardous Wastes

Some wastes cannot be recycled easily or safely. A waste product that may cause illness is a **hazardous waste.** Plastics, paints, and the chemicals used to kill insects are examples.

These wastes are dangerous in a number of ways. Some easily catch fire and can produce dangerous smoke. Some are so powerful they can eat away at the containers that hold them. The chemicals then enter the water supply and harm plants, animals, and even people. Some can sit in the soil for years and poison the food supply.

Solid Wastes

Every year we throw out enough trash to fill five million large truck trailers. Plastic wrappers, newspaper, and soft drink cans clutter roads. This waste is called **litter.** Litter makes public places and communities look unattractive. People who use trash cans to hold solid waste help prevent litter.

Some of this trash is burned, but the amount is not large. Much of this waste, though, could be **recycled.** This means that it could be changed in some way and used again. Paper can be broken down and made into new paper. Aluminum cans and glass bottles can be crushed and used to make new cans and bottles.

People And Pollution

People have caused many pollution problems. Now people are themselves becoming a pollution problem. That means that we harm the environment by our presence, especially as our number increases. Naturally, the more people there are, the more garbage and natural waste there are. More people means more cars on the road and more exhaust pollution. More people means the depletion of many natural resources, including marine life, forests, soil, and animals that are sources of food.

Becoming a caring person means caring for more than yourself, your family, and your friends. Being a caring person means caring about all of life, that which is alive now and that which will be alive after we are gone. It means caring for your environment so that more and more people can live healthy lives.

Caring For The Environment

Caring for the environment is another way of caring for yourself and your family. It is a way of caring for your friends and your community. It is also caring for those persons not yet born. Caring for the environment is not only an investment in today. It is an investment in tomorrow as well.

The earth is our home. It is the only home we have. It is the only home future generations are likely to have. Nothing on earth is limitless.

When we abuse the environment, we are abusing ourselves. It is a big step to start thinking about the consequences of small choices that you take for granted every day. This is how we can work together to save our earth. Here are some small choices that make a big difference:

- **practice responsible pet care.** When people who have pets do not clean up after them, the mess draws flies, smells, and becomes a breeding place for diseases. When dogs and cats keep reproducing and there is nobody to care for the pets, they are often abandoned. Turning animals loose is sentencing them to inhumane treatment. Another way to care for the earth is to have your dogs or cats spayed or neutered so they can't reproduce.
- **recycling garbage.** Many teens make an important contribution to the environment by learning how to recycle cans, glass, or paper. Saving and recycling garbage may be a lot of trouble, but being responsible isn't always easy.
- **find the garbage can.** It takes self-discipline and a sense of responsibility to break the litter habit. But caring people don't litter. They hang on to their garbage until they are near a trash can. If every teen in America were to begin to throw away trash, it would make a visible difference! The whole country would begin to look better. Younger children would see teens and start to follow their example. And adults would see them and begin to remember their manners, too!

SECTION REVIEW
STOP AND REFLECT

1. Make a list of the major sources of environmental pollution where you live. Next to each one, list the kinds of pollutants that come from that source.

2. Make a list of those groups or organizations in your community that are fighting environmental pollution. What kinds of pollution are they working against?

3. If everyone in your community shared your attitude about the environment, what kind of community would it be?

CHAPTER 19 REVIEW

Putting Your Values To Work

STRENGTHENING YOUR VALUES

Reread *A Teen Speaks* **on page 423. Then answer the following questions.**

1. When Tyrone drops the candy wrapper on his grandmother's front lawn, what kind of message is he sending to his grandmother?
2. Why do you think Tyrone's grandmother gets on his case so quickly about things? Why does the candy wrapper bother her so much?
3. Is there any difference between Tyrone's dropping the candy wrapper by the school fence with the other trash, and dropping it on his grandmother's lawn? Explain your answer.
4. Should the trash that has collected by the school fence be picked up? Who is responsible for picking it up?
5. When people throw their own trash away and pick up the litter they see, what values are they exhibiting?

INTERPRETING KNOWLEDGE

1. With a group of friends, plan and make a video that delivers this message: people need to protect their environment. If you don't have video equipment, write a one-act play and act it out for your class.
2. Find items that would normally be thrown away. Create a sculpture with the materials.

SHARPENING YOUR THINKING SKILLS

1. In what ways is it possible for people to alter nature? Give at least four examples.
2. Explain why a threat to one kind of life would be a threat to all other kinds of life.
3. Name several ways air is important to people.
4. How does air pollution affect the earth's water supply? How, in turn, does this affect food sources for people?
5. Is it possible for teens to make a difference in protecting the environment? Explain your answer.
6. In what way are people pollutants (not polluters)?
7. Explain why protecting the environment is a way of showing that you care not only for yourself, but for family and friends as well.

APPLYING KNOWLEDGE

1. With your class, organize a clean-up day for the whole school or for your community.
2. Start a recycling campaign: collect newspapers, bottles, and cans, and take your collections to the recycling centers in your community.

CHAPTER 19 REVIEW

Putting Your Values To Work

PRACTICING DECISION-MAKING SKILLS

Read about each situation. Then answer the questions.

Situation A: One afternoon after school, Maria, Amy, and Devon are sitting and talking on the stairs in the front of Devon's apartment building. Devon tells his friends he's had a hard day and lights up a cigarette. Amy tells Devon he shouldn't smoke. She explains that the smoke from his cigarette is polluting her air. Devon asks how his cigarette could be responsible for polluting Amy's air when they're living on a street heavily traveled by buses, cars, and trucks. Devon says that even if his cigarette is polluting the air—and he doesn't believe it is—he doesn't care, because cigarettes allow him to relax after a hard day.

1. What values are reflected in Devon's response to Amy's request that he stop smoking?
2. Compare the pollution from cars with the pollution from cigarettes. Is one a more serious problem than the other? Explain your answer.

Situation B: Same as Situation A. Amy says that Devon is just being selfish and that he should take up a cleaner habit. Why not chew gum? It doesn't pollute the air, and it calms her nerves, Amy tells him, as she puts a piece in her mouth. Suddenly Devon jumps up and calls Amy a hypocrite, pointing at the ground. There, by Amy's feet, is the gum wrapper she just dropped. Devon explains that Amy has no right to accuse him of polluting the air when she pollutes the ground by littering. It's just one gum wrapper, Amy tells Devon. If there were a trash can around, she would have used it. Besides, if it bothers someone, let him or her pick it up.

Devon is really angry now. He asks Amy who she thinks is going to pick up the gum wrapper. Devon explains that his father is the super of the building, and he has to pick up the trash around the apartment. Amy says that is his job.

3. If you were in Amy's position, without a trash can in sight, what would you have done with the wrapper?
4. After Devon tells Amy that his father will have to pick up the wrapper, what values does her response reflect?

Situation C: Same as Situation B. During Devon and Amy's argument, Maria is quiet. At one point, she picks up the wrapper and puts it in her pocket. Amy and Devon turn to Maria and ask her who is right.

5. Why do you think Maria picked up the wrapper?
6. What should Maria say to Devon and Amy? Explain your answer.

CHAPTER 20

Building A Caring Community

A TEEN SPEAKS

I volunteered to babysit after school, so teen mothers could learn more about managing their time, taking care of their homes, and preparing nutritious meals. I really loved watching those little kids, especially this little girl named Jessie. Jessie always cuddled up on my lap and would stop crying the minute I picked her up. One day I was singing to Jessie when her mom came in after class. Her mom said she liked the way I took care of Jessie and asked me to come over and visit with them. I said I would. I really wanted to go because Jessie's mom, Angela, is about my age, and I was curious to see what it was like to be a teenage mom.

Now sometimes I go over and watch Jessie while her mom gets supper. It feels really good to help out, but I realize what having Jessie has done to change Angela's life. She's trying to balance school with the job of taking care of a baby. She has to work weekends to get enough money to pay for the things she and Jessie need. When she comes home at night, she's all tired out. But Jessie doesn't know that and demands a lot of attention. Angela's mom helps out as much as she can, but she works, too. The baby's father has disappeared.

I used to be a little wild. But I'm a lot more careful and particular than I used to be. Babysitting is fun—and work, too. But I can go home afterwards. I can go out with my friends. I don't have to have two jobs in addition to school. I realize I want time just to be me. The rest will come when I'm ready. That's the lesson Jessie taught me.

Melinda

SECTION 1

HABITS OF THE HEART

have you ever heard of Alexis de Tocqueville? He was a young Frenchman who came to America in 1831 to see at first hand how this new nation worked and to write about it. What struck him most of all in America was the way people—especially young people—joined together on their own to work for the greater good. The babysitter at the beginning of this chapter is an example of what de Tocqueville was referring to. He called these volunteer efforts, "habits of the heart." It's a good term to remember as you think about how you can help make every group you belong to a part of a caring community.

How Do We Develop Habits Of The Heart?

Habits are patterns of behavior that have become nearly or completely involuntary. Some are good and some are bad. When we act out of habit, we act without thinking. As you know, habits are difficult to break. You may have watched someone work very hard to stop smoking, for example. People get "hooked" on cigarettes, drugs, or alcohol because they contain substances that are addictive or habit-forming. Putting off doing your homework and leaving the house before anyone can ask you to help with the dishes are also habits.

All habits are formed by repeating certain actions over and over. Good habits, "habits of the heart," can become part of your character through the exercise of your values. You can build your arm muscles by regularly, repeatedly exercising your arms. You can build "habits of the heart" by exercising the values you want to be known by. You start by taking responsibility for yourself—your health, your decisions, your actions, your future. And then, when self-responsibility becomes a habit, it's much easier to become a willing volunteer at home, at school, and in the community.

If Alexis de Tocqueville came to America today, what would he find? Young people concerned only with themselves? What some people call the "me generation"? Or would he find young people with deeply held "habits of the heart"? And where would he find you?

John's Story — "Okay, where do I start?"

John knew there were some "special" (physically or mentally handicapped) students in his school, but he had never met any of them. In fact, he tried to stay away from them. Some of them looked different. Some had funny ways of speaking. He wondered why they were in his school at all. Then he met Artie.

Artie sat down at his table in the lunchroom one day. He was one of those "special" students. And, to John's discomfort, he introduced himself and wanted to be friendly. "Hi, I'm Artie," he said. And then without waiting for an answer, Artie said, "I play basketball."

As it happened, John did, too. In fact, he was on the varsity team. He wondered what team Artie played on. He finally said, "I'm John. Yeah, I play basketball, too." Then he finished his lunch fast and left the table.

At basketball practice the next week, John noticed a group of kids shooting baskets at the far end of the gym. They weren't very good. A teacher he knew was coaching them. Then he saw that one of them

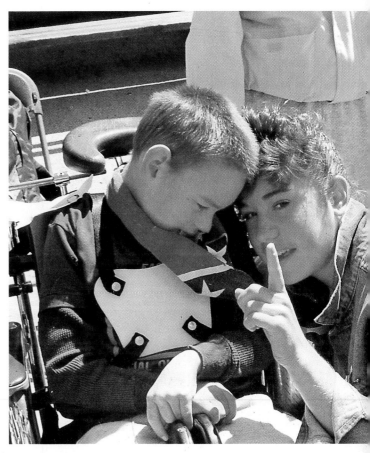

"Habits of the heart" are shown by volunteering to help others who are less fortunate.

was Artie. He left practice and wandered over to watch. "Hey Artie," he called, "keep your eye on the basket."

The teacher came over. "I didn't know you knew Artie," she said. "This is our Special Olympics team. They're getting ready for their tournament. Want to help me out coaching?"

John didn't know what to say. He'd always been a little scared of the "special" kids because he thought they were

Volunteers have fun and know the satisfaction of helping others — while increasing their feelings of self-worth.

"different." But now he'd met one of them, Artie, who liked basketball just as he did. "Okay, I guess," he said. "When do I start?"

Soon John learned that all the "special" kids had names and likes and dislikes and a love of sports just as he did. His coaching made a big difference, and his team won its division at the Special Olympics basketball tournament. At the school assembly at the end of the year, they called him up to help pass out the awards to the Special Olympics team. Later, they all went out for pizza.

This is not a made-up story. It's true. It's happening in hundreds of schools all over America. Could it happen to you?

SECTION REVIEW
STOP AND REFLECT

1. Why was John scared of the "special" students?

2. What "habits of the heart" did John need to develop before he became their coach?

3. How do you think John felt when he gave out the awards?

4. How do you think Artie felt having John as his coach and friend?

5. Is there a chance at your school to volunteer for a "helping" activity like Special Olympics?

6. If so, would you volunteer? Why? Why not?

SECTION 2

WHAT IS GOOD CITIZENSHIP?

A *citizen* is a resident of a city, a town, a state, or a country. A citizen is entitled to the rights, privileges, duties, and responsibilities that come with being a resident of that community. People usually speak of citizenship in terms of one's country. But you can be a citizen of a family, a school, a classroom, a congregation, a team.

Some people think that being a good citizen is pretty boring—like something older people have to do. They think that it doesn't affect you until you're old enough to vote. They're wrong. *Good citizens* are people who take part in making their community a better place to live.

Citizenship Starts At Home

Your home is the first caring community you have a chance to help build. That's where "habits of the heart" get their start. There you have a chance to exercise your values before taking them out into the world.

By helping out willingly around the house, you show other members of the family that you care. You show that you feel a responsibility for making it a good home. You show that you trust and respect your family members. Think of it this way: If you offered to help your mom get dinner ready on a night when she was really busy, do you think she'd realize how much you care about her? Of course, she would.

If your father asked you to turn down the stereo because he had a headache, would you do it willingly? What if you wanted to go out with your friends, but your mother asked you to babysit your little brother? How would you respond?

It's not always easy, but being a good citizen often means making some sacrifice for a larger good. It sometimes means putting off something you want to do now in favor of something that will be better for everyone in the long run. Being willing to wait for a reward later, rather than expecting all the benefits right now, is called **delayed gratification.** When you were a child you weren't capable of delayed gratification. You wanted or needed things right away. Since you

Helping a neighbor with a special project can make you feel good and of worth.

didn't understand time, you didn't understand that you could wait for what you wanted. Being able to delay gratification is one of the ways you know you are growing up!

Sometimes delaying gratification is tough. But it's the only way you can build a caring community. It's like giving the right-of-way in heavy traffic. If everyone wanted to get through the intersection at the same time, there would be a terrible crash and no one would get through.

Most of the time, if you are willing to wait and do what your values and not your wants tell you should come first, the benefit will follow. And you will feel even better than if you had crowded through first.

Making A Difference At Home — "I'm ready to help!"

No matter whether you live in a house in the suburbs, on a farm, or in an apartment in the city, the place where you live becomes a home only when the people in it decide to become caring citizens of a family. Doing what you are asked to do helps make life easier for everyone. When you help out without being asked, you are exercising your value of caring and developing "habits of the heart."

How To Be A Caring Citizen At School

If your school is not already a caring community, you can help make it one. Many students think school is a place that belongs only to adults. Students may also think school is there only to teach subjects that are supposed to be useful in later life. But even though your classwork is your first responsibility, school offers many opportunities to exercise your values and help build a caring community. Here are some examples:

- You can volunteer to be a peer tutor for students who need help.
- You can run for student government. Then you can help set rules

When teens help willingly around the house, things usually go smoother for everyone.

and policies to make school a more caring place.
- You can become a volunteer in Special Olympics or become a coach and friend to a special student.
- You can get together with other students to help keep the school grounds free of litter.
- You can join with others to ask the principal to start a community service program if your school does not already have one.

You might even start a "Teens Care" program that could focus on any problem that concerns young people in your area. In Fort Wayne, Indiana, a group of teens decided that they wanted to help educate others about AIDS. They help at community workshops and provide office assistance to the local AIDS Task Force. You could do the same. Or, you could select another issue.

Showing That You Care at Home

Here are a few caring exercises you can do at home:
- Help set the table for dinner.
- Volunteer to cook one meal a week.
- Clean up the kitchen after mealtime.
- Pick up your things around the house and put them away.
- Keep your own room or space clean and orderly.
- Offer to babysit a younger brother or sister.
- Read to an elderly relative.
- Keep the noise level down when others are around.
- Take the dog for a walk.
- Take out the trash.

Because you know what you can do to make the life of your own family more comfortable, you can add to this list and then do the things you've added.

Peace Corps—The Hardest Work You'll Ever Love

A new water well in the Dominican Republic, a handicraft business for handicapped artisans in Malawi, a newly built school in Gabon, food for children in Peru, maternal health care in Asia—these are among the worldwide projects where Peace Corps volunteers have made a difference!

Created in 1961 by President John F. Kennedy, the Peace Corps is composed of volunteer Americans from every walk of life. Its purpose is to promote world peace and friendship, to provide skills that help people in other countries, and to help Americans and citizens of other nations better understand one another.

Today thousands of Peace Corps volunteers serve in Africa, Asia, Latin America, the Near East, and the Pacific region. Working mainly with agriculture, education, health, conservation, and small business projects, volunteers serve two years. They live among the people with whom they work, so all broaden their understanding.

Participate In The Community

In his inaugural address in 1960, President John Kennedy sent a message to the young people of America. He said, "Ask not what your country can do for you—ask what you can do for your country." Those words formed the basis of the Peace Corps and hundreds of other volunteer programs in communities across America.

Many teenagers think they have to be old enough to vote or to be out of school to make a contribution to their community. They are just making excuses for doing nothing. You can become a valuable member of your community while you're still a teen. It's all a matter of getting involved. But before you take on a project outside of your home or school, the first thing to do is to become well-informed. Imagine how embarrassing it would be if you worked for months raising money to buy new swings for the children's park and then found that the Parks Department was already planning to buy a swing set. If you'd read the papers and kept up with what was going on in your community, you might have found another project that would benefit more from your time and energy.

Staying Informed — "What's going on?"

There are ways to know what's going on. Here are some things you might do:
- Read the local paper every day.
- Watch the local news on television or listen to it on radio.
- Attend public meetings on topics of interest. Students are always welcome.
- Read the bulletin board in the library and other public places.
- Listen to public service announcements on local radio and television. They often call for volunteer help with special projects.
- Work with adults to organize a community meeting to air ideas and opinions about issues in your neighborhood.
- Ask your parents or other adults about the community, church or synagogue, service club, or other volunteer programs they are involved in. This is one good way of involving yourself in your parents' interests and activities.

Can you think of other ways to find out what's happening and what's needed in your community?

Politics And You — "I'm only one person, but I count!"

Voting for elected officials is another way citizens become involved in their community. Elected officials make decisions about how a community will work and what services will be available there. If your community doesn't have enough social, recreational, and educational facilities for teens, this might be a good place to start learning how the political process works and how you can influence it.

But what difference does it make if I can't vote?

You've probably asked yourself this question. Even though the voting age is 18, only a very small percentage of those teens who can vote actually do vote. Regardless of your age now, you can make a promise to yourself that you are going to make your vote count as soon as you are old enough to go to the polls. Right now, even before you are old

There are many ways to find out where your help is needed most.

enough, you can keep informed on the issues and think about what candidates and what positions you prefer. Don't just follow blindly what others think your political preferences should be. Make voting one of your "habits of the heart."

Even before you can vote, you can help support the candidate of your choice by passing out flyers, volunteering to stuff envelopes, lettering signs and posters, or arranging to have the candidates from all parties come to the school and address a special assembly. Finding out about the issues now, understanding the political process, making up your own mind about what is important—all of these will help you become a more responsible citizen in the future and make yours a more caring community.

SECTION REVIEW
STOP AND REFLECT

1. Think of a time when you (or someone you know) wanted something in the worst way and refused to wait. Tell who was involved and what happened.

2. Why didn't you (or someone you know) delay gratification?

3. Describe a problem at home that might be solved if you and your family were all good citizens at home.

4. What could you do to help your family be good citizens at home.

5. Tell about a class at school that needs some good citizenship. Why do you think so?

6. What could you do to be a good citizen in this class? What if nobody else will be a good citizen?

7. Describe a problem that you know about in your community, state, country, or the world. Tell how volunteers might help solve the problem.

8. How might you help to solve the problem?

SECTION 3

WHY IS CITIZENSHIP IMPORTANT?

talk is cheap. You can talk about caring all you want to. No matter how much you'd like to help build a caring community, no matter how much you'd like to develop caring "habits of the heart," if you don't put your thoughts and words into action, they don't mean a thing. And sitting back moaning or complaining about how bad things are is not going to change them. *Involvement*—that is, taking part and making an effort—does accomplish something. Taking action can make a difference—especially when many citizens get together to get a project done. There's real strength in approaching a difficult project as a team.

Responsible Citizens Help Stop Crime

Apathy, which means not caring and not trying, is the opposite of involvement. People who show apathy see what's going on around them as someone else's responsibility. They don't care enough to get involved. Think what would happen if everybody in your community had an attitude of apathy. People could take whatever they wanted—no matter to whom it belonged—and no one would stop them. When someone got hurt, he or she would just be left lying there. No one would bother even to get help.

Have you ever heard the story of Kitty Genovese? She was attacked outside her apartment by a man with a knife who wanted her money. Dozens of people stood at their windows and watched while she tried to escape her attacker. They heard her call for help. No one answered her call. No one even called the police from the safety of their apartment. No one wanted to get involved. Finally, she was stabbed to death and the attacker got away.

People who care about others do not steal, vandalize, destroy property or otherwise harm anyone else. It doesn't matter where you live—on a ranch in Montana or in a city like Los Angeles. Crime moves into a community, a neighborhood, a housing project, or a single block when the idea gets around that no one cares. Here are some signs of a community where no one cares:

These teens are working to rehabilitate their community, so it will be a better place for everyone. What could you do in your community?

- trash in the streets or on the roads,
- drugs sold openly on street corners,
- graffiti on fences and building walls,
- broken, unused recreational equipment in local parks or playgrounds,
- overgrown grass or weeds in vacant lots,
- old refrigerators or abandoned cars in vacant lots or on streets,
- absence of any law enforcement agents,
- broken windows and doors in apartments and buildings,
- homeless people living on the streets,
- people going hungry.

These are just some of the signs of an uncaring community. They are an open invitation to crime. Many communities have started Neighborhood Rehab projects and Crime Watch teams to give constant, visible proof that their community does care.

With a Neighborhood Rehab project, families get together and raise money to plant flowers, repair and paint buildings, contact civic agencies responsible for enforcing codes, write letters to local newspaper editors, and contact the elected officials to get problems solved.

Where a Neighborhood Watch exists, neighbors watch one another's property and report possible problems to the police.

Are there such groups where you live? If not, what would it take to get such programs started? What could students, parents, and teachers do together to improve the neighborhood around the school?

Serving Others

Serving or helping people is a special way of showing that you care. It's a "habit of the heart" that lasts you all your life. Doing things to benefit others is known as **service**.

Careers In Public Service

Some people choose careers as public servants. They work for various agencies of government, or for public institutions such as schools, hospitals, recreation departments, and other institutions devoted to helping people.

In some way, you are involved with public servants every day. Among them are teachers, custodians, librarians, police officers, firefighters, clergy, park employees, judges, social workers, counselors, school nurses, park and recreation employees, elected and appointed public officials.

People such as these make citizenship part of what they do every day. Have you ever considered a service career for your future?

Service In Private Business

Many private businesses also value citizenship. They encourage their employees to volunteer for community projects. They make contributions to local and national charities. Have you heard of the Westinghouse Science Awards? This is one of the oldest and most effective programs for encouraging young people to go into science as a career.

McDonald's has a very caring program, the Ronald McDonald Houses, in which the families of chronically ill children can stay while their children are undergoing treatment.

At the community level, most stores and businesses support projects that benefit the community through the United Way. The National Football League also helps the United Way by paying for television announcements during its games. If you are planning a career in the business or professional world, there is no reason why you can't combine that work with caring and service.

In fact, by being a caring citizen at work you can help the place you work become interested in citizenship. Business organizations that are honest and caring are needed in our society.

Public servants combine their ideals of service with their careers.

Volunteer Service Organizations

In every community, organizations are formed to meet specific needs of people within that community. These are known as volunteer service organizations or civic clubs. In addition, there are churches and synagogues that work closely with people to meet their needs. In your town or city there are probably organizations such as Rotary, Kiwanis, Civitan, Junior League, and others. These groups help raise money and support causes in their local area. They also undertake special projects of national and even international service.

For example, Shriners hold circuses in various parts of the country. Most of the proceeds go to support hospitals for burn victims. Lions Clubs raise funds to provide glasses and eye operations for those who need them but can't pay for such services. Jaycees frequently plan events to buy food for people who might otherwise go without. Civitans send volunteers to Special Olympics.

Some youth organizations like Boy Scouts and Girl Scouts work on projects that benefit the community, too. They have taken leadership in programs to clean up streams and riverbanks and to combat environmental pollution.

Alexis de Tocqueville would be pleased to discover, if he returned to the United States today, that the "habits of the heart" are still very much a part of our communities. Do you know what groups like these have done to improve your community?

SECTION REVIEW
STOP AND REFLECT

1. Take a look around your community. Does it look like a community that cares? Describe why you think it is or isn't

2. Discuss some ways your community could be more caring.

3. Refer back to page 445. Why do you think no one helped Kitty Genovese? How do you think those people felt about themselves later?

4. Think of someone you know who has a career in public service. Tell who the person is and what he or she does. Also tell how you happen to know that person.

5. Tell about someone you know who works in a store or shop and seems to be a good citizen. Explain why you think so.

6. What do you plan to do when you finish school and are on your own? Tell about your plans for the future.

7. How will you be able to be a good citizen through your work in the future?

SECTION 4

VOLUNTEERING: CARING ENOUGH TO SHARE YOURSELF FREELY

Volunteers are people who offer themselves or their services for free. Many elderly people in nursing homes have no one to visit them, talk to them, or read to them. In some communities, students in middle schools and junior high schools are given special training so they can visit nursing homes. These students give the companionship that would otherwise be missing. In other schools, teenagers are recruited as volunteers to tutor young children. What kind of volunteer program do you think you are best suited for? There is probably just the right organization in your community waiting for you to join.

Rewards Of Volunteering

Why would anyone work for nothing? When people have the "habit of the heart" of caring, they gain things that are far more valuable than money. When you volunteer, whether it's just to do the dishes when your mom and dad are tired or to coach a Special Olympics basketball team, you feel good about yourself. You respect yourself, and that's a strong, satisfying feeling that money can't buy. Best of all, that good feeling is contagious. Other people see how you feel about your commitment to be a caring citizen, and soon they want to help out too.

Giving also involves thinking about others beyond yourself—what they would like, what makes them happy or sad, what they could really use that you can provide. Part of the joy of giving of yourself is matching what you have to give with what the other person needs. When you make a good match, you feel proud of "caring well done."

Volunteering also offers unique opportunities to find new friends. When you volunteer, you become involved with others who have similar interests. These new friends are not competing with you. They are working with you to help others. Try it out and see. Some of those new people you meet just might become friends you'll treasure for the rest of your life.

Some young people actually learn job skills while volunteering. Susan

She realized that being a nurse was really what she wanted. When she graduated from nursing school, she went to work as a nurse in the same hospital.

What It Takes To Be A Good Volunteer

It takes a lot to be a good volunteer. Even though you don't punch a time clock or collect a paycheck, people count on you. They trust you, and you can't let them down.

When programs count on volunteers, it's because they cannot afford to pay for very important work. So it's up to the volunteers to make the program a success. Those who don't show up when they are supposed to or are always trying to do things their way, make it tough on everyone else. They can even give a whole program a bad name.

If you are going to volunteer, practice the "habits of the heart" that make you a giver instead of a taker. Be responsible and dependable. Let people know they can count on you to do what you say you will do. In other words, earn their trust. Work hard to get along with the other people you work with. Here's a place to put into practice the leadership and conflict resolution skills you've learned.

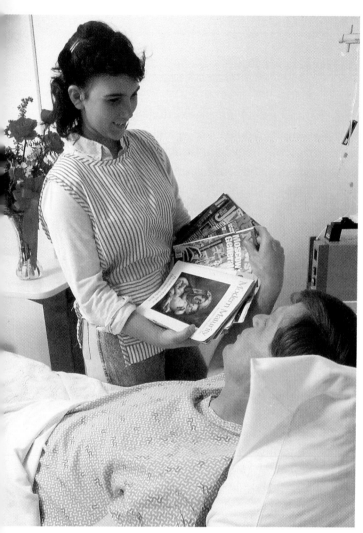

Learning job skills is often a side benefit for many teen volunteers.

thought she wanted to become a nurse, but she didn't know where to begin. She went to the local hospital and got an afterschool volunteer job in the gift shop. There, she made friends with doctors and nurses. They helped her get another volunteer job as a "candy striper" on a surgical floor. She got to know many patients, and had the chance to do things for them that made their recovery easier.

Can One Person Really Make A Difference?

One volunteer can make a big difference in the lives of many people. For example, when Louis Braille was only 12 years old, he decided to simplify a special kind of writing that could be read in the dark. He spent hours working to create an alphabet that could be read with the fingers. Many times Braille was tempted to give up. He often wondered if he'd ever come up with a system that would make reading easy for blind people. When he was 15, three years after he started, he had what he wanted, a system of raised, coded dots that has given millions of blind people the chance to read.

Another young man, James Phipps, volunteered to help Dr. Edward Jenner develop a vaccine against smallpox. He allowed Dr. Jenner to inject matter from the scars of cowpox under the skin of his arm. People who had cowpox were known not to get smallpox. In about a week, the boy's arm got very sore. But he didn't get smallpox. Today, children all over the world can thank James Phipps and Dr. Jenner for developing the vaccination.

You don't have to wait until you are an adult to start building a caring community where you live and work. As shown by the stories of John, Louis Braille, and James Phipps, teens can make a difference in the lives of others.

Build a Team to Meet a Need

Can you think of ways volunteers can work together to help meet a need in your community? Working together, people can accomplish what no individual can do alone.

- Provide homes for students visiting from a foreign country as part of an exchange program or just invite the exchange students to dinner or out for an evening.
- Tutor students who are having difficulty with school work.
- Visit elderly people who are unable to leave their homes.
- Provide help for families who care for severely disabled relatives at home.
- Organize clean-up teams around an apartment building or housing project.
- Organize a team for Special Olympics and help coach it or play on it.
- Read to blind persons who need companionship.

Can you list five other projects you could undertake as a team of volunteers?

Special Olympics provides many rewards for participants and volunteers — regardless of their ages.

Becoming a responsible citizen, one who has put his or her values on the line, who is not afraid to volunteer when there is work to be done, can begin at any age. The values you choose now will become your "habits of the heart" as you exercise them by putting them into practice. You will strengthen your family by helping to create a caring community at home. You will make your school a more caring place by looking around to see where you are needed and doing it. And you will improve the quality of life in your neighborhood or community by being a good and caring citizen, respectful of the rights of others and recognizing that for every right that is granted, there is an obligation that must be paid.

Moving Forward

You can help to build a stronger home, a more caring school, a better community, a peaceful world by developing "habits of the heart" and practicing them. This sounds like a lot to ask from teens! But it is both real and practical. As bad as the headlines might seem, for every act of violence in the world there are thousands of acts of kindness. For every example of terrorism, there are a million examples of caring, comforting, nurturing.

As you've seen in earlier chapters of this book, you learn caring by practicing it, starting with yourself and then expanding your practice of caring to include your family, your friends, your school, your community, and finally your country and the world.

You have been thinking about accepting the values of caring, trust, respect, and family responsibility, as "habits" to cultivate. Now you can start looking for ways to put these values into practice in your own life and in the lives of others. Growing up caring is a process that never ends. As long as you live you will be faced with new challenges and opportunities. There will be new ways to show that you care as a person, a mate, a parent, as a worker, and as a citizen. The important thing is to start now.

Good luck!

Every good thing a person does helps make a better community for all.

SECTION REVIEW
STOP AND REFLECT

1. Describe a problem that you know about. It can be one that involves just your community or state, or it can be a problem for this country or for the world.

2. Tell how volunteers might help solve the problem you have described.

3. Think of one person who has made a big difference in your life. Describe this person. What would you say are his or her "habits of the heart"?

4. Think of one person for whom you could make a big difference. Describe this person. Tell how you could make a difference by being more caring.

5. What "habits of the heart" have you developed in your life?

6. What "habits of the heart" would you like to develop? What has kept you from developing these habits in the past?

7. What actions might you take to develop the "habits of the heart" that you have listed?

CHAPTER 20 REVIEW

Putting Your Values To Work

STRENGTHENING YOUR VALUES

Reread *A Teen Speaks* **on page 435. Then answer the following questions.**

1. What do you think were some of the personal reasons Melinda volunteered to babysit after school?
2. Why did Melinda agree to go to Jessie's house and visit with her and her mom, Angela? What did she learn as a result of getting to know Angela?
3. How did Angela's life change after Jessie was born?
4. In what ways do you think Angela's values changed after she became a mother?
5. What are some reasons why Jessie's father may have disappeared? What does his disappearance tell you about his values?
6. In what ways have Melinda's values changed because of her friendship with Jessie and Angela?

INTERPRETING KNOWLEDGE

1. Design a mural to illustrate ways of exercising "habits of the heart" at home, at school, or in your community.
2. Visit several different places where volunteers are at work. Take photographs or videotapes of volunteers in action. Share your results with the rest of the class.

SHARPENING YOUR THINKING SKILLS

1. Why did Alexis de Tocqueville refer to volunteer efforts as "habits of the heart?"
2. What are some habits that you have formed? Which habits would you like to keep? Which habits would you like to break? Explain your answer.
3. Which of your habits have become part of your character through the exercise of your values? How have you exercised those values?
4. Who are some good citizens in your school? In your community?
5. In what ways have you been a good citizen at home, at school, and in your community?
6. Give an example of how delayed gratification relates to good citizenship.

APPLYING KNOWLEDGE

1. Spend some time as a volunteer in your school, community, or religious organization. Summarize your experience in a paper, a poem, or a skit.
2. Make a list of questions to ask about the needs of your community. Use your questions as a guide to interview one or more elected community leaders. Try to find out how the political process works and how you can influence it to help make your community a more caring one.

CHAPTER 20 REVIEW

Putting Your Values To Work

PRACTICING DECISION-MAKING SKILLS

Read about each situation. Then answer the questions.

Situation A: Refer to *A Teen Speaks* on page 435. Jessie's father, Nick, has contacted Angela, asking to see his baby. He dropped out of school and moved to another town when he learned that Angela was pregnant with Jessie. Jessie is 18 months old now, and Nick has never seen her.

1. What options are available to Angela?
2. What are the possible consequences of each option?

Situation B: Same as Situation A, but now Nick is looking for a full-time job and has asked Angela to marry him. He wants her to quit school and to stay home with Jessie.

3. What advice would you give Angela? What advice would you give Nick?
4. Would your advice to them be any different if Nick had a full-time job? Explain your answer.

Situation C: Dana would like to join the Peace Corps when she finishes college. She's a junior in high school and plans to major in world history. Someday, after serving in the Peace Corps, she'd like to be a teacher. Dana is making plans for the summer, and her friends are pressuring her to "be lazy" and to "just let the summer happen." She'd like to have a couple of months for fun and relaxation, but, deep down, she would also like to find an interesting job or volunteer activity.

5. How would you advise Dana to deal with the peer pressure she is experiencing?
6. What are some of Dana's values? How do they seem to differ from those of her friends?

Situation D: Gil and Juan are walking home from the movies one night when they spot several guys they know from school. The guys are spraying graffiti on the outside of the community center and all over the sidewalk. They have a reputation for being pretty tough, and they are older and bigger than Gil and Juan.

7. What options do Gil and Juan have?
8. What are the possible consequences of each option?

Situation E: Sid, who is 16 years old, entered a radio contest and won $1,500. The money will be delivered to him tomorrow, and he is really excited. His mind is full of all kinds of ways to spend the money. He'd love to get some video equipment and some new clothes. Also, he'd like to put some money down on an old car that Mr. Smith, his auto mechanics teacher, has put up for sale. On the other hand, Sid knows his mother, who supports the family, is having trouble making ends meet. She works long hours, but there is still seldom money for anything but food, rent, and Sid's school expenses.

9. What are some of Sid's options?
10. What are the possible consequences of each option?
11. Would your advice change if Sid had a part-time job? Explain your answer.

Glossary

A

Abuse The use of drugs in ways that are illegal and unhealthy.

Acid rain Polluted rain.

Acquaintance Someone you see and talk with but do not know very well.

Addiction The condition of being chemically dependent or "hooked."

Adolescence The teen years.

Aerobic activity Rhythmic, nonstop, vigorous activity that exercises the heart.

Aggressive Overly forceful, pushy, hostile, or destructive.

Alcohol A widely abused drug that depresses, or slows down, all of the body's systems.

Alcoholic A person who is addicted to alcohol.

Alcoholism A lifelong but treatable disease in which a person's life becomes unmanageable to some degree due to the use of alcohol.

Alienated Feeling cut off from others; feeling lonely.

Anorexia An eating disorder involving the extreme fear of becoming overweight, which results in severe weight loss from self-starvation.

Apathy Not caring and not trying.

Aptitude The tendency to learn to do something easily.

Assault A violent attack.

Assertive Standing up for yourself and for what you believe in firm but positive ways.

Axillary hair Hair that grows in the armpit.

B

Balanced diet A diet that gives you the variety of foods your body needs.

Body language The way you express yourself through movement, posture, and facial expressions.

Bulimia An eating disorder that involves consuming a large quantity of food in a short period of time and then vomiting to get rid of it.

C

Calories The measure of the energy food gives you and the energy your body burns.

Citizen A resident of a city, a town, a state, or a country.

Cliques Exclusive friendship groups.

Codependents People who live with or are closely involved with chemically dependent people.

Commitment A special kind of caring that includes a long-term promise.

Communication A two-way process involving a sender, a receiver, and a message; occurs when the sender and the receiver understand the message in about the same way.

Compassion The ability to feel with another person.

Compatibility A sense of harmony and agreement.

Compatible Having similar values and personalities that blend well.

Conception The process in which a sperm from a male fertilizes an egg and the fertilized egg goes to the uterus, attaches itself to the uterine wall, and becomes an embryo, the beginning of a baby.

Conflict Disagreement; a struggle or fight.

Conflict resolution The process of solving a conflict by problem solving and cooperation.

Conform Make yourself like other people to fit in.

Conformity Acting exactly the way others in a group act.

Consequences Results.

Contraceptives Products designed to prevent pregnancy.

Coordination The ability to move your muscles in a harmonious way.

Cross-addicted Addicted to more than one drug.

Culture The beliefs and social behavior of a racial, religious, or social group.

Custom What others usually do.

D

Decisions Choices.

Dehydration Excessive water loss.

Delay gratification Wait for something you want.

Dependence A problem with drug abuse that occurs over time; the body gets used to having the drug in its system.

Depressants Drugs that slow down your body systems; examples include tranquilizers and barbiturates.

Designer drugs Drugs that are made in a lab from substances not currently on the federal government's list of controlled substances.

Detoxification A period of "drying out" or becoming otherwise drug-free.

Drugs Substances other than food that change the way your mind and body work; examples of drugs include alcohol, marijuana, cocaine, tobacco, sleeping pills.

E

Ejaculation Emission of sperm from the penis.

Emotions The feelings that all people have, such as joy, sorrow, love, hate, fear, frustration, guilt, grief, pride, happiness.

Empathetic Feeling with others; trying to understand and support them.

Empathy The ability to put yourself in another person's place.

Empty — calorie foods Foods that provide calories but very few — or even no — nutrients.

Enabling Protecting a drug user from suffering the consequences of his or her drug use or addiction.

Endurance The ability to handle prolonged physical stress or exercise.

Environment Everything (except your heredity) that surrounds you and affects the way you develop or behave.

Envy Wanting something that someone else has and that you can't or don't have.

Ethical Moral or right.

Exercise Vigorous physical activity.

Extended family Your relatives who do not live with you.

F

Family A group of two or more people who live together and/or are related by blood or marriage.

Family council Family meeting where every member has a right to express his or her feelings.

Feminine Womanlike.

Fidelity Treating your partner in a special way that you treat no other person.

Flexibility The ability to move your muscles — stretch, bend, and twist — to their fullest extent.

Friend Someone you know well and like to spend time with.

G

Gateway drugs Drugs that lead to experimentation with even harder drugs.

Goal Something you want to do or be.

Good citizens People who take part in making their community a better place to live.

Good nutrition The habit of eating a variety of healthy foods.

Growth spurts Times when your body begins to make big changes.

Guilt A feeling of responsibility when something goes wrong or when you know you've done something that isn't right.

H

Habits Patterns of behavior that have become nearly or completely involuntary; what you usually do.

Hallucinations Imagined sights and sounds that seem real at the time.

Hallucinogens Drugs that cause temporary confusion of mental images; examples include LSD and PCP.

Hangover The physical effects of withdrawal the day after drinking.

Hazardous waste A waste product that may cause illness; examples include plastics, paints, and the chemicals used to kill insects.

Healthy family A family that is loving and caring.

Heredity Traits that are inherited from your parents.

Hormones Body chemicals.

I

Ideals Standards of perfection.

Individualism The belief that each person needs to look out for individual or self-interests and well-being.

Inhalants Fumes or chemicals such as lighter fluid, paint thinner, glue, or typewriter correction fluid that produce a brief high when inhaled.

Interdependence Choosing to depend on one another.

Internal rules Rules within you.

Intervention A meeting in which close friends and/or relatives — people who care — confront a chemically dependent person.

Interview A meeting with an employer who wants to ask you questions and explain a job to you.

Intimacy A feeling of closeness and familiarity.

Invincibility The feeling you have when you believe "It can't happen to me."

Involvement Taking part and making an effort.

J

Jealousy Feeling hostile or suspicious of another's possessions or good fortune.

L

Litter Waste, such as paper bags, plastic wrappers, newspaper, or empty cans, that has been thrown onto roadsides, parks, and other inappropriate places.

Look-alike drugs Drugs that are made from legal substances to look like illegal or street drugs.

Loyalty Being faithful to someone.

M

Manipulate Try to control others to get what you want.

Marijuana A drug that is harmful when inhaled into the body; it has dangerous effects on growth and learning.

Masculine Manlike.

Masturbation The touching of one's genitals for pleasure.

Menstruation A monthly discharge of bloody fluid released from the uterus.

Mental health The state of being comfortable with yourself, with others, and with your surroundings.

Modeling Patterning yourself after someone you admire.

Multiplier effect The harmful and even deadly effect that drugs have in combination.

N

Narcotics Drugs that relieve pain and bring on sleep; examples include heroin, morphine, codeine, and opium.

Natural environment Those parts of the environment produced by nature, as well as our yards, our parks, the rivers, and manmade lakes.

Network All of the people you know and all of the people they know.

Nicotine a stimulant that speeds the heartbeat and raises blood pressure; a highly addictive drug found in tobacco.

Nocturnal emission *(wet dream)* Ejaculation or emission of sperm from the penis during sleep.

Nurturing Caring for.

Nutrients Tiny substances that are found in food and that are necessary for the body to function, to grow, to provide energy, and to repair itself.

Nutrition Eating the foods the body needs to grow, to develop, and to work properly.

O

Obesity The condition of being 20 percent or more above your best weight.

Overdose Severe illness or death resulting from a single dose of a drug.

Ovulation The process in which an egg leaves the ovary.

P

Passive Giving up, giving in, or backing down without considering what's best for you.

Passive smoking Breathing in smoke from another person's cigarette, pipe, or cigar.

Peacemaker A person who works to bring about peace through positive conflict and through conflict resolution.

Peer pressure The control and influence people your own age have over you.

Peers People close to your own age.

Platonic Without romantic or sexual attraction.

Prejudice A feeling toward someone or something that is not based on experience.

Premature resolution A solution that is made too quickly, without considering all the issues involved in the conflict.

Prescriptive rules Rules that tell you what you should do.

Puberty The beginning of adolescence.

Pubic hair Hair that grows in the pelvic area.

R

Rape Forced sexual intercourse.

Recycled Changed in some way and used again.

References The names of people who can tell about your responsibility and your skill as a worker.

Relationship A bond between people who share some of the same interests, who exchange information, who discuss their feelings.

Resources Money, materials, and time.

Restrictive rules Rules that set limits.

S

Sedentary Having an inactive way of life that is mostly spent sitting down.

Self-identity Your sense of who you are and what you want to become.

Self-image (self-concept) How you see yourself.

Self-worth (self-esteem) The value or importance you place on yourself.

Service Doing things to benefit others.

Sexual abuse Any sexual activity that involves a child or young person and an adult.

Sexual attraction The desire to be physically close.

Sexual identity How you see yourself as a sexual human being.

Social norms What your community expects of you.

Society A group of people living together in a community or country.

Spirituality The part of yourself that allows you to know you are alive and to feel the wonder of being alive.

STDs (sexually transmitted diseases) Diseases that are spread through oral, genital, or anal intercourse.

Stereotype An exaggerated belief about a category such as race, religion, jocks, or intellectuals.

Steroids Hormones; legal drugs that are often misused.

Stimulants Drugs that speed up your nervous system; examples include cocaine and amphetamines.

Stress The strains or tensions that can be caused by changes in your life.

Superficial On the surface.

Support groups People with similar problems or concerns who join together to help each other.

T

Tact Communicate something difficult without offending or hurting.

Tolerance A problem with drug abuse that occurs over time; the body requires larger and larger amounts of the drug to get the same high.

Trends The direction in which things seem to be going.

U

Unethical Not moral; wrong.

Unhealthy family A family that is hostile and uncaring.

V

Values Those things that you prize or cherish most.

Volunteers People who offer themselves or their services for free.

W

Wellness The process of becoming and staying physically, mentally, emotionally, and socially healthy.

Withdrawal The reaction, sometimes violent, of the body and mind when the drug is taken away from an addicted person's body.

Index

A

AA (Alcoholics Anonymous), 122, 196
Abuse
   drug, 150-175, 180-197
   of environment, 427, 431
   in a family, 213, 391-393
   sexual, 392
Acceptance, 28-29, 310-311
Acid rain, 429
Acne, 64, 162
Acquaintances, 281-282
Actions
   consequences of, 132-133
   and dealing with feelings, 115
   and decisions, 55
   and setting goals, 17-18
   and values, 13-14
   and words, 238
Addiction, 154, 165, 169
Adolescents, 205. *See also* Teens
Adolescent sexuality, 303-311
   physical changes in, 303-307
   psychological changes in, 308-311
Adults, 206
Advertising, 63, 66, 302
Aerobic activity, 98, 99
Affection, 110, 284
Age, 209
Aggressiveness, 136-137
AIDS (Acquired immunodeficiency syndrome)
   disease, 345-349
   and education about, 441
   and needles, 160, 162
   and sexual freedom of past, 412
AIDS Task Force, 441
Air, 423-424, 427-428
Alateen, 193, 196
Al-Anon, 195-196
Alatot, 196
Alcohol, 77, 169, 152, 155-157
Alcoholics Anonymous (AA), 122, 196
Alcoholics and alcoholism, 152, 155, 157, 188-196
Alcohol rehabilitation center, 183
Alertness, 86

B

Alienation, 113
America
   and drugs, 185, 189
   education in, 218
   and exercise, 87
   hunger in, 66-67
Amphetamines, 158
"Angel dust," 160
Anger, 115-118, 389
Animals, 422-427, 429-430
Anorexia nervosa, 68
Anxiety, 118, 389
Apathy, 445
Appearance, 85, 240, 303
Aptitude, 208-212
Assault, 391
Assertiveness, 139
Athletes, 161-162. *See also* Exercise
Attraction, 315, 316
Axillary hair, 305

Babies
   and AIDS, 346, 349
   born of addicts, 160, 165-166
   and pregnant woman's choices, 358
   and risks of teen mothers, 344
Balanced diet, 62-64
Barbituates, 159
Behavior, 7-8, 13-15
Bias, Len, 150, 194
Bias, Louise, 194
Birth, 341, 344, 347. *See also* Pregnancy
Birth control, 344-345
Bill of Rights
   and democratic values, 58
   nonsmokers, 165
Blackmail, 130
Blood, and AIDS, 346, 349
Blood pressure, 77
Blood sugar, 85
Body language, 238, 240, 242
Braille, Louis, 451
Bread, 75
Breaking up, 318
Bribery, 130
Bulimia, 68
Businesses, 447
Bicycles, 100

C

Caffeine, 161
Calisthenics, 87, 99
Calories, 71, 75, 76, 85
Cancer, 71, 75, 162, 164
Carbohydrates, 70, 75
Careers, 210-212, 447
Caring about yourself
   decisions, and, 40-56, 453
   your health and, 62-77, 82-101, 106-123
   and recovery from addiction, 196-197
   threats to, 31-34, 171
   as a value, 7-8, 19, 24, 30, 35, 453
Caring about others
   communication and, 233-234, 263
   your community and, 436-453
   the environment and, 430-431
   your family, and, 356-371, 376-393
   friendship and, 276-293
   in sexual relationships, 316-317, 405
   threats to, 31-34, 171
Carbon dioxide, 423-424, 428-429
Carbon monoxide, 428
Challenge, and decision-making, 50-52, 55
Change
   and families, 383-384
   and marriage, 409
   physical, 303-307
   psychological, 308-311
Chemical dependency, 180, 184, 187, 189, 194-197
Child abuse, 188
Child care, 398-399
Childhood, 408
Children, 211-212, 413
Child Welfare Services, 392
Chlamydia, 347
Chloride, 77
Choices. *See also* Decisions
   and decision-making, 4, 46-47
   and drugs, 167-174
   for drug treatment, 182-186
   about the future, 207-208
   food, 62-65
   freedom and responsibility of, 46-47, 52-53
   and values, 4, 12-13
Cholesterol, 69, 74

Churches, 7, 93, 448. *See also* Clergy members
Citizens and Citizenship
   and the community, 442-443
   at home, 439-440
   practicing habits of good, 452-453
   at school, 440-442
   and serving others, 447-453
   and stopping crime, 445-446
Clergy members, 122, 169, 214, 378
Cliques, 279
Closeness, 402-403
Clothes, 100-101
Cocaine, 152, 156-159
Codeine, 160
Codependents, 187-197
Coke, 159
Colds, 158
Cold turkey, 154
College, 217-218
Coma, 159
Commitment, 86, 90, 336, 400-401
Communicating, 228-245
   and body language, 238, 240
   and relationships, 111, 404-405, 409-410
   and resolving conflicts, 265-271
   ways of, 237-245
Community, 258, 436-453
   building a caring, 436-438
   conflicts in a, 258
   and crime, 445-446
   at home, 439-441
   participation in the, 442-444
   at school, 440-441
   service in the, 447, 453
Compassion, 35
Compatibility, 316, 407
Competition, 263-267
Conception, 306
Condoms, 346-347
Confidence, 87, 94-95, 264-267, 286-287
Conflict, 250-271
   and communication, 240-241, 243
   dealing with, 259-262
   and family problems, 381
   recognizing kinds of, 10, 14-15, 250-254
   resolution of, 263-271, 409
   sources of, 255-258
Conformity, 287
Confrontation, 259-262

Consequences, 8
   and decision making, 45-46, 53-54, 132-135
   kinds of, 131-135, 142-143
Constitution, U.S., 5, 8
Contraceptive foam, 347
Cool down, 100
Cooperation
   in conflict resolution, 263, 268-271
   in friendship, 280, 290
   in sports and exercise, 86
Co-op study program, 217
Coordination, 83-84
Counselors, guidance
   for drug problems, 169, 183-184, 196
   for emotional problems, 68, 119, 122
   for family problems, 378-379
   for pregnancy, 215
   for school problems, 215, 217, 219
   for work, 108, 210
Counseling, 184, 193, 195-196, 391
Crack, 159
Crime, 445
Crime Watch teams, 446
Crisis, 391. *See also* Families, troubled
Crisis hotline, 188, 192
Criticism, 241
Cross-addiction, 165
Culture and cultural differences
   and choices, 52, 65
   in communicating, 120, 138, 242, 257, 337-338
   in relationships, 26, 284, 407
   and society, 10, 300
Customs, 40
Cycling, 83, 100

D

Dance, 83-84
Dating, 314-319
Deadlocks, 271
Dealers, drug, 170-171
Deals, 130
Death
   and AIDS, 346, 349
   and drugs, 150, 154, 159, 170-171
   and grief, 121
   in a family, 389-390
   and suicide, 170, 386-388
   and teen pregnancy risks, 344

Decisions
   ethical, 48-49
   sexual, 324-349
   snap, 134
   and values, 4, 44
Decision making
   and consequences of, 132-135
   about drugs, 150, 172-173
   about food choices, 62-66
   and your future, 202-203, 221
   in marriage, 410-414
   on your own, 40, 44
   risks in, 45-47
   and saying no, 136-145, 172-173
   steps in, 45-55, 172-174
   and values, 4, 44, 174
Declaration of Independence, 5
Dehydration, 101
Delayed gratification, 439-440
Democracy, 5, 8
Dependability, of friends, 284
Dependence
   of babies for care, 358-359
   on drugs, 153, 180
Depressants, 159
Depression, 71, 86, 118-119, 385-386, 389
Designer drugs, 161
deTocqueville, Alexis, 436, 448
Detoxification ("detox"), 184
Dextrose, 76
Diabetes, 71, 75
Dietary guidelines, 69-77
Dietary Guidelines for Americans, 69
Diet, balanced, 62-64
Disaster, 382-383
Diseases
   drugs and, 156-166, 192
   heart, 74, 77, 82, 159, 162
   and inactivity, 82
   and obesity, 72
   sexually transmitted, 341, 345-349
Divorce, 50-55, 119, 121, 388-389
Doctors, 63, 379, 391
Dreams
   and actions, 220-221
   and your future, 203-207, 221
Driving, 132-135, 150

Drug-free, 184
Drugs and drug abuse, 149-175, 179-197
   dealing, 170-171
   decision-making about, 167-175
   definition of, 152-154
   diseases and, 318-319
   facts about, 150-166
   nations and, 153, 185, 189
Drugs, recovering from, 179-197
   helping friends and family, 187-197
   treatment choices and, 182-186
"Drying out," 184

E

Earth, 24, 422-426, 429, 431
Eating
   for fitness, 69-77
   and food choices, 62-68
Education. *See also* School
   in America, 218
   at drug rehab centers, 184-185
   and marriage, 407
Electives, 217
Emergency, 382, 391
Emotions, 68, 115-123, 308, 383
Empathy, 9, 110, 231, 245, 280
Empty-calorie foods, 76
Enabling, 190-191
Endurance, 84
Energy, 423
Engagement, 406
Environment, 299-302, 422-431
Envy, 113-114
Ethical decisions, 48-49
Ethnic background, 407
Evaluation, 55, 143
Exercise
   motivation for, 95-101
   nations and, 87
      obstacles to, 92-96
   and teens, 75
   and values, 82-91
   and weight, 71, 85
Expectations, 408-414
Exploitation, 285
Extended family, 371, 378

F

Facts and myths
   about foods, 63-64
   about sexual relations, 329-330, 341-344
Family. *See also* Families, troubled; Families, understanding
   building a strong, 398-415
   marriage and, 401-415
   parenting decisions for a, 413-414
   values and trends for, 398-401
Families, troubled
   admitting problems in, 378-381
   balance in, 377-379
   common problems of, 381-384
   depression in, 385-386
   and drug problems and treatment, 183, 187-197
   loss in, 386-390
   needs and teen dropouts in, 213, 215
   sources of help for, 378-379
   violence in, 385-393
Families, understanding, 356-371
   accepting your, 356-365
   conflicts at home, 256-257, 259-262
   in different cultures, 10, 360
   and emotional support, 108-109, 121-122, 278
   and influences on food choices, 62-64, 77
   pattern, traditional, 399
   recognizing healthy families, 365-371
   and role responsibilities, 312
   and sharing activities, 87, 91
   and values, 6-8, 12, 19, 331, 335
Fantasies, 204, 329
Fat, 69-70, 70-71, 74-75
Fear, 95, 115, 118
Feedback, 243-244
Feelings, 115, 167-168, 308
Femininity, 298
Fetal alcohol syndrome, 165-166
Fiber, 69, 75
Fidelity, 412
Fighting, in and for a family, 369-371
Finances, 411
Fitness, 66, 69-77, 101

"Flashing lights," 134, 144
Flattery, 130
Flexibility, 84, 87
Food, 62-77
   and choices, 62-65
   and dietary guidelines, 69-77
   and emotions, 68
   groups, 70, 72-73
Foreign exchange students, 283
Forgiveness, 280
Friends and friendship, 276-293
   barriers to friendship, 285-291
   building friendships, 285, 292-293
   definition of, 16, 276-284
   and exercise, 81, 90-91
   in love and marriage, 400, 404-405
   and manipulation, 130
   stages of friendships, 281-285, 291-293
   with both sexes, 310-311, 314-319
   and values, 16, 276
Frostbite, 100
Fruits, 75
Frustration, 115, 120
Fuel, 425, 428
Future. *See also* Plans and planning
   planning for your, 202-212
   and your education, 213-221

G

Garbage, 429, 431
Gases, 427-429
Gateway drugs, 164-165
Glucose, 76
Goals
   and decisions, 17-18, 52
   definition of, 18
   and dreams, 202-204, 207, 221
   and jobs, 207-212
   and staying in school, 213-221
Gonorrhea, 347
Gossip, 234, 287
Government, 7-8, 251, 444
Grass, 161
Grief, 115, 121, 389
Groups,
   and culture, 10, 63
   dating in, 315
   food, 70, 72-73
   friendship, 279
   support, 184, 186

Growing, 6, 8-11, 123
Growth spurts, 304
Guidance counselors. See Counselors, guidance
Guidelines, for behavior, 7-8
Guidelines, dietary, 69-77
Guilt, 115, 119-120, 130, 389

H

Habits, 436
"Habits of the heart," 436-439
Hallucinations, 158-160
Hallucinogens, 160
Handicaps, 93-94
Hazardous waste, 430
Health
   mental, 106-123
   and nutrition, 62-77
   physical, and fitness, 82-101
   problems from drugs, 150-166
   and values, 16, 51
Heart, 83
Heart disease
   and drugs, 159, 162
   and lack of exercise, 82
   and nutrition, 74-75, 77
Heart rate, 98-99
Help, sources of. See also Counselors; Doctors
   for drug treatment, 182-186
   for troubled families, 378-379, 384, 387-393
Hemorrhage, 344
Hepatitis, 160
Heredity, 299
Heroin, 163
Herpes simplex virus (herpes II), 347
High school, 217-221
Hiking, 83-84, 100
HIV (AIDS virus), 162, 346-349
Home
   as a caring community, 439-440
   and cultural values, 414
   earth as, 425, 431
Honesty, 232-233, 280
"Hooked" on drugs, 154, 168-170
Hormones, 161-162, 305
Housekeeping, 399
Housing, 410
Humor, 232
Hunger, 66-67
Hypothermia, 100

I

"Ice," 159
Ideals, 206
Identity, 27-28, 298-302, 313
Illiteracy, 300
"I messages," 233-234, 237, 241
Independence, 51, 361-364, 325
Individualism, 10
Inhalants, 160-161
Injury, and safety precautions, 98-100
Insults, 130, 234
Intercourse, 345-347, 392
Interdependence, 358, 364
Interests, 210-211, 217
Interruption, 236
Interview, 208
Intervention, 194
Intimacy, 281, 284, 359
Invincibility, 13
Involvement, 445

J

Jealousy, 113-114, 289
Jenner, Dr. Edward, 451
Jobs
   getting, 208-210
   and marriage, 410-411
   part-time, 209-211
   selection of, 313
   volunteer, 449-450
Johnson, Ben, 161
Joy, 115, 119
Junior high school, 217

K

Kennedy, President John F., 442
Kids Anonymous, 188
Kindness, 280

L

Labor, U.S. Department of, 210
Laws, 7-8, 258
Legal agreement, 400-401
Lies, 234, 280
Life
   and care, 24-25
   liberty and pursuit of happiness, 5
   interrelatedness of, 422, 430
Listening, 244-245
Litter, 430-431
Loneliness, 113
Look-alike drugs, 161
Loss, 386-390
Love. See also Family
   and emotions, 115, 117, 404-405
   and your future family, 413-415
   and support, 309
   true, 398-405, 408, 415
Loyalty, 36
LSD, 160

M

Manipulation, 130
Marijuana, 152-153, 157-158, 164
Marriage, 401-415
   building a strong, 408-415
   customs and trends of, 360, 398-399
   or living together, 401
   preparing for, 406-408
   tasks of, 408-415
   teenage, 404-405
Masculinity, 298
Mate selection, 317
Maturity, 6, 406
Meaning
   and communication, 229-231, 242
   and religion, 107-108

Mediator, 271
Medicines, 152
Menstruation, 305
Mental health, 106-123
Mental retardation
   of babies, 344, 346
   and Special Olympics, 437-438, 448-449, 451
Messages
   in communication, 231-238, 243
   media, 63, 70, 302
   sexual, 337-338, 340
Methamphetamine, 159
Minerals, as nutrients, 70
Mistakes
   and friends, 277, 280, 290
   and parents, 362-363
Mocking, 130
Modeling, 6-7
Money
   and food choices, 66-67
   and school dropouts, 213-214
   and sports equipment, 96
   and work-study programs, 217
Moods, 115, 118-119. *See also* Emotions
Morphine, 160
Motivation, for exercise, 95
Multiplier effect, of drugs, 165
Muscles, 83-87, 98-100
Myths and facts
   about foods, 63-64
   about sexual behavior, 329-330, 341-344

N

Narcotics, 160
Nations, and drugs, 185, 189
Natural environment, 422-431
   caring for the, 427-431
   understanding the, 422-426
Natural resources, 425, 430
Nature, 24, 107. *See also* Natural environment
Needles, 160, 162, 346, 348-349
Needs, 284, 308-309
Negative ways to say no, 136-137
Negative values, 15-16
Neighborhood Rehab projects, 446
Neighborhood Watch, 446
Network and networking, 282, 289

Newspapers, 66
Nicotine, 153, 163-164
Nitrogen, 424, 428
Nocturnal emissions, 307
Non-enabling, 190-192
Nonsmokers' Bill of Rights, 165
Nuclear family, 360
Nurse, school, 68, 379
Nurturing, 110
Nutrients, 62, 70
Nutrition, 62-63, 160

O

Obesity, 71, 76, 87
Olympics, Special 90, 94, 437, 448-449, 451
Openness, 265-266, 268, 269, 280
Opinions, 236
Opium, 160, 185
Options, family, 415
Overdose, 154, 159
Oxygen
   during exercise, 83-84, 86, 99
   in the natural environment, 423-424, 428-429
Ozone layer, 429

P

Parents, 7, 362-364, 378, 391
Parents Anonymous, 188
Passiveness, 137
Passive smoking, 164
Payoffs, 171
PCP, 158, 160
Peace Corps, 442
Peacemaker, 263, 267
Peer mediation, 271
Peer pressure, 42, 64, 129, 169, 289
People, 430-431
Pets, 431
Phipps, James, 451
Pills, 152

Plans and Planning. *See also* Goals
   for careers, 208-211
   and decision-making, 50-53, 55
   for fitness, 66, 97
   for marriage and family, 406-415
   and romance, 329-332, 335
Plants, 422-430
Platonic relationship, 278
Police, 188
Politics, 443-444
Pollution, 427-431
Popularity, 288
Positive ways to say no, 138-139, 142-143
Possessiveness, 289
Posture, 85
Power plays, 290
Pregnancy
   and consequences, 143, 341-344
   and drugs, 165-166
   teen, 214-215, 305
Prejudice, 285-286
Pressure
   inner, 128-129
   peer, 42, 64, 129, 169, 289
Privacy, 10, 311, 381, 414
Problems
   and communication, 240-241
   and conflict, 259-263
   and families. *See also* Families, troubled
Property rights, 9
Protein, 70
Pryor, Richard, 158
Psychologists, 122
Puberty, 305
Pubic hair, 305
Public service, 447-448

R

Rape, 339, 390-391
Rape crisis center, 391
Radiation, 429
Rain, 429
Reactions, from drugs, 150-165
Reactions, emotional, 11. *See also* emotions
Reality, 204, 206
Recovery, 196-197
Recreation departments, 93-94
Recycled waste, 430-431
Red Cross, 379, 383
References, 208
Rehabilitation center, 183-184

Rejection, 112-113
Relationships. See also Friends; Family
    building, 110-114, 281, 284
    and mental health, 110-114
    between people, 19
    sexual, 325-326
Relaxation, 86-87
Religion
    and marriage, 399-400, 407
    and meaning of life, 107-108
Reproductive organs, 305, 307
Resources
    definition of, 42
    for a family, 381
    natural, 425, 430
    at school, 219
Respect
    in communication, 233
    in friendship and love, 280, 405
    and recovery from addiction, 197
    and sexual identity, 309
    as a value, 7, 19
Responsibility
    in communication, 243
    in a family, 312-313, 381
    and guilt, 119-120
    in marriage, 405
    as a value, 7, 19
Risks
    for babies and teen mothers, 344
    and consequences, 131-135
    in decision-making, 45-47
    in exercise, 100
    in sharing friendship, 282
    of STDs, 347
Roles,
    and jobs, 313
    of women and men, 300-301, 312-313
Romance, and myths, 329, 344
Romantic attraction, 402-403
Ronald McDonald Houses, 447
Rules, 7-8, 10, 359
Rumors, 266, 280
Running, 83-84, 100

S

Safe sex, 346
Salt, 77
Sarcasm, 236
Saving face, 257
Saying NO
    consequences and, 131-135
    pressures and, 128-129, 140-145
    reasons for, 163
    and sexual decisions, 335, 341-342, 349
    steps in, 136-139, 142-143

Saying YES
    to your future, 349
    by saying no, 145
    and sexual consequences, 327, 341
    and snap decisions, 134
School
    as a caring community, 440
    competition and cooperation in, 264-271
    getting the most out of, 216-221
    and sports participation, 93-94
    and teen drop outs, 213-214
    why teens stay in, 214-216
Secrecy, 265, 380-381
Sedentary, 98
Self-centeredness, 287
Self-concept, 107
Self-confidence, 284
Self-control, 64, 90
Self-discipline, 62-63
Self-doubt, 318
Self-esteem,
    definition of, 106-107
    and peer pressure, 64
    and fitness, 89
    and saying no, 139
Self-identity, 27-29
Self-image, 71, 107
Self-indulgence, 32
Selfishness, 31, 280
Self-protection, 32
Self-reliance, 52
Self-respect, 27, 197
Self-righteousness, 32-34
Self-worth, 16
    and decisions, 42, 51
    and emotions, 115-123
    and relationships, 282
    and values, 109-110
Sensitivity, 233, 292
Service, 447-448
    private business and, 447
    public, 447
    volunteer organizations and, 448
Sex
    and abuse, 188, 392-393
    and disease (STDs), 341, 345-349
    and sexual identity, 298-302, 312-313
Sexual decisions, 324-349
    attraction and, 311, 324
    consequences of, 341, 349
    and feelings, 324, 403, 405, 412
    and options, 331-340
    and saying yes and no, 140-145, 324-330
    and values, 311, 340

Sexually transmitted diseases (STDs), 341, 345-349
Sharing, 8, 31, 280-282
Shock, 382-383
Shyness, 111-112
Silence
    and abuse, 392
    code of, in troubled families, 380-381
    instead of communication, 238-239, 245, 253
    in friendships, 280
Singles, 415
Skills, 215
Sleep, 86, 159
Smoking, 163-165
Snap decisions, 134
Snow, 159
Socializing, 414
Social norms, 258
Social relationships, 51
Social Security card, 209
Society, 4, 7-8, 19, 58, 67
Sodium, 77
Sorrow, 115, 119
Special Olympics, 90, 437-438, 448-449, 451
Speed, 159, 161
Spirituality, 106-108
Sports, 84, 93. See also Exercise
Starch, 69, 75
STDs (sexually transmitted disease), 341, 345-349
Stealing, 172
Stereotype, 286
Sterility, 162
Steroids, 161-162, 172-174
Stimulants, 158-159
Stress, 68
Strokes, 77
Substance abuse, 142. See also Drugs
Sucrose, 76
Sugar, 69, 76-77
Sugar substitute, in drugs, 161
Suicide, 170, 386-388
Sun, 423
Sunburn, 101
Support
    emotional, 121-123, 309
    groups, for becoming drug-free, 169, 184, 192
    in a healthy family, 409
Surgeon General Report, U.S., 163
Swimming, 101
Synagogues, 7, 93, 448
Syphilis, 347

T

Tact, 233
Talents, and jobs, 209
Talking, 231-237, 308, 370
Teachers, 63, 68, 169, 215, 378-379
Teaching, direct, 6-8
Teams and teammates, 86, 88-90
Teasing, 130
Technology, 399
Teens
   and conflict, 256-258
   and drugs, 168-171
   and friendship patterns, 278-279
   and ideals and reality, 204-207
   and school, 213-221
   and sex, 324, 341, 344
   and teen mothers, 344
   volunteer work for, 442-443
Television, 7, 63, 66, 75
Tension, 261
Threats, 130
Time, 42, 66-67
Tobacco, 163
Tolerance, drug, 153
Tooth decay, 76
Toxemia, 344
Treatment center, 196. *See also* Rehabilitation
Treatment, drug, 182-186
Trends, 398-399
Trial and error, 6, 8
True love, 398-405, 408, 415
Trust, 7, 19, 86, 233, 280, 405
Truth, 7

U

United Nations, 153, 348
United States. *See* America
United Way, 447
Users, and using behaviors, 332-333, 335, 342

V

Values, 4-19
   and where they come from 4-11
   finding your own, 12-19
Values, positive, 15-16
   of caring, 7-8, 19, 24-30, 35, 453
   of family, 7, 19, 356-365. *See also* Family
   of respect, 7, 19, 197, 233, 280, 309, 405
   of responsibility, 7, 19, 243, 312, 381, 406
   of trust, 7, 19, 86, 233, 280, 405
Vegetables, 75
Violence, 171, 267-268, 390-393, 452
Vitamins, 64, 70
Voice, tone of, 238-239, 245
Volunteering, 113, 208, 440-444, 448-451
Volunteer service organizations, 448
Voting, 443-444
Vows, marriage, 399, 400, 406

W

Walking, 87, 89, 92, 98-99
Warm-up, 98-99
Wastes, 430
Water
   as a nutrient, 70, 101
   as a resource, 424
   pollution, 427, 429-430
   sports, 101
Weddings, 399-401
Weight, 69-71, 85
Wellness, 82
Westinghouse Science Awards, 447
Wet dreams, 307
WHO (World Health Organization), 348
Withdrawal, 154
Workout, 83, 89, 98-100
Work permits, 209
Work-study programs, 217, 221

Y

YMCA, 94, 379
YMHA, 94, 379
"You messages," 235
Youth worker, 379

Acknowledgements

Amateur Softball Association: Page 95; American Red Cross: Page 393 (©1989); Arnold and Brown: Pages 7, 15 (L-R), 25, 32, 33, 35, 51 (L-R), 54, 63, 67, 71, 74, 75, 77, 83, 86, 89, 96, 97 (L-R), 98, 99 (R-L), 100, 108, 109, 110, 129, 132, 133, 135, 139, 141, 142, 157, 159, 160, 162, 164, 165, 168, 170, 171, 173, 174, 189, 190, 195, 197, 203, 211, 212, 216, 217, 220, 229, 233, 234 (L-C-R), 238, 239, 240, 243, 244, 253, 256, 261, 265, 270, 277, 286, 287, 291 (T-B), 330, 332, 336, 337 (T-C-B), 343, 345, 366, 367, 381, 386, 388, 399, 407, 410, 411, 441; Special thanks to the following individuals, schools and businesses for their help with Arnold and Brown photographs. In Peoria, IL: Sandra M. Sallee, Production Coordinator; E.N. Woodruff High School; MaryLou Schlesinger, Woodruff Counseling; Peoria Central High School; Mrs. Nan Kleffman; Irving Grade School; The Alfredo Cisneros Family; The Rigoberto Diaz Family; The Dr. Sam Pittmon Family; Bogard Drug Store; K's Merchandise; Kroger North Point; Peoria Police Department: Officers Bill Calbow and Clint Wilson; Peoria Police Explorer Troop; Proctor Community Hospital; River City Athletic Club; Rocky's Hitching Post; The Silver Bullet Saloon; Thompson Food Basket at Candletree; University Ford; Walters Brothers Harley-Davidson Sales; Westminister Presbyterian Church. Illini Bluffs High School, Glassford, IL; Beau Didley's, Peoria Heights, IL; Royal Haeger Lamp Company, Macomb, IL. James L. Ballard: Cover, Pages II, 2, 3, 6, 14, 22, 23, 27, 28, 29 (L-R), 30, 34, 38, 39, 41, 43, 47, 48, 60, 61, 65, 80, 81, 98, 99 (L-R), 104, 105, 107, 112, 119, 126, 127, 138, 142, 145, 148, 149, 178, 179, 182, 191, 200, 201, 226, 227, 230, 231, 235, 241, 248, 249, 251, 252, 254, 257, 262, 266, 269, 274, 275, 278, 282, 296, 297, 307, 310, 312, 319 (T-B), 322, 323, 325, 334, 338, 355, 374, 375, 377, 380, 396, 397, 401, 402, 418-419, 420, 421, 434, 435, 438, 444, 446; Roger B. Bean: Page 390; Black Star: Pages 166 (Joseph Rodriguez), 242 (Spiegel), 429 (©Dennis Capolongo); California Angels: Page 204 (John Cordes); Tim Campbell: Page 9; David Cunningham (Illustrations): Pages 144, 151, 181, 304, 424; Duomo: Page 116 (Dan Helms); Lynn Fitzgerald: Page 340; James Gaffney: Page 359; Marilyn Gartman Agency: Pages IV (Tom McCarthy), 13 (Spencer Grant), 18 (Bannock/Photri), 101 (Tom McCarthy), 260 (©Diane Schmidt), 313 (©Ellis Herwig), 316 (©Ellis Herwig), 358 (©Mark E. Gibson), 404 (©Lee Balterman), 413 (©Brent Jones), 440 (©Brent Jones); Ann Garvin: Pages 72 (T-B), 73 (T-C-B), 110; Girl Scouts of the U.S.A.: Page 117; ©Ken Graham: Page 123; Graphics Associates: Page 76; ©Chip Henderson: Pages XIII-1; The Image Bank: Pages 58-59 (Janeart Ltd.), 90 (Butch Powell), 91 (Luis Castaneda), 113 (©Marc Romanelli), 224-225 (Kaz Mori), 264 (©Michael Melford), 293 (Butch Martin), 326 Mike Maas), 328 (Butch Martin); Index Stock International, Inc.: Pages 10, 91, 240, 314; Magnum Photos, Inc.: Pages 17 (Bruce Davidson), 196 (Michael Nichols), 207 (©Gilles Peress), 288 (©Michael Nichols), 363 (©Hiroji Kubota),; McDonald's Corporation: Page 209; ©Abraham Menashe: Page 360; Nawrocki Stock Photo: Page 426 (Steere); Peace Corps: Page 442; H. Armstrong Roberts, Inc.: Pages 168 (P. Buddle), 428L (R. Krubner), 428R (Camerique), 450 (Camerique), 453 (Murphy/Stills); Rotary International: Pages 283, 437; ©Charles Schabes: Page 120; Jeanne Seabright: Page 301; Sovfoto/Eastfoto: Page 425 (Tass); Special Olympics International: Pages 88, 94, 452 (©Ken Regan); Stock, Boston: Pages 26 (Patricia L. Pfister), 93 (Leonard Harris), 121 (©Bob Daemmrich), 122 (©Deborah Kahn Kalas), 137 (©Gale Zucker), 191 (Mike Mazzaschi), 193 (Rick Brown), 196 (©Bill Horsman), 281 (Chris Brown), 288 (©Bill Horsman), 299 (©Nancy Dudley), 348 (©Lawrence Migdale), 368 (©Billy E. Barnes), 370 (©Bob Daemmrich), 400 (Billy E. Barnes), 408 (©Willie L. Hill, Jr.), 414 (Robert Caputo); U Haul International Inc.: page 383 (Copyright 1989-all rights reserved); UNICEF: Pages 153 (#1914), 218 (#1425), 379 (Aslak Aarbus); Uniphoto, Inc.: Pages I (Uniphoto), 85 (©Daemmrich), 185 (Daemmrich), 346 (Uniphoto), 357 (©Rick Brady); United Nations: Pages 52 (#154322-Gayle Jann), 67 (#164652-John Isaac), 87 (#137158-J.P. Laffont), 300 (#153455-John Isaac), Carlos Vergara: Page 369; Viewfinders: Pages 123 (Multimedia), 215 (Multimedia), 240 (©Wayne R. Johnson), 240 (Multimedia); 403 (Multimedia); Walgreens: Page 154; West Stock: Pages 346 (©Jeffrey W. Myers), 352-353 (Ed Bock); The Yale University Art Gallery: Page 5 (The Declaration of Independence by John Trumball. American, 1756-1843).

While the examples and stories in this text are true, models and fictional names have been used to portray them.